Substrate Integrated Antennas and Arrays

Substrate Integrated Antennas and Arrays

Yu Jian Cheng
SCHOOL OF ELECTRONIC ENGINEERING
UNIVERSITY OF ELECTRONIC SCIENCE AND TECHNOLOGY OF CHINA
CHENGDU, PEOPLE'S REPUBLIC OF CHINA

CRC Press
Taylor & Francis Group
Boca Raton London New York

CRC Press is an imprint of the
Taylor & Francis Group, an **informa** business

CRC Press
Taylor & Francis Group
6000 Broken Sound Parkway NW, Suite 300
Boca Raton, FL 33487-2742

First issued in hardback 2019

ISBN-13: 978-1-4987-1453-2 (hbk)

Visit the Taylor & Francis Web site at
http://www.taylorandfrancis.com

and the CRC Press Web site at
http://www.crcpress.com

Contents

Preface

Theoretically speaking, high-frequency antennas can be developed from the existing designs that work well in the low-frequency band. However, the design and structure implementation may be complicated by shorter wavelengths. Conventional printed transmission lines, such as microstrip lines and coplanar waveguides, are popular for integrated designs of circuits and antennas, but they cannot be easily applied in the millimeter-wave frequency band. Conventional nonplanar transmission lines, such as rectangular metallic waveguides and coaxial lines, are still widely used for the development of millimeter-wave and terahertz components even though they are expensive, bulky, and nonintegrated. The most attractive features of those nonplanar structures are their high-quality, high-power handling capability and completely shielded waveguiding channel.

As a new generation of guided-wave structure, substrate integrated circuits (SICs) have gained wide attention since their inception. The principle of SICs is to convert nonplanar structures into the corresponding planar form, enabling the planar fabrication processing of those nonplanar and three-dimensional structures. In this case, SICs inherit merits of those nonplanar counterparts and are promising for high-frequency antennas, arrays, and wireless systems. This research area is of great worth in wireless communication base station antennas, portable microwave point-to-point systems, collision avoidance radars, conformal antennas, satellite antennas, etc.

This book is a research monograph about substrate integrated antennas and arrays. I hope it is able to provide all the necessary information about substrate integrated antennas and arrays without readers having to collect bits and pieces from different sources and papers. This book systematically describes a new antenna technique based on the concept of SICs, which has the potential to replace those existing antenna technologies. After introduction to the substrate integrated feeding network, different types of substrate integrated feeding networks, antennas, and arrays are discussed with respect to theoretical and experimental results, including substrate integrated slot array antennas, substrate integrated traveling-wave antennas, substrate integrated feeding antennas, substrate integrated monopulse antennas, and substrate integrated multibeam antennas.

I hope this book will help antenna researchers have some insight into SIC antennas and arrays. Other targeted readers are antenna engineers who want to introduce SIC antennas to the wireless communication industry. Another goal of this book is to provide a shortcut for electronic and system engineers who do not know a lot about antennas but want to learn something about SIC antennas without the hassle of learning from various sources. The comprehensive contents are helpful to further understand SIC antenna technology and overcome difficulties, such as how to meet all the bandwidth, gain, and efficiency specifications, for antenna engineers already working in the field.

I spent more than seven years researching substrate integrated antennas and arrays. During these years, with great pleasure, I acknowledge many people who have influenced my thinking and contributed to my knowledge, in particular Prof. Wei Hong

(Southeast University, China) and Prof. Ke Wu (University of Montreal, Canada). I also acknowledge support from the National Natural Science Foundation of China (NSFC) under grant 61001028, the Foundation for the Author of National Excellent Doctoral Dissertation of PR China under grant 201338, and the Program for New Century Excellent Talents in University under grant NCET-13-0089.

Author

Yu Jian Cheng earned his BS degree from the University of Electronic Science and Technology of China in 2005 and his PhD degree in 2010 without going through the conventional master's degree at Southeast University, Nanjing, China. Since 2010, he has been with the School of Electric Engineering, University of Electronic Science and Technology of China, and is currently a professor. From 2012 to 2013, he was part of the research staff in the Department of Electrical and Computer Engineering, National University of Singapore. His current research interests include microwave and millimeter-wave antennas, arrays, and circuits. He has authored or coauthored more than 70 papers in journals and conferences.

Dr. Cheng was the recipient of awards from the One Hundred Person Project of UESTC in 2014, New Century Excellent Talents in University in 2013, National Excellent Doctorate Dissertation of China in 2012, and the Jiangsu Province Excellent Doctorate Dissertation in 2011. He is a senior member of the Institute of Electrical and Electronics Engineers and a senior member of the Chinese Institute of Electronics. Cheng has served on the editorial board of the *International Journal of Antennas and Propagation*, and review boards of various technical journals, including *IEEE Transactions on Antennas and Propagation*, *IEEE Microwave and Wireless Components Letters*, *IEEE Antennas and Propagation Letters*, and *IEEE Antennas Propagation Magazine*.

1 Substrate Integrated Circuits

The development of substrate integration antenna technology is based on the concept of substrate integrated circuits (SICs). The principle of SICs is to convert nonplanar structures into their corresponding planar form, enabling the planar fabrication processing of those nonplanar and three-dimensional structures. In this case, conventional planar transmission lines, such as microstrip line and coplanar waveguide (CPW), can be integrated with these SIC structures on the same substrate with the same processing technology. In fact, any nonplanar guided-wave structure can be converted into its planar counterpart. SICs inherit the merits of nonplanar guided-wave structures and is very attractive for high-frequency applications.

In this chapter, incomplete but typical substrate integrated technologies are introduced, including substrate integrated waveguide (SIW), substrate integrated image guide (SIIG), and substrate integrated nonradiative dielectric (SINRD) guide.

1.1 SUBSTRATE INTEGRATED WAVEGUIDE

As the most mature and popular SIC structure, an SIW is synthesized by two rows of metalized vias or metalized slot trenches embedded in a dielectric substrate with the conductor claddings on the top and bottom broad walls, as shown in Figure 1.1. It can be considered a dielectric-filled rectangular waveguide.

Similar structures were proposed earlier than the inception of SICs under different terminologies, i.e., laminated waveguide [1–5] and post-wall waveguide [6–11]. Quite naturally, specific advantages of the conventional rectangular waveguide, such as high-quality factor and self-consistent electrical and mechanical shielding, are maintained by these similar structures. The most attractive feature is that a full-scale integration can be expected, involving passive and active circuits as well as antennas on the same substrate with the same processing technique. This is crucial for millimeter-wave applications because antennas and circuits cannot be separately designed and handled at such high-frequency bands. The single-layered or multilayered SIW has been studied extensively for the development of different circuits and antennas with various processing techniques, including the printed circuit board (PCB) process, low-temperature cofired ceramic (LTCC) technique, photo-imageable process, etc.

1.1.1 Design Rule

Except for the substrate thickness h and the relative permittivity ε_r, there are three main parameters for an SIW, i.e., the SIW width a, the diameter of the metalized via d, and the distance between neighboring metalized vias s. After taking the effect of

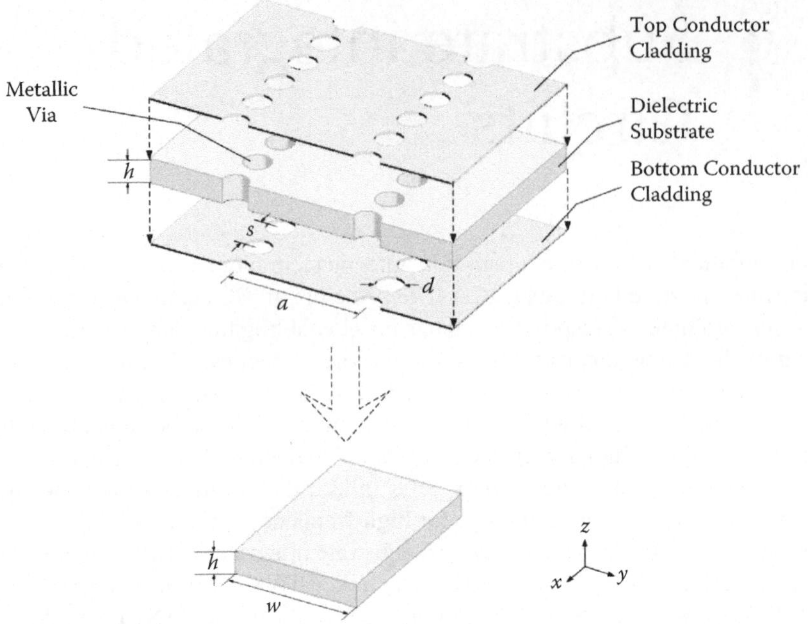

FIGURE 1.1 Configuration of an SIW.

the dielectric material into consideration, the distance between the two rows of periodic metalized vias determines the cutoff frequency. This nonstandard waveguide only allows the propagation of TE_{n0} (transverse electric) modes, not supporting the propagation of transverse magnetic (TM) modes. The reason is that the current flow along the two sidewalls does not exist.

There are several works focusing on how to design an SIW. The simplest method is described in [12]. Combined with the BI-RME method and the Floquet theorem, dispersion properties of an SIW were rigorously investigated and recognized. This early work shows that the SIW basically has the same transmission characteristics as the conventional rectangular waveguide filled with the same dielectric material. After that, empirical equations were fitted from the calculated dispersion curves so as to estimate the cutoff frequencies of the first two dominant modes of the SIW. To validate the analysis results, the propagation constant was extracted by measuring the phase difference of two SIWs with different length. Then, the SIW can be analyzed as a rectangular waveguide filled with the same dielectric material just by using the equivalent width, w, of the SIW, which can be obtained by

$$w = a - \frac{d^2}{0.95 \times s} \tag{1.1}$$

However, Equation (1.1) does not include the effect of diameter-to-width ratio, i.e., d/a. When d increases, the error of Equation (1.1) will appear. To overcome this weakness, a multimode numerical calibration procedure was proposed in [13] combined with a commercial software package based on a full-wave finite-element method (FEM) for extracting complex propagation constants of the SIW structure accurately. In this case, the complex propagation constant can be accurately extracted for each SIW mode. A more accurate empirical equation can be given:

$$w = a - 1.08 \times \frac{d^2}{s} + 0.1 \times \frac{d^2}{a} \tag{1.2}$$

In [14], another rigorous full-wave approach on the basis of the method of lines (MoL) was presented to analyze propagation characteristics of the SIW. A generalized matrix eigenvalue equation was derived to replace the conventional transcend equation. Thus, the computational efficiency can be improved much more. Furthermore, an efficient Z-transform absorbing boundary condition (Z-ABC) is used for further improving the calculated accuracy of the propagation constants. By means of this proposed technique, a series of useful empirical equations were obtained to design SIW components.

$$w = a \times \left(\xi_1 + \frac{\xi_2}{\frac{s}{d} + \frac{\xi_1 + \xi_2 - \xi_3}{\xi_3 - \xi_1}} \right) \tag{1.3}$$

In Equation (1.3),

$$\xi_1 = 1.0198 + \frac{0.3465}{\frac{a}{s} - 1.0684} \tag{1.4}$$

$$\xi_2 = -0.1183 - \frac{1.2729}{\frac{a}{s} - 1.2010} \tag{1.5}$$

$$\xi_3 = 1.0082 - \frac{0.9163}{\frac{a}{s} + 0.2152} \tag{1.6}$$

To verify the validity of these design equations, some full wave simulations for SIWs at different frequencies are implemented to compare with the calculated results obtained by Equations (1.1) to (1.3). As shown in Figures 1.2 and 1.3, the accuracy of Equation (1.3) is better than others. The accuracy of all design equations is improved with the increase of a/λ. Besides, these equations have a better approximation at a lower-frequency band.

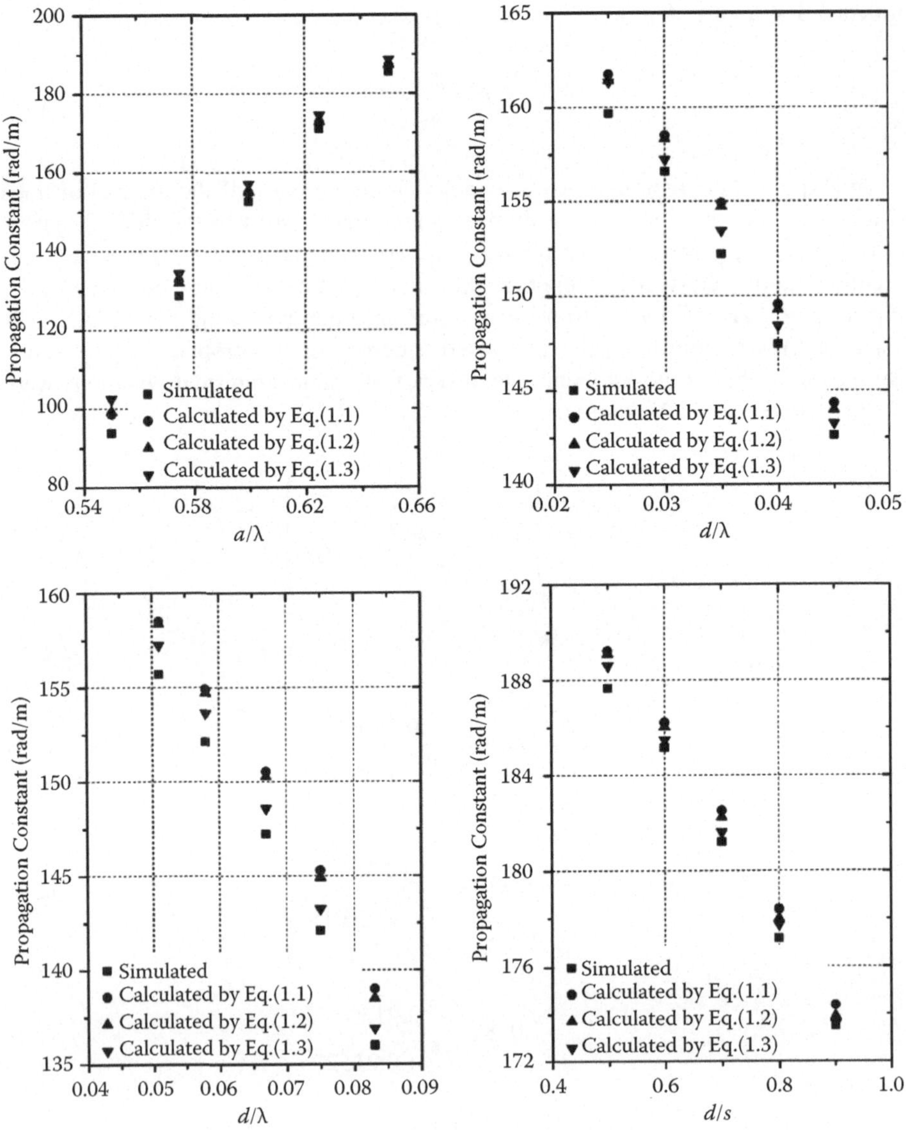

FIGURE 1.2 Propagation constants of a 10 GHz SIW calculated by different design equations.

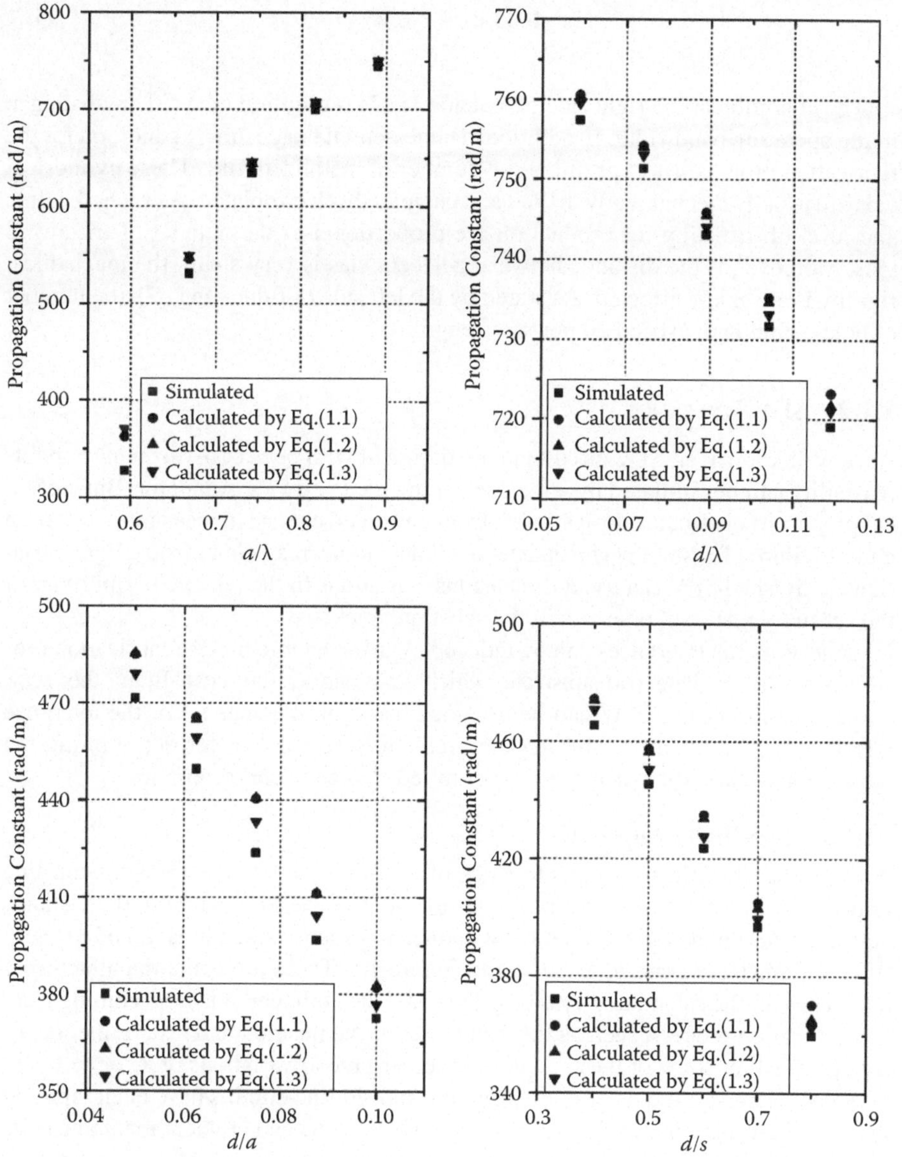

FIGURE 1.3 Propagation constants of a 30 GHz SIW calculated by different design equations.

There is a design rule defined in [15] that the SIW can be equivalent to a conventional rectangular waveguide that does not present any bandgap over the band of interest. That is:

$$0.05 < \frac{s}{\lambda_c} < 0.25 \tag{1.7}$$

The condition at the right side of Equation (1.7) is required to avoid any bandgap in the operating bandwidth. In addition, a nonessential but desirable condition for the fabrication process is to minimize the number of metalized vias. These cylindrical vias are usually mechanically drilled and chemically electroplated. As such, the time and level of difficulty for production are proportional to the number of metalized vias. Moreover, if the distance between adjacent vias is very small, the mechanical rigidity is adversely affected. As stated by the left side of Equation (1.7), the number of holes should not exceed 20 per wavelength.

1.1.2 SIW Tolerance

As is well known, an SIW circuit can be designed with perfect performance and its feasibility can be validated by a mature commercial software simulator. But electrical properties of circuits always suffer from unwanted change in measurements, even a total failure of system performance. Such deviations may come from slight variations in finished SIW electrical dimensions, variations in the relative permittivity of the substrate material, or errors in the substrate thickness.

In general, the tolerances can be reduced by using a more precise fabrication process and a better dielectric substrate, which also lead to high cost. In another way, a tolerance-insensitive SIW can be designed based on the analysis of the tolerance effect on SIW structures. The inverse problem is also useful for determining the fabrication accuracies to achieve the permitted change in characteristics.

1.1.2.1 Sensitivity Approach

As described before, the characteristics of an SIW, such as the transmission line impedance Z_0 and phase constant β, are functions of the SIW width a, the distance between adjacent metalized vias s, the diameter of the metalized via d, and the substrate parameters ε_r and h, as shown in Figure 1.1. There are some manufacturing tolerances for the substrate properties. There are also influential factors coming from the fabrication process, such as the accuracies of the position and size of the metalized post. All or some of these parameters contribute to variations of Z_0 or β.

Some tolerance studies of microstrip line and slotline circuits have been reported in [16, 17] through the sensitivity approach. This simple and efficient method can be used to evaluate the effect of tolerances on SIW performance as well. The worst case can be calculated by the first-order effects of parameter variations [18]. Different from other analytical methods, like the Monte Carlo analysis, only the maximum absolute values of tolerances are needed for the sensitivity approach, rather than the detailed distributions.

The sensitivity of a parameter A with respect to a parameter B, S_B^A, is defined as

$$S_B^A = \frac{B}{A}\frac{\partial A}{\partial B} \tag{1.8}$$

The sensitivity as defined in Equation (1.8) can be utilized to determine the deviation in circuit characteristics for a given tolerance in a parameter. Sensitivity analyses for Z_0 and β of the SIW are presented in the following section. The effects of tolerances on SIW's performances, such as voltage standing wave ratio (VSWR) and phase shift, are evaluated as well.

1.1.2.2 Tolerance on SIW Impedance

The impedance of an SIW, Z_0, can be calculated by

$$Z_0 = \frac{h}{w}\frac{\eta}{\sqrt{1-\left(\frac{\lambda}{\lambda_c}\right)^2}} \tag{1.9}$$

In Equation (1.9),

$$\eta = \frac{120\pi}{\sqrt{\varepsilon_r}} \tag{1.10}$$

The change in Z_0 is related to tolerances in different parameters, and its worst-case behavior can be calculated as follows:

$$\frac{|\Delta Z_0|_{\max}}{Z_0} = \left|\frac{\Delta a}{a}S_a^{Z_0}\right| + \left|\frac{\Delta h}{h}S_h^{Z_0}\right| + \left|\frac{\Delta \varepsilon_r}{\varepsilon_r}S_{\varepsilon_r}^{Z_0}\right| + \left|\frac{\Delta s}{s}S_s^{Z_0}\right| + \left|\frac{\Delta d}{d}S_d^{Z_0}\right| \tag{1.11}$$

In Equation (1.11), Δa, Δh, $\Delta \varepsilon_r$, Δs, and Δd are the maximum errors for a, h, ε_r, s, and d, respectively.

Two SIWs, which have the center operating frequencies of 10 and 30 GHz, respectively, are investigated. Each one only permits the propagation of the dominant mode TE_{10}. A complete set of graphs for sensitivities of Z_0 with respect to a, h, ε_r, s, and d as functions of width-to-height ratio a/h, width to diameter ratio a/d, diameter-to-distance ratio d/s, and working wavelength λ_0 for a 10 GHz SIW are given in Figure 1.4. As shown in Figure 1.5, four similar abscissas are defined to plot sensitivity curves of Z_0 for a 30 GHz SIW.

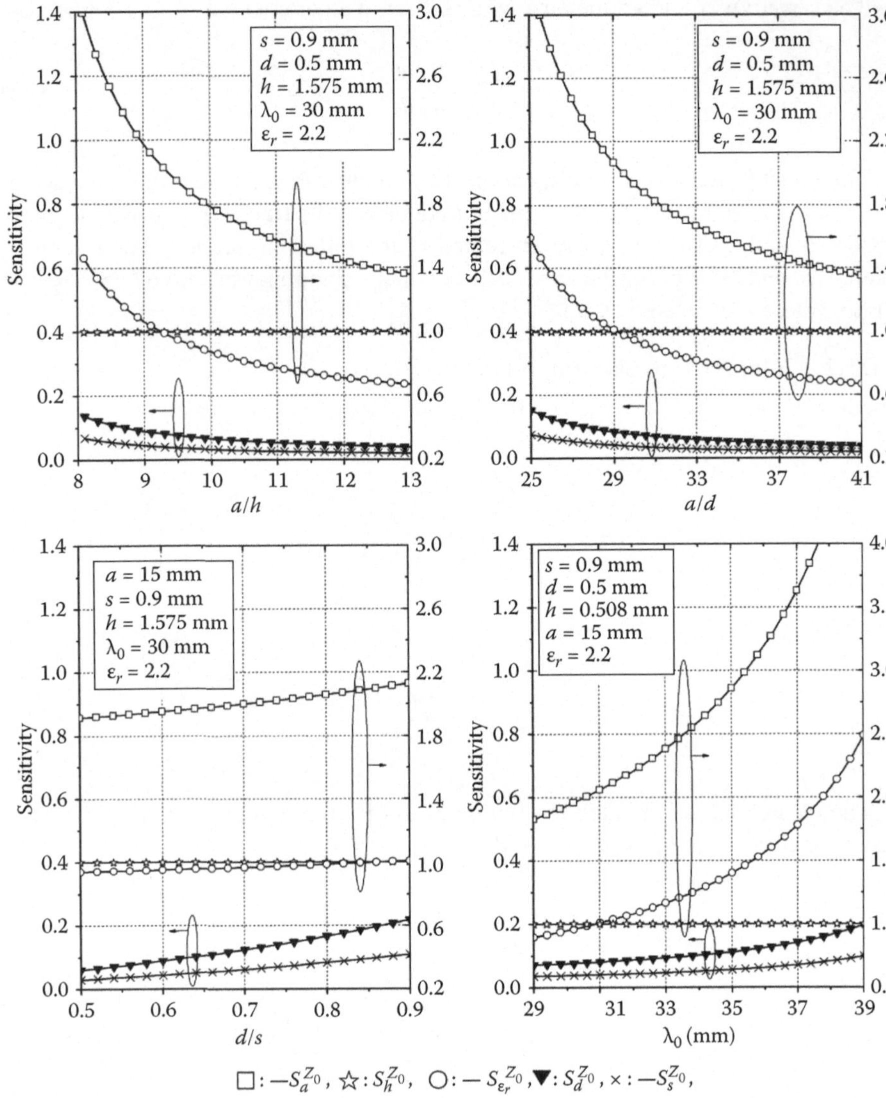

FIGURE 1.4 Impedance sensitivity curves of a 10 GHz SIW impedance with respect to *a*, *d*, *s*, ε_r, and *h*.

Some interesting conclusions are observed from these sensitivity curves:

1. All of the sensitivity curves are smooth and monotone except $S_h^{Z_0}$.
2. All of the sensitivity curves have the same amplitude-varying tendency except $S_h^{Z_0}$. The changing trend of the tolerance effect of each parameter is roughly the same, but the changing extent is different.
3. $S_h^{Z_0}$ equals to 1 at all times. That means the tolerance effect of the substrate thickness remains unchanged.

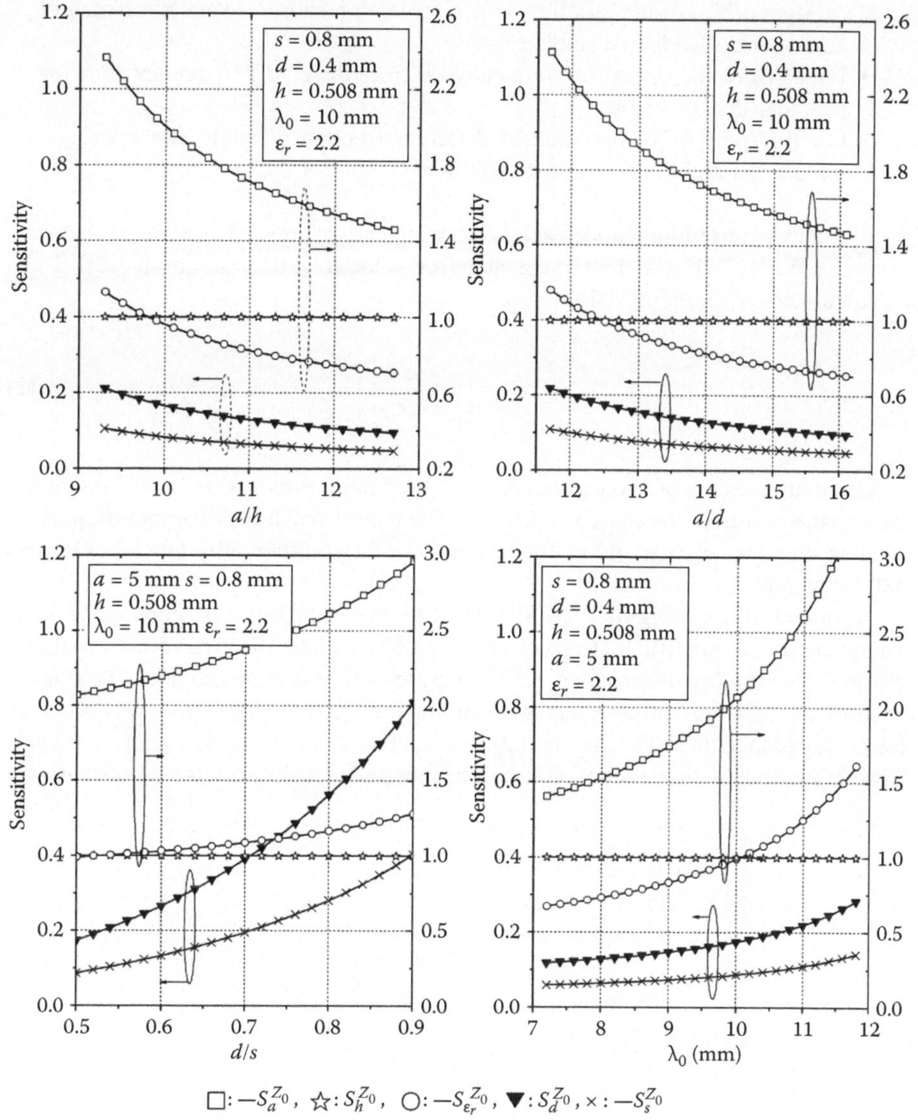

FIGURE 1.5 Impedance sensitivity curves of a 30 GHz SIW impedance with respect to a, d, s, ε_r, and h.

4. $|S_d^{Z_0}| = 2\,|S_s^{Z_0}|$. That means the tolerance effect of d is twice than that of s all the time.
5. $S_a^{Z_0}$, $S_{\varepsilon_r}^{Z_0}$, and $S_s^{Z_0}$ are negative. This is because the SIW impedance decreases with an increase in a, ε_r, and s.
6. $S_h^{Z_0}$ and $S_d^{Z_0}$ are positive because Z_0 increases with h and d.
7. Generally speaking, the tolerance effects of a, h, and ε_r on the SIW impedance are greater than the effects of s and d.

8. A larger a/h leads to a smaller error, a larger a/d leads to a smaller error, and a smaller d/s leads to a smaller error.
9. For an SIW, the operating frequency far from the cutoff frequency contributes to a smaller error.
10. For different SIWs operated at different frequency bands, the tolerance effects of s and d increase with frequency.

Tolerances cause impedance variations, which are expressed in terms of the maximum VSWR [17]. Using the largest changing value ΔZ_0, the maximum VSWR can be calculated by Equation (1.12):

$$\text{VSWR} = \left[1 - \frac{|\Delta Z_0|}{Z_0}\right]^{-1} \tag{1.12}$$

Calculated results of two examples are plotted in Figures 1.6 and 1.7. The fabrication accuracy was assumed to be 0.01 mm. The tolerances in the substrate properties, such as h and ε_r, correspond to the tolerances of a commercially available Rogers 5880 substrate.

As shown in these figures, an SIW with higher impedance is more sensitive to tolerances and more difficult to reach the low VSWR under the given fabrication precision. Thus, the low-impedance SIW structure will be a better choice. The impedance of the SIW varies inversely with the operating frequency. In order to obtain lower impedance, the SIW are supposed to operate away from the cutoff frequency. In addition, the low-impedance SIW can be integrated easily with other conventional

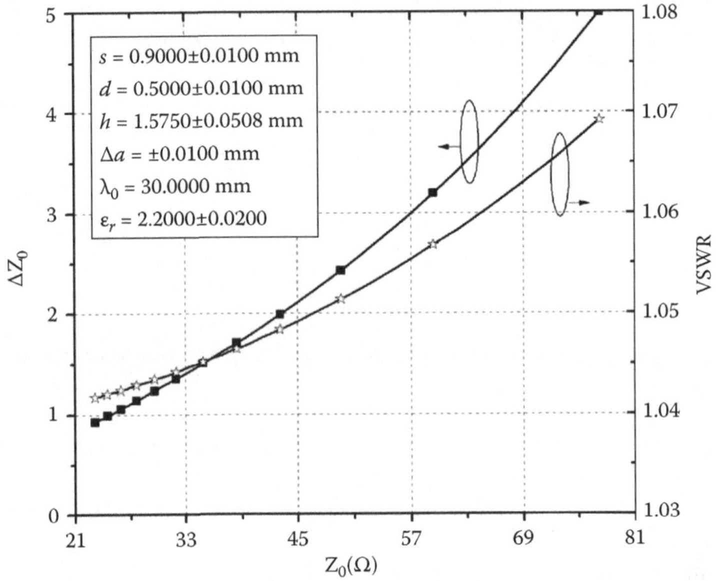

FIGURE 1.6 VSWR due to the tolerances for a 10 GHz SIW.

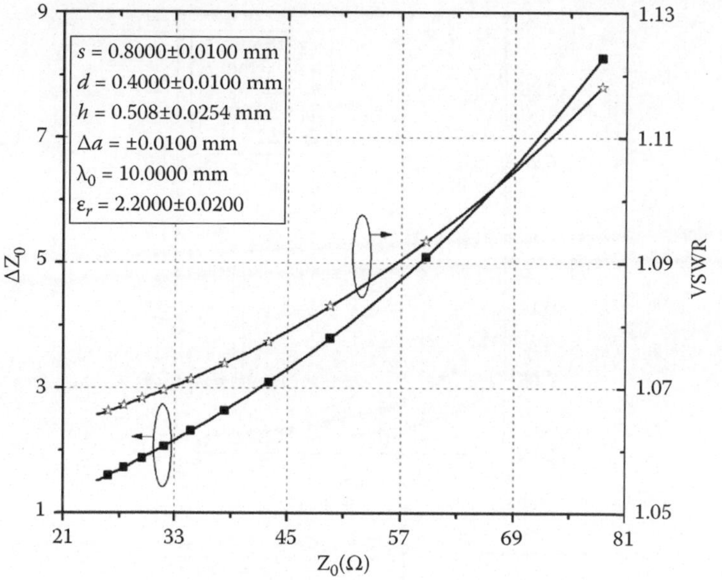

FIGURE 1.7 VSWR due to the tolerances for a 30 GHz SIW.

planar circuits. Compared with the results shown in Figures 1.6 and 1.7, the SIW operated in the high-frequency band is more sensitive to tolerances and causes a higher VSWR under the given fabrication precision.

1.1.2.3 Tolerance on SIW Phase Constant

Phase constant β is of great importance for SIW designs as well. Variations in β result in a change of wave phase, which is a serious problem to those phase-sensitive applications. If a dielectric material with low tan δ is used, the SIW phase constant β will be calculated as

$$\beta = \sqrt{\left(\frac{2\pi\sqrt{\varepsilon_r}f}{300}\right)^2 - \left(\frac{\pi}{w}\right)^2} \tag{1.13}$$

In Equation (1.13), the unit of f is GHz and the unit of w is mm. Tolerances cause β drifting from the designed value. With the same definition of sensitivity, there is

$$\frac{|\Delta\beta|_{\max}}{\beta} = \left|\frac{\Delta a}{a} S_a^{\beta}\right| + \left|\frac{\Delta\varepsilon_r}{\varepsilon_r} S_{\varepsilon_r}^{\beta}\right| + \left|\frac{\Delta s}{s} S_s^{\varphi}\right| + \left|\frac{\Delta d}{d} S_d^{\varphi}\right| \tag{1.14}$$

A complete set of graphs for sensitivities of β with respect to a, ε_r, s, and d as functions of width-to-diameter ratio a/d, diameter-to-distance ratio d/s, and working wavelength λ_0 for a 10 GHz SIW are given in Figure 1.8. Four similar abscissas are defined to plot sensitivity curves of β for a 30 GHz SIW, as shown in Figure 1.9.

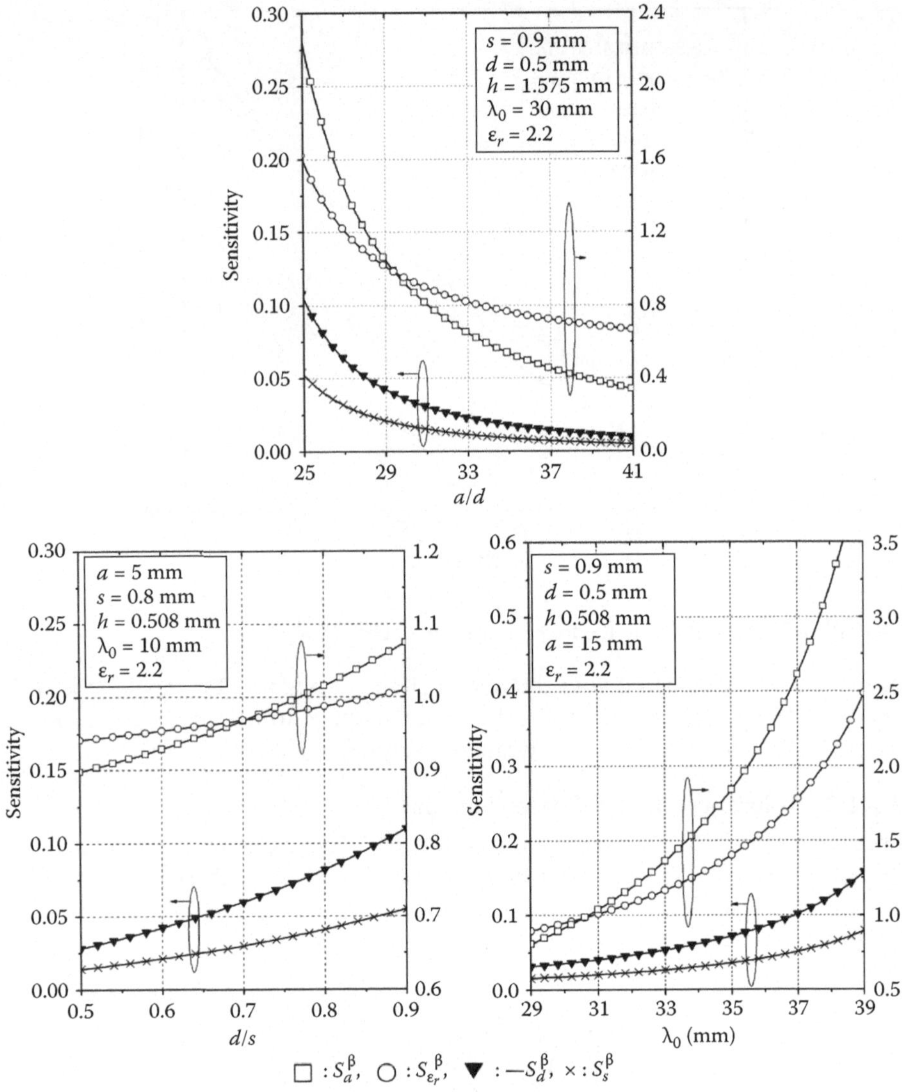

FIGURE 1.8 Phase constant sensitivity curves of a 10 GHz SIW with respect to a, d, s, and ε_r.

Some conclusions obtained from these sensitivity curves are listed as follows:

1. All of the sensitivity curves are smooth and monotone.
2. Absolute values of the sensitivity curves have the same varying tendency; thus, the tolerance effects of different parameters have the same change in the trend.
3. The tolerance effect of d is also twice than that of s all the time.
4. $S_a^{Z_0}$, $S_{\varepsilon_r}^{Z_0}$, and $S_s^{Z_0}$ are positive; thus, β increases with a, ε_r, and s.

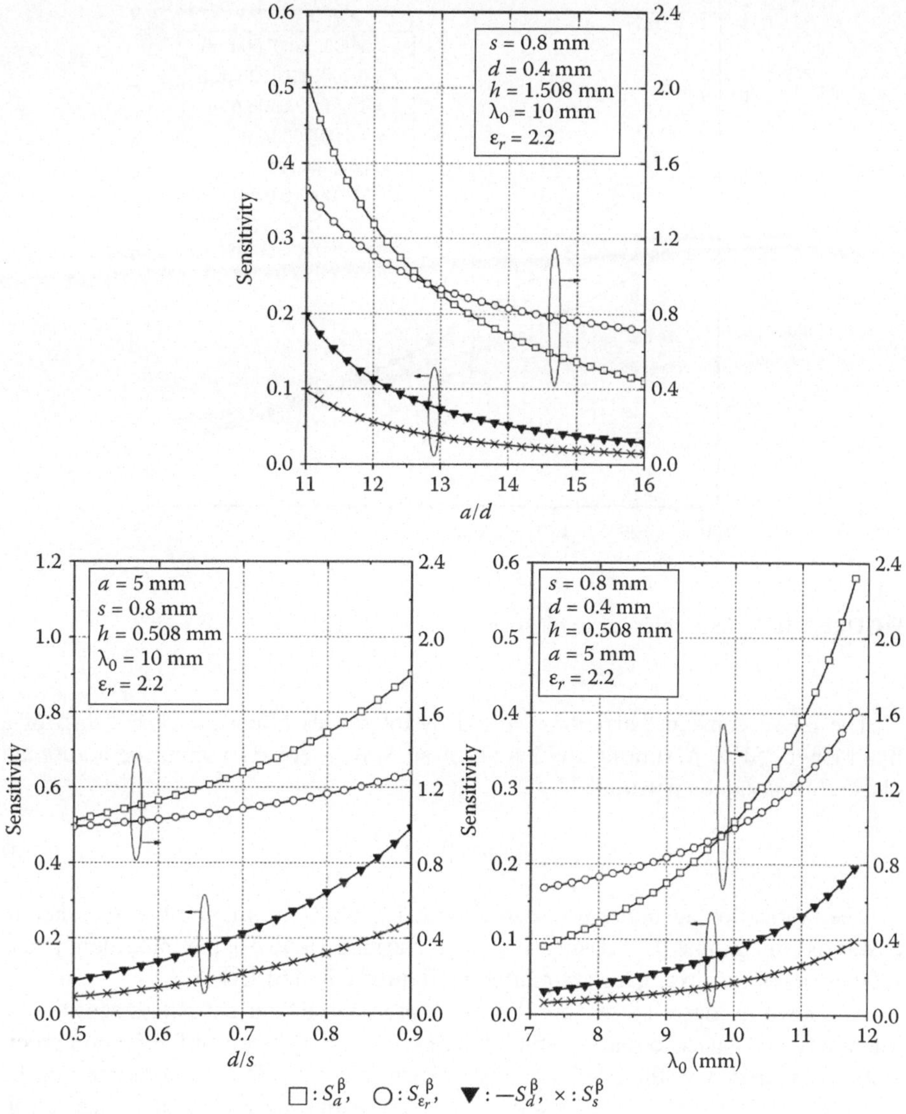

FIGURE 1.9 Phase constant sensitivity curves of a 30 GHz SIW with respect to a, d, s, and ε_r.

5. S_d^β is negative; thus, β decreases with d.
6. Generally speaking, the phase constant of an SIW is more sensitive to tolerances in a and ε_r than those in s and d.
7. A larger a/d contributes to a smaller error, while a smaller d/s contributes to a smaller error.
8. For an SIW, increasing operation frequency contributes to a smaller error.
9. For different SIWs operated at different frequency bands, the tolerance effects of s and d increase with frequency.

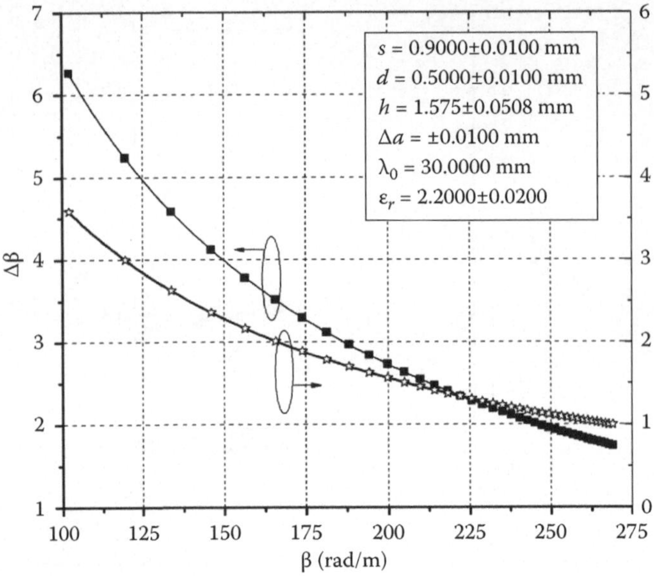

FIGURE 1.10 Phase shift due to the tolerances for a 10 GHz SIW.

The phase constant variations caused by tolerances affect the phase shift of a fixed-length SIW. Assuming the length of an SIW, l, equals 1 mm, the additional phase shift φ can be calculated by

$$\varphi = \beta \times l \tag{1.15}$$

The fabrication accuracy was also assumed to be 0.01 mm, and the tolerance in ε_r was assumed to be 0.02. Using the largest varying value $\Delta\beta$, the maximum phase shift errors of two examples are plotted in Figures 1.10 and 1.11.

As shown in these figures, an SIW with large phase constant is less sensitive to tolerances and easier to reach a small phase error under the given fabrication precision. Compared with the results in Figures 1.10 and 1.11, the SIW operated in the high-frequency band is more sensitive to tolerances and causes a greater phase error.

1.1.2.4 Practical Tolerance Consideration

The tolerances in the substrate thickness and dielectric permittivity are specified by the manufacturer. It is desirable to know the permitted tolerance in dimensions of an SIW for specified values of ΔZ_0 (or VSWR) and $\Delta\beta$ (or phase shift). The alternative problem of the tolerance analysis can be used for determining the required fabrication accuracy.

The accuracy of a given parameter (such as the SIW width a) can be obtained by Equation (1.16) derived from Equation (1.11):

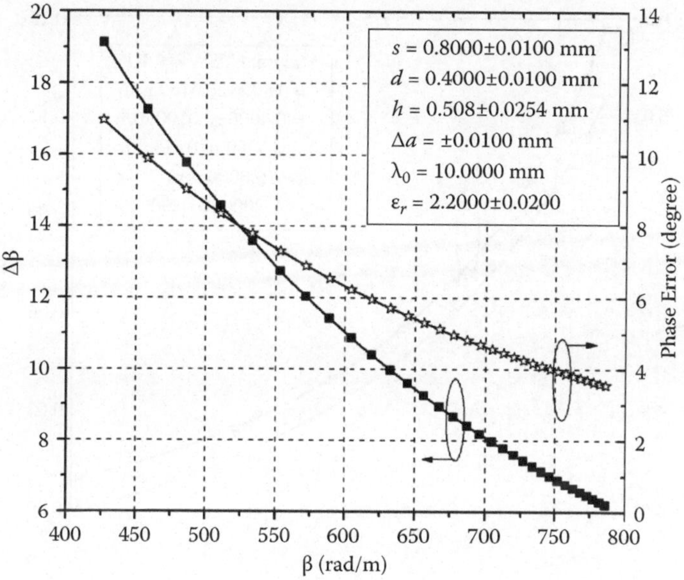

FIGURE 1.11 Phase shift due to the tolerances for a 30 GHz SIW.

$$\frac{|\Delta a|}{a}\left|S_a^{Z_0}\right| = \frac{|\Delta Z_0|}{Z_0} - \left|\frac{\Delta h}{h}S_h^{Z_0}\right| - \left|\frac{\Delta \varepsilon_r}{\varepsilon_r}S_{\varepsilon_r}^{Z_0}\right| - \left|\frac{\Delta s}{s}S_s^{Z_0}\right| - \left|\frac{\Delta d}{d}S_d^{Z_0}\right| \tag{1.16}$$

Here, Δh, $\Delta\varepsilon_r$, Δs, and Δd are the same as provided before. The maximum value of VSWR is set to be 1.1. The required fabrication accuracy of Δa for a 30 GHz SIW is obtained by Equation (1.16) and plotted in Figure 1.12 versus Z_0. It also indicates that an SIW with higher impedance needs a more accurate fabrication and a better substrate to achieve the same VSWR. The required fabrication accuracies of h, ε_r, s, and d can be determined in a similar manner.

Using Equation (1.17), a similar analysis can be used to evaluate the required fabrication accuracy of Δa for a 10 GHz SIW, which keeps the relative phase error below 3%. The result is illustrated in Figure 1.13.

$$\frac{|\Delta a|}{a}\left|S_a^{\beta}\right| = \frac{|\Delta \beta|}{\beta} - \left|\frac{\Delta \varepsilon_r}{\varepsilon_r}S_{\varepsilon_r}^{\beta}\right| - \left|\frac{\Delta s}{s}S_s^{\beta}\right| - \left|\frac{\Delta d}{d}S_d^{\beta}\right| \tag{1.17}$$

The SIW with small β needs a more accurate fabrication processing to achieve the same $\Delta\varphi/\varphi$. If β is below 136.46 rad/m, $\Delta\varphi/\varphi$ will exceed the limit of 3% even Δa equals zero.

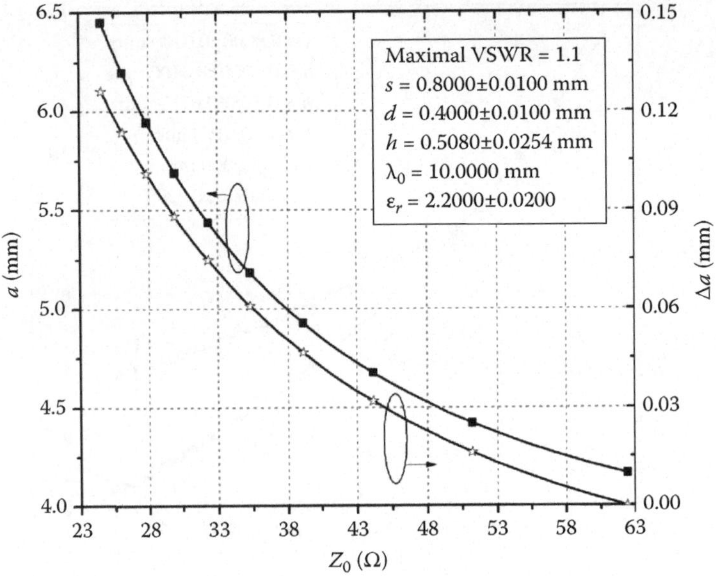

FIGURE 1.12 Tolerance in SIW width allowed for a VSWR value of 1.1.

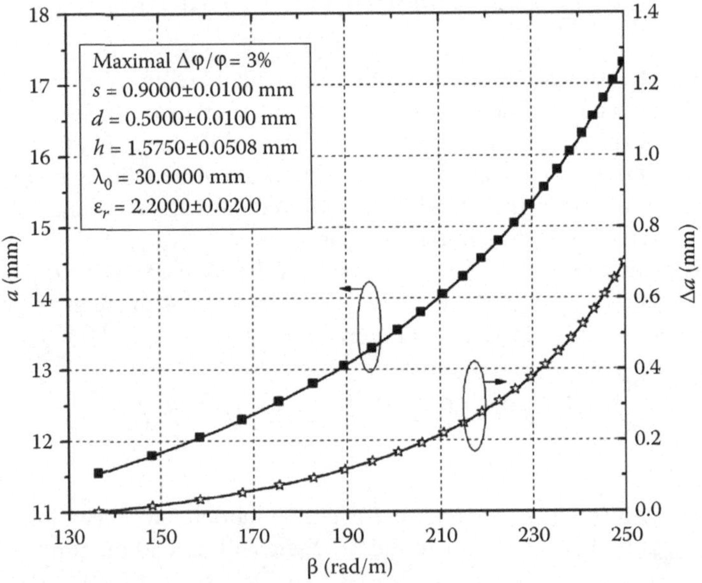

FIGURE 1.13 Tolerance in SIW width allowed for a 3% relative error of phase shift.

1.1.3 SIW Power Handling Capacity

The reliability of transmission lines is affected by the thermal effect, which is a fundamental design issue particularly for planar transmission lines [19]. As a new guided-wave structure, the SIW is investigated in terms of its power handling capability (PHC), including average power handling capability (APHC) and peak power handling capability (PPHC).

To investigate the APHC and PPHC problems at different frequencies, the equivalent width of the SIW, w, is set to be consistent with the operating frequency, f. Equation (1.18) presents the relationship between f and equivalent width w:

$$f = \frac{225}{w\sqrt{\varepsilon_r}} \tag{1.18}$$

It is assumed that each SIW operates at the middle point of each dominant mode zone.

1.1.3.1 Average Power Handling Capability of SIW

The APHC of conventional planar transmission lines has been investigated, such as microstrip line [17], multilayer microstrip line [20], thin-film microstrip line (TFML) [21], and CPW [22]. The conclusion is that these structures are appropriate for relatively low average power applications. Since the SIW has become a popular choice in circuit and antenna applications, its APHC also needs to be studied.

There are several key factors that have effects on the accuracy in characterizing the frequency-dependent APHC of an SIW, which is primarily determined by the temperature rise of the dielectric substrate. They are:

1. Ohmic attenuation
2. Dielectric attenuation
3. Cross-sectional shape
4. Melting temperature T_m or glass transition temperature T_g
5. Ambient temperature T_∞

The frequency-dependent ohmic and dielectric attenuation of the SIW should be considered first. Analytical equations given in [23] for the ohmic and dielectric losses, named α_c and α_d (Np/m), in a rectangular metallic waveguide can be rewritten as

$$\alpha_c = \frac{R_m}{\eta h\sqrt{1-\left(\frac{\lambda_0}{2w\sqrt{\varepsilon_r}}\right)^2}}\left(1+\frac{\lambda_0^2 h}{2w^3\varepsilon_r}\right) \tag{1.19}$$

$$\alpha_d = \frac{\pi\lambda_g}{\lambda^2}\tan\delta = \frac{\pi}{\lambda_0}\sqrt{\varepsilon_r}\frac{1}{\sqrt{1-\left(\frac{\lambda_0}{2w\sqrt{\varepsilon_r}}\right)^2}}\tan\delta \tag{1.20}$$

In Equations (1.19) and (1.20), w can be calculated by Equation (1.1), tan δ represents dielectric loss tangent, σ is metal conductivity, and λ_0 is the wavelength in vacuum. Besides,

$$R_m = \sqrt{\frac{\omega\mu_0}{2\sigma}}$$

$$\eta = \sqrt{\frac{\mu_0}{\varepsilon_0\varepsilon_r}}$$

The propagation properties of the quasi-TE_{10} mode within an SIW are similar to those of the TE_{10} mode within a conventional waveguide. Thus, these attenuation factors can be obtained directly according to these well-established closed-form formulae. Neglecting the radiation loss, the total power attenuation constant α per unit length of the SIW can be divided into two parts: the ohmic and dielectric components.

SIW transmission lines are able to operate over the frequency range of 10–300 GHz on the basis of the existing low-cost dielectric materials and fabrication technology. Dielectric substrates used in these frequency bands have a dielectric loss tangent ranging from 10^{-3} to 10^{-4}. The relative permittivity of these substrates varies from 2 to 13. Furthermore, the dielectric substrate to construct SIW circuits is generally covered with copper claddings having a conductivity $\sigma = 5.8 \times 10^7$ Ω/m.

In order to validate Equations (1.19) and (1.20), two SIWs were fabricated on the Rogers 5880 substrate having a relative permittivity of 2.2 and a thickness of 0.508 mm. One operates at the center frequency of 10 GHz with $a = 15.21$ mm, and the other operates at the center frequency of 25 GHz with $a = 6.21$ mm. A thru-line (TL) calibration technique was used to extract attenuation constants from measurements of two SIWs with different lengths. Results calculated by Equations (1.19) and (1.20) are compared with simulated and measured results. The agreement between the calculated, simulated, and measured results is acceptable. The ohmic and dielectric attenuation constants of a series of SIWs were also calculated over 7 to 50 GHz as described in [19]. It implies that the difference between the calculated and simulated results is small. Besides, the dielectric attenuation is dominant in these cases.

Now, there is a possibility of estimating the APHC of an SIW. The APHC is mainly determined by the temperature rise of the supporting substrate and the metallic cladding because of different losses. Unlike conventional transmission lines, such as microstrip line, the SIW is a uniconductor guided-wave structure, which has the same electric potential for each metallic surface. Thus, the methods described in [17, 20–22] cannot be simply repeated. The way used here is to find out the maximum average power, which makes the temperature of the substrate reach its glass transition or melting point, through a heat-transfer analysis method as described in [24, 25].

Before the analysis, four assumptions are made:

1. The melting point of the metallic cladding is usually much higher than that of the dielectric substrate. For example, the melting point of the copper can be up to 1093°C. Therefore, the glass transition or melting temperature of the substrate is assumed to be responsible for the APHC limitation of an SIW.

2. On the cross section of an SIW, the absorbed power is treated as a uniform heat source. The SIW usually has a large width-to-height ratio and a relatively small transverse size; thus, good heat uniformity can be guaranteed.
3. The temperature almost remains unchanged over the cross section of the SIW. That means the temperature at the sidewalls of the SIW is approximately equal to the temperature at the center.
4. The temperatures of the substrate and the metal cladding are assumed the same. In general, this hypothesis can be satisfied because the thermal conductivity of the metal is excellent.

Heat transfer is a science that aims to predict the energy transfer that takes place between material bodies with a temperature difference [24]. As described in this classical theory, the heat conduct is equal to the sum of the thermal radiation and convection.

$$q_{cond} = q_{rad} + q_{conv} \tag{1.21}$$

Thermal radiation is defined as the electromagnetic radiation that is emitted by a body because of its temperature [24]. Thermodynamic theory shows that energy emitted by an ideal radiator is proportional to the fourth power of the absolute temperature of the body.

Such a heat exchange can be expressed by the Stefan-Boltzmann law of the thermal radiation [24] when an object is enclosed by a very large surface:

$$q_{rad} = h_r A(T_w - T_\infty) \tag{1.22}$$

In Equation (1.22), A (m^2) is the surface area of the object, T_w (K) is the temperature of the dielectric substrate, T_∞ (K) is the temperature of the environment (usually 20°C), and the quantity h_r is called the radiation heat-transfer coefficient.

$$h_r = \xi\tau\left(T_w^2 + T_\infty^2\right)(T_w + T_\infty) \tag{1.23}$$

Constant τ is equal to 5.669×10^{-8} W/m^{2}K^4, and ξ is the emissivity of the radiating surface. The emissivity of copper cladding is equal to 0.65.

To express the effect of convection, Newton's law of cooling is used [24]:

$$q_{conv} = h_c A(T_w - T_\infty) \tag{1.24}$$

The quantity h_c is called the convection heat transfer coefficient. For a horizontal heating plate, the heated plate facing upward or the cooled plate facing downward has

$$h_c^u = 1.32\left(\frac{T_w - T_\infty}{w}\right)^{1/4} \tag{1.25}$$

On the other hand, the heated plate facing downward or the cooled plate facing upward has

$$h_c^d = 0.61\left(\frac{T_w - T_\infty}{w^2}\right)^{1/5} \tag{1.26}$$

That means the top and bottom conductor claddings of an SIW have different convection heat-transfer coefficients.

When a temperature gradient exists in a body, energy will be transferred by conduction from a higher-temperature region to a lower-temperature one. The propagating power is supposed to be P_0 at $z = 0$. After propagating along the guided-wave structure, the power decays as

$$P = P_0 e^{-2\alpha z} \tag{1.27}$$

After first-order derivation calculus for z, the power loss per unit length is equal to the propagating power decrement along the z-direction. That is,

$$\Delta P = \frac{\partial P}{\partial z} = -2\alpha P \tag{1.28}$$

Here, a very useful relationship between q_{cond} and P can be obtained:

$$q_{cond} = \Delta P = 2\alpha P \Delta z \tag{1.29}$$

Considering the large width-to-height ratio of an SIW, the heat conduction through lateral walls of the SIW can be neglected. Therefore,

$$q_{cond} = \left(h_c^u + h_c^d + 2h_r\right) A(T_w - T_\infty) = 2\alpha P \Delta z \tag{1.30}$$

The equality relationship presented in Equation (1.30) can be rewritten as

$$P = \frac{\left(h_c^u + h_c^d + 2h_r\right) w(T_w - T_\infty)}{2\alpha} \tag{1.31}$$

Based on the above detailed mathematical description, numerical calculations can be performed for the APHC discussion of an SIW. Substituting the attenuation values calculated by Equations (1.19) and (1.20) and full-wave simulations into Equation (1.31), the maximum APHCs of SIWs over 7 to 14 GHz and 18 to 34 GHz can be estimated as shown in Figures 1.14 and 1.15. There exist differences between the APHC values owing to different attenuation constants obtained from different methods. The calculated result can be viewed as the limiting value of APHC.

Figure 1.16 illustrates the calculated maximum APHC for different common dielectric materials, the properties of which are given as in [17, 26]. The relative

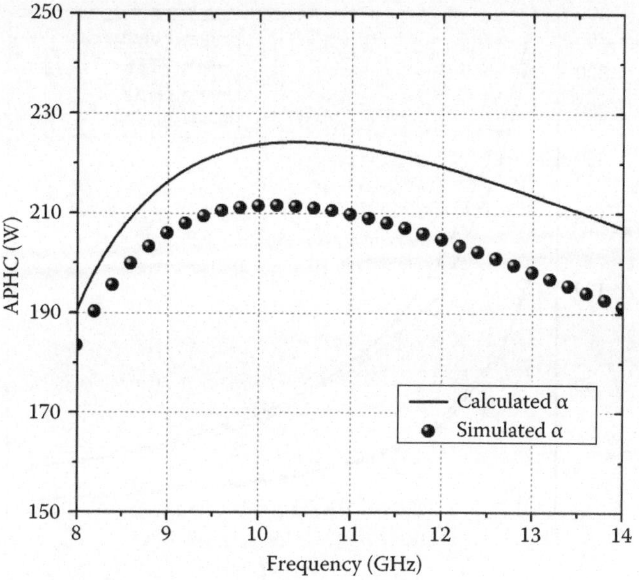

FIGURE 1.14 Calculated and simulated APHCs of an SIW with $a = 15.21$ mm.

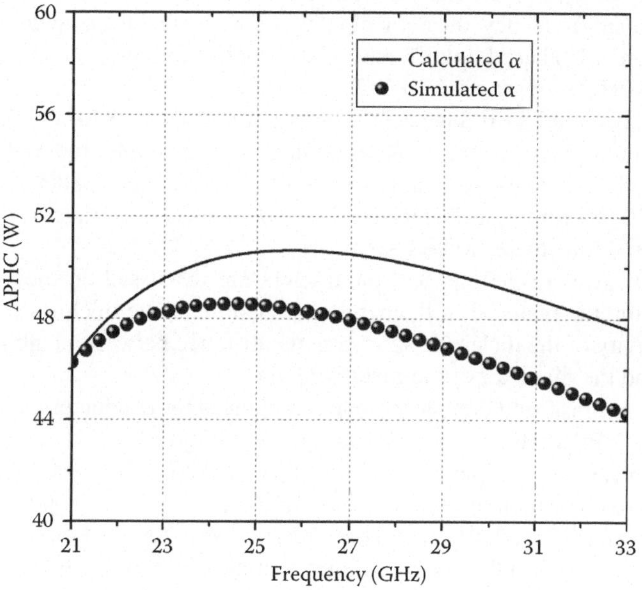

FIGURE 1.15 Calculated and simulated APHCs of an SIW with $a = 6.21$ mm.

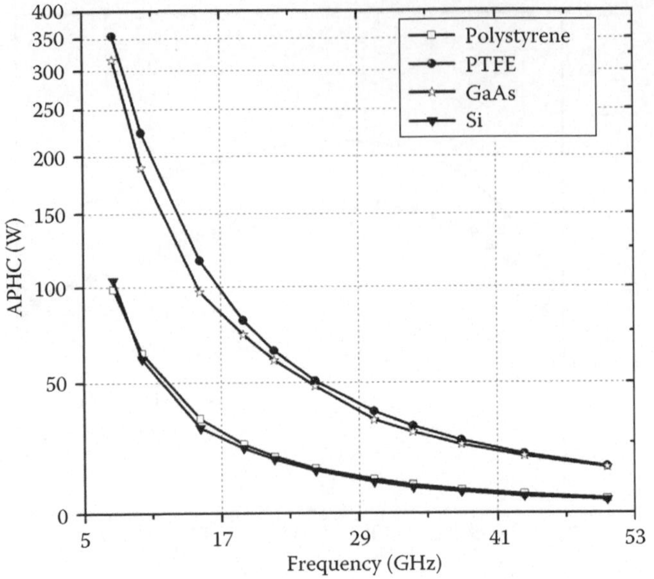

FIGURE 1.16 Comparison of APHCs for SIWs based on different dielectric materials ($s = 0.8$ mm, $d = 0.4$ mm, $h = 0.508$ mm, $\varepsilon_r = 2.2$).

permittivities of polystyrene, polytetrafluoroethylene (PTFE), GaAs, and Si are 2.53, 11.7, 11.7, and 12.3, respectively. The loss tangents of polystyrene, PTFE, GaAs, and Si are 0.00047, 0.0005, 0.00016, and 0.0005, respectively. The heat distortion temperatures of polystyrene, PTFE, GaAs, and Si are 100, 200, 300, and 200°C, respectively. At a given frequency, the equivalent width w should be changed in compliance with Equation (1.18) based on different relative permittivities. It can be seen that the maximum APHC is observed for PTFE.

As a widely used PTFE dielectric substrate, the Rogers 5880 substrate is able to support medium power in the SIW technology. For example, the SIW can handle 225 W at 10 GHz or 37 W average power at 30 GHz on Rogers 5880 substrate with a thickness of 0.508 mm. It is able to satisfy the demands for average power in general microwave and millimeter-wave systems.

Next, influences of geometrical parameters are discussed on the APHC of the SIW, including the material of the conductor claddings, the thickness, and the relative permittivity of the dielectric substrate, the distance between neighboring metalized vias, and the diameter of the metalized via.

Aluminum, brass, and copper are popular choices for conductor claddings. Their influence on SIW APHC is discussed as shown in Figure 1.17. The conductivities of aluminum, brass, and copper are 3.72×10^7 Ω/m, 1.57×10^7 Ω/m, and 5.80×10^7 Ω/m, respectively. The emissivities of aluminum, brass, and copper can be found in [27]. They are 0.765, 0.610, and 0.650, respectively. As illustrated in Figure 1.17, the copper cladding handles the maximum average power and the aluminum comes second. Thus, the remaining APHC problems are investigated based on the copper cladding.

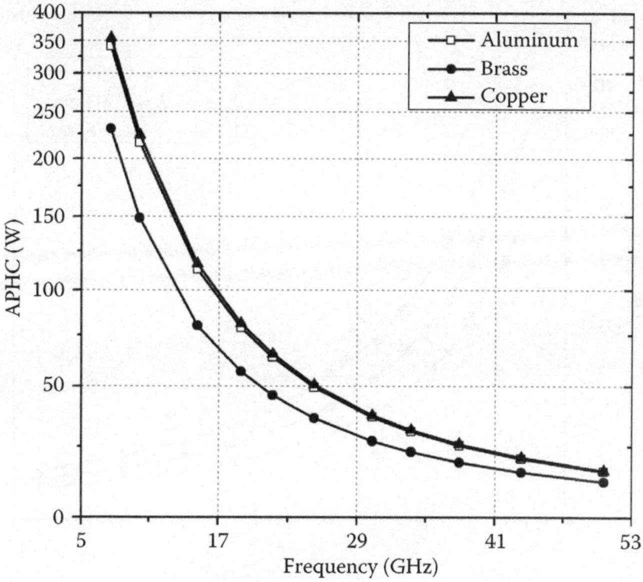

FIGURE 1.17 Comparison of APHCs for SIWs with different metal claddings ($d = 0.4$ mm, $s = 0.8$ mm, $h = 0.508$ mm, $\varepsilon_r = 2.2$).

The effects of the substrate thickness h and the relative permittivity ε_r on the APHC of the SIW are shown in Figures 1.18 and 1.19. The conclusion is that the thinner SIW is, the smaller APHC is.

For example, at $f = 10$ GHz:

$$\frac{P_{APHC}\big|_{h=0.254\,\text{mm}}}{P_{APHC}\big|_{h=0.508\,\text{mm}}} \approx 0.673 \tag{1.32}$$

Besides, at $f = 30$ GHz:

$$\frac{P_{APHC}\big|_{h=0.254\,\text{mm}}}{P_{APHC}\big|_{h=0.508\,\text{mm}}} \approx 0.742 \tag{1.33}$$

Meanwhile, increasing relative permittivity ε_r lowers the SIW APHC because of an effective increasing of w.

For example, at $f = 25$ GHz:

$$\frac{P_{APHC}\big|_{\varepsilon_r=4.4}}{P_{APHC}\big|_{\varepsilon_r=2.2}} \approx 0.646$$
$$\frac{P_{APHC}\big|_{\varepsilon_r=10.2}}{P_{APHC}\big|_{\varepsilon_r=6.6}} \approx 0.727 \tag{1.34}$$

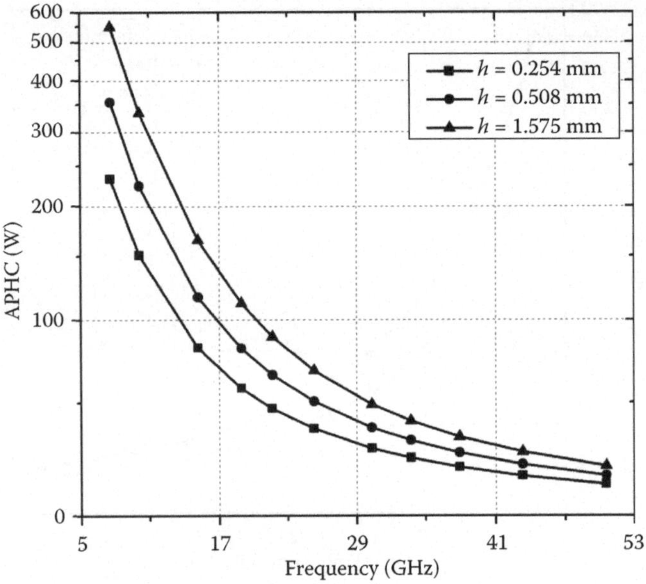

FIGURE 1.18 Comparison of APHCs of SIWs with different substrate thicknesses h (d = 0.4 mm, s = 0.8 mm, ε_r = 2.2).

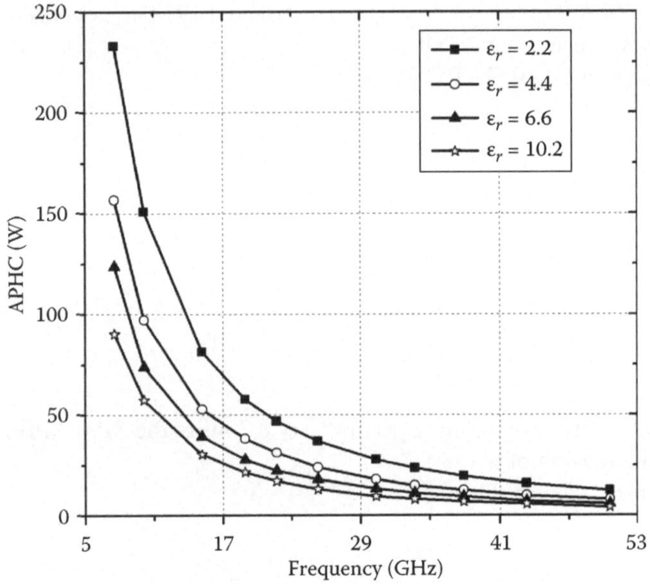

FIGURE 1.19 Comparison of APHCs of SIWs with different relative permittivities ε_r (d = 0.4 mm, s = 0.8 mm, h = 0.254 mm).

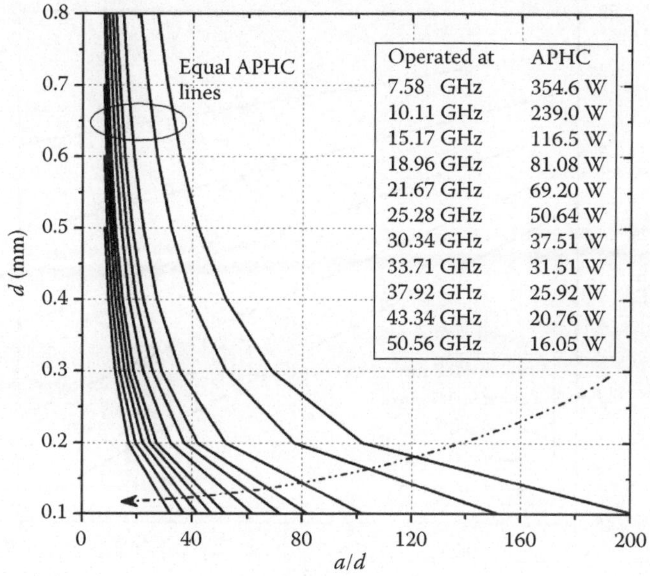

FIGURE 1.20 Comparison of APHCs of SIWs with different a/d and d ($d/s = 0.5$, $h = 0.508$ mm, $\varepsilon_r = 2.2$).

Moreover, a/d has to be larger than 8.0 and $0.4 < d/s < 1.0$ after considering limited manufacturing techniques for SIW circuits [13]. Influences of the distance between neighboring metalized vias s and the diameter of the metalized vias d on SIW APHC will be investigated within these limitations. Figure 1.20 gives a series of equal APHC lines. Each one shows an APHC value at a specific frequency. The APHC decreases with the operating frequency, following the arrow. When a/d is equal, a larger d increases the APHC. When d is equal, a larger a/d increases the APHC. Figure 1.21 shows the APHC versus d/s with different diameters of the metalized via. These SIWs were designed at 25 GHz with $a = 6.21$ mm. Two conclusions can be drawn:

1. A larger d/s is accompanied by a smaller APHC at a given d.
   For example, when $d = 0.4$, with different d/s,

$$\frac{P_{APHC}\big|_{d/s=0.5,s=0.8\text{mm}}}{P_{APHC}\big|_{d/s=0.4,s=1\text{mm}}} \approx 0.988$$
$$\frac{P_{APHC}\big|_{d/s=0.6,s=0.67\text{mm}}}{P_{APHC}\big|_{d/s=0.5,s=0.8\text{mm}}} \approx 0.988 \tag{1.35}$$

   This conclusion can be restated as the APHC increases with s.

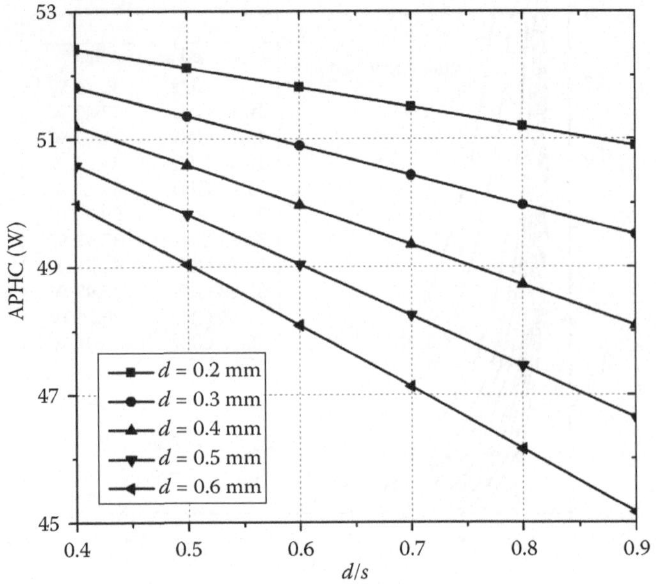

FIGURE 1.21 Comparison of APHCs of SIWs with different d/s and d (a = 6.21 mm, h = 0.508 mm, ε_r = 2.2, operated at 25 GHz).

2. When d/s is unchanged, the APHC increases monotonously by reducing d. For example, when $d/s = 0.5$,

$$\frac{P_{APHC}\big|_{d=0.2\text{mm},s=0.4\text{mm}}}{P_{APHC}\big|_{d=0.3\text{mm},s=0.6\text{mm}}} \approx 1.015$$

$$\frac{P_{APHC}\big|_{d=0.3\text{mm},s=0.6\text{mm}}}{P_{APHC}\big|_{d=0.4\text{mm},s=0.8\text{mm}}} \approx 1.015 \tag{1.36}$$

It indicates that the APHC decreases with s.

Such conclusions seem to disagree with each other. Decreasing s reduces the APHC, while decreasing d increases the APHC. However, the influence of d is greater than that of s. Therefore, the influence of s cannot be considered alone. The effects of d and s should be combined together to ascertain any change in APHC.

1.1.3.2 Peak Power Handling Capability of SIW

There exists a similar limitation for peak power consideration. When the electric field within the dielectric exceeds a specific critical value, dielectric breakdown will happen. The APHC discussed above is the issue of the continuous wave power, while the PPHC can be considered the issue of the pulsed-wave power. As the SIW systems may be used in microwave and millimeter-wave high-pulse-power applications, it seems necessary to discuss the PPHC of an SIW as well.

Similar to the conventional rectangular waveguide, it is necessary to relate the maximum electric field E_0 within an SIW to the transmitted power so as to determine the maximum peak power, which can be handled before breakdown happens. The electric field within the SIW operating in the dominant TE_{10} mode is

$$E_y = E_0 \sin\frac{\pi x}{w} \tag{1.37}$$

In Equation (1.37), E_0 is the peak value of the electric field. The transmitted power is

$$P = \frac{1}{2Z}\int_0^h \int_0^w E_y^2 dxdy \tag{1.38}$$

with impedance Z being

$$Z = \frac{\sqrt{\dfrac{\mu_0}{\varepsilon_0 \varepsilon_r}}}{\sqrt{1-\left(\dfrac{\omega_c}{\omega}\right)^2}} \tag{1.39}$$

The transmitted peak power can be calculated by combining Equations (1.37) to (1.39):

$$P = \frac{\sqrt{1-\left(\dfrac{\omega_c}{\omega}\right)^2}}{480\pi} wh\sqrt{\varepsilon_r} E_0^2 \tag{1.40}$$

When E_0 reaches the critical value, the dielectric substrate is punctured and this value is PPHC. The PPHC of an SIW can be calculated by Equation (1.40) with E_0 replaced by the dielectric strength of the corresponding materials. For air, its dielectric strength is 3 kV/mm. This value for a dielectric is much higher. For polystyrene, its dielectric strength is 24 kV/mm. For PTFE, its dielectric strength is 17 kV/mm. For GaAs, its dielectric strength is 24 kV/mm.

As shown in Equation (1.40), the PPHC of an SIW is mainly determined by the dielectric strength of the used dielectric material and the substrate thickness at a given frequency. Absolutely, a thicker substrate has the higher PPHC. Besides, an SIW should be designed to operate in the higher region of the dominant mode zone. That is because the PPHC is inversely proportional to the factor ω_c/ω as described in Equation (1.40). Full-wave simulations have been used to validate Equation (1.40). Simulated results agree well with calculated results.

It implies that the SIW has really high peak power capability in both microwave and millimeter-wave bands. As shown in Figure 1.22, the SIW can handle 807.23 kW at 10 GHz or 269.08 kW peak powers at 30 GHz with a thickness of 0.254 mm.

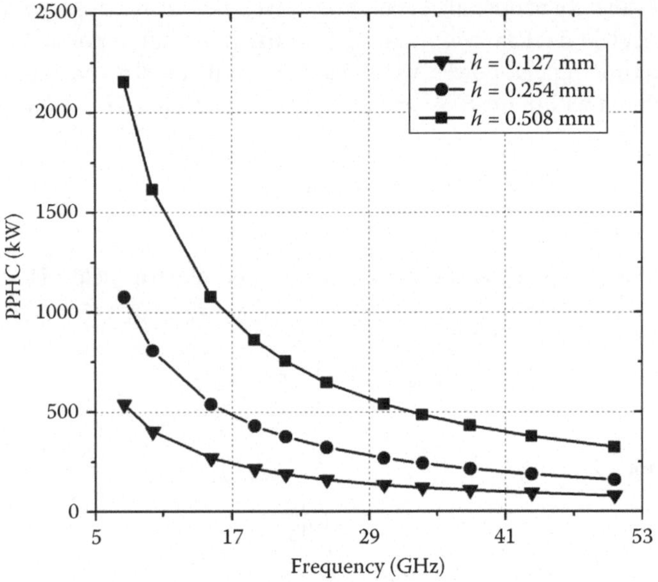

FIGURE 1.22 Comparison of PPHCs of SIWs with different substrate thicknesses h ($s = 0.8$ mm, $d = 0.4$ mm, $\varepsilon_r = 2.2$).

There is an interesting phenomenon, as shown in Figure 1.23. It is that SIW PPHCs are almost independent of different relative permittivities. A larger relative permittivity will increase the PPHC, while a larger relative permittivity associated with a smaller w will decrease the PPHC. These interactions are almost canceled out.

We also examine the SIW PPHC in connection with influences of a, d, and s within the same limitations. Figure 1.24 gives a series of equal PPHC lines. Each one shows a given PPHC value at a specific frequency. The arrow points to the smaller PPHC and the higher frequency. When d becomes equal, a smaller a/d will decrease the PPHC. When a/d becomes equal, a smaller d will decrease the PPHC as well. Figure 1.25 shows the PPHC as a function of d/s with different diameters of the metalized via. These SIWs were designed at 25 GHz with $a = 6.21$ mm. It can be concluded that a smaller d/s leads to a larger PPHC with the same d. It also means that a larger s leads to a larger PPHC. Meanwhile, a larger d leads to a smaller PPHC when d/s remains unchanged.

1.1.3.3 Power Handling Capability of SIW Bends

Until now, the investigation of the SIW PHC has been presented for matched lines. If the transmission line is not matched, the power distribution will be nonuniform along the line because of standing waves. The APHC of the transmission line is decreased as a result of the nonuniform heat dissipation. When the SIW is used in the design of matching networks and passive components, the APHC is determined at the input point of where the signal enters and is strongest.

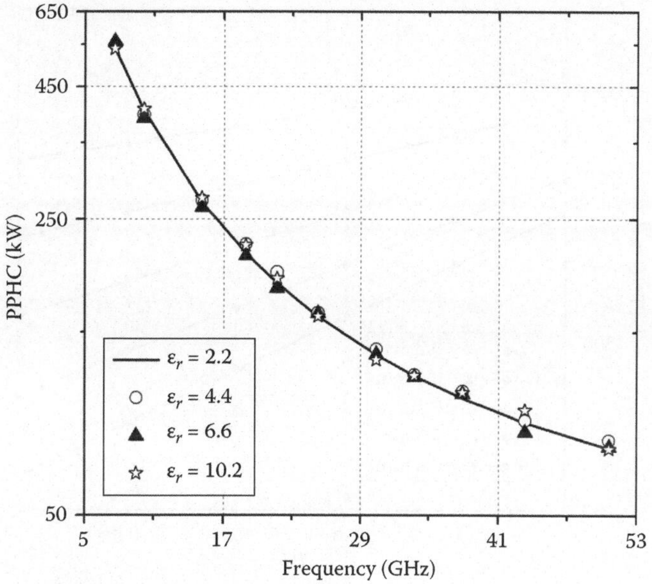

FIGURE 1.23 Comparison of PPHCs of SIWs with different relative permittivities ε_r ($d = 0.4$ mm, $s = 0.8$ mm, $h = 0.127$ mm).

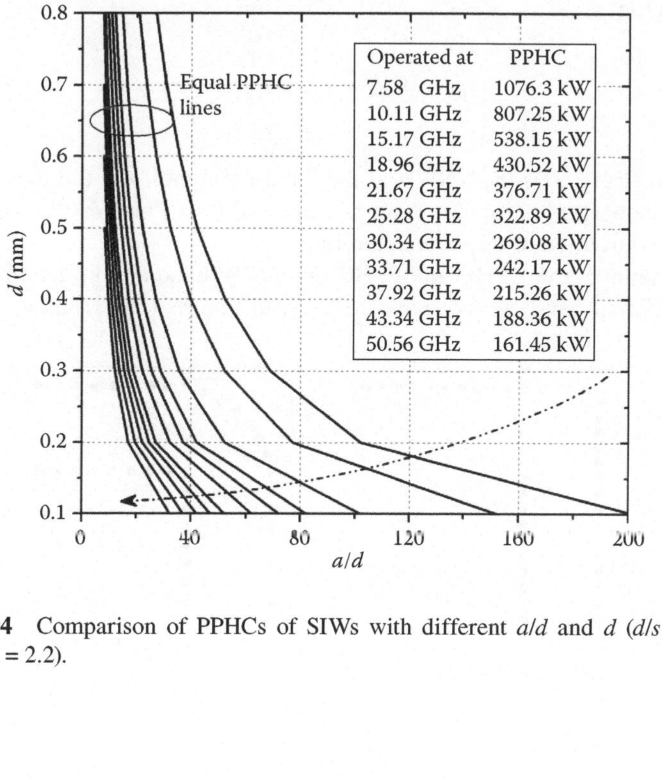

FIGURE 1.24 Comparison of PPHCs of SIWs with different a/d and d ($d/s = 0.5$, $h = 0.254$ mm, $\varepsilon_r = 2.2$).

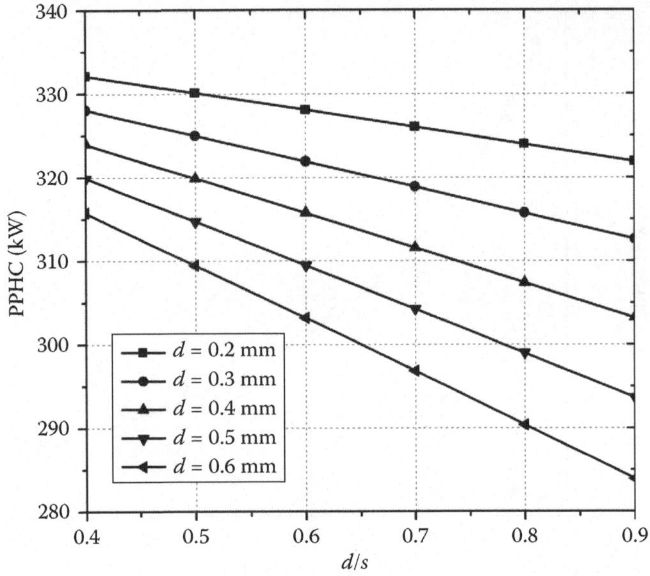

FIGURE 1.25 Comparison of PPHCs of SIWs with different d/s and d (a = 6.21 mm, h = 0.254 mm, ε_r = 2.2, operated at 25 GHz).

To analyze this issue, the attenuation constant, α_m, is defined in [20] for the unmatched line:

$$\alpha_m = 2\alpha \frac{\left(S_i^2 + 1\right)}{\left(S_i + 1\right)^2} \tag{1.41}$$

In Equation (1.41), S_i is the VSWR at the input point and α is the total attenuation constant. The best α_m is equal to α for the matched line. The worst α_m is equal to 2α for the short-circuited or open-circuited line.

As an example, the APHC and PPHC of an SIW bend are discussed here. There are two different bend configurations, shown in Figure 1.26. One has a metalized

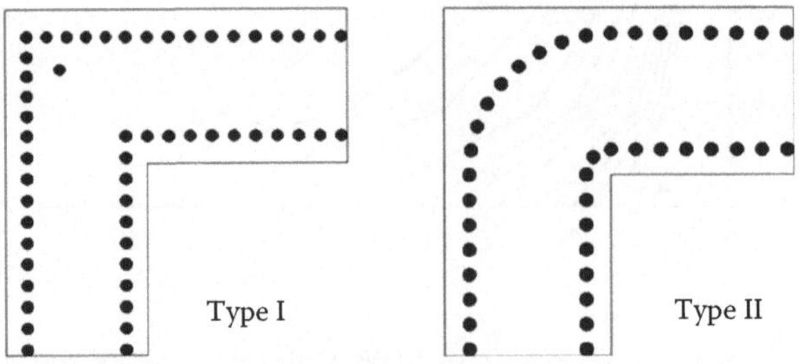

FIGURE 1.26 Different SIW bend structures.

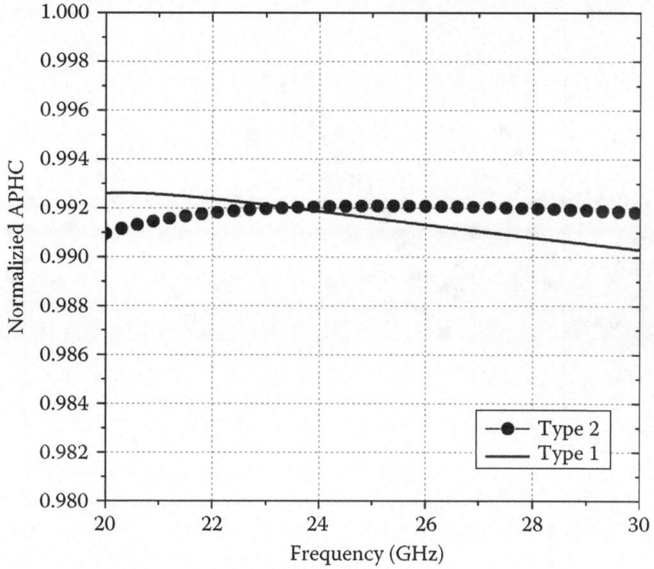

FIGURE 1.27 Normalized APHCs of two types of SIW bends.

via at the corner of a right sidewall, and the other is a round bend. The SIW width is unchanged throughout the bends. Both of them were designed to operate at 25 GHz on a substrate with $\varepsilon_r = 2.2$. Thus, Equation (1.31) can be used to calculate the APHC of the bend by simply substituting α_m for α.

Based on the simulated S-parameters over 20 to 30 GHz, Figure 1.27 shows the related normalized APHCs. It can be observed that the APHCs of them are almost the same. The APHC of a straight SIW is used for reference. The discontinuities lower the maximum average power slightly because of the good matching of these SIW bends.

It is known the PPHC is primarily determined by the peak electric field inside the SIW. The discontinuities distort the electric field and then affect the PPHC. The same SIW bends shown in Figure 1.26 are used to analyze this problem. As shown in Figure 1.28, such discontinuities have remarkable effects on the maximum peak power. Besides, there are some fluctuations appearing in PPHC curves. The results depicted in Figure 1.28 also demonstrate that two types of bend have a varying tendency similar to that of PPHC.

1.1.4 Temperature Effect on SIW

Similar to those conventional planar transmission lines, the characteristic of an SIW is also sensitive to the change of environment temperature. Recently, SIW technology has been applied in various fields, and some of them had to work in the environment of high or low temperature. As such, it is essential to carefully consider the temperature effect on SIW in order to evaluate a system's performance and reliability.

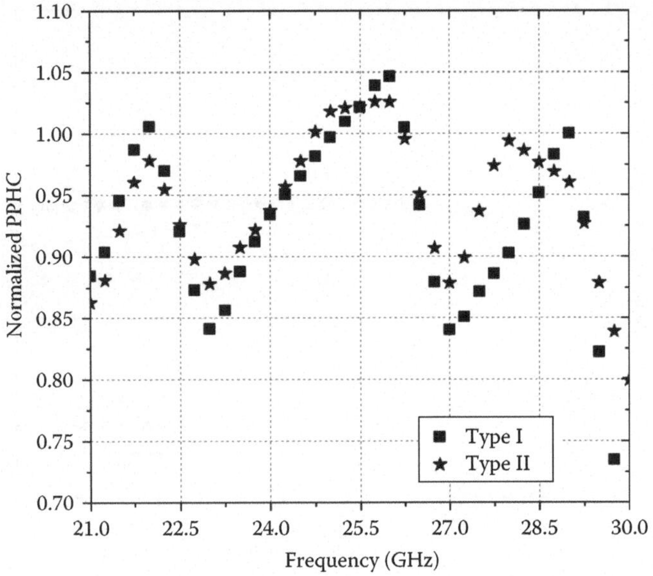

FIGURE 1.28 Normalized PPHCs of two types of SIW bends.

The thermal coefficient of relative permittivity (TCK) is a measure of how ε_r is affected by the change in temperature [29]. The change of ε_r can be expressed with TCK:

$$\varepsilon_r(T+\Delta T)=\varepsilon_r(T)(1+TCK\times\Delta T) \tag{1.42}$$

Changes in the relative permittivity affect the characteristics of SIWs, thereby impacting the performance of circuits.

Several measured (1 + TCK × ΔT) curves for common commercial dielectric substrates are depicted in Figure 1.29, which reveals the discrepancies between actual TCKs and those quoted in the manufacturer's data sheet [26]. Therefore, materials should be measured in a specialized laboratory to obtain accurate temperature properties. This step is very important in the practical design.

As shown in Figure 1.29, ε_r of dielectric laminates may increase, decrease, or remain nearly unchanged with temperature depending on the used dielectric material, such as PTFE glass, hydrocarbon ceramic, and PTFE ceramic.

Here, the investigated substrates are divided into three categories with different variation tendencies of ε_r versus temperature. All of them have the relative permittivity of 2.2 and the loss tangent of 0.0009 at 23°C. Substrate A is a commonly used PTFE laminate, i.e., RT/duroid 5880, having the measured TCK shown in Figure 1.29. That means its ε_r decreases with temperature. To make a comparison, two other substrate materials are presented. Substrate B (TCK(B) = –TCK(A)) is just the reverse of class A, which is characterized by ε_r, whose value increases with temperature. Substrate C is ideal material whose ε_r keeps unchanged over the temperature range of concern, i.e., TCK(C) = 0 ppm/°C.

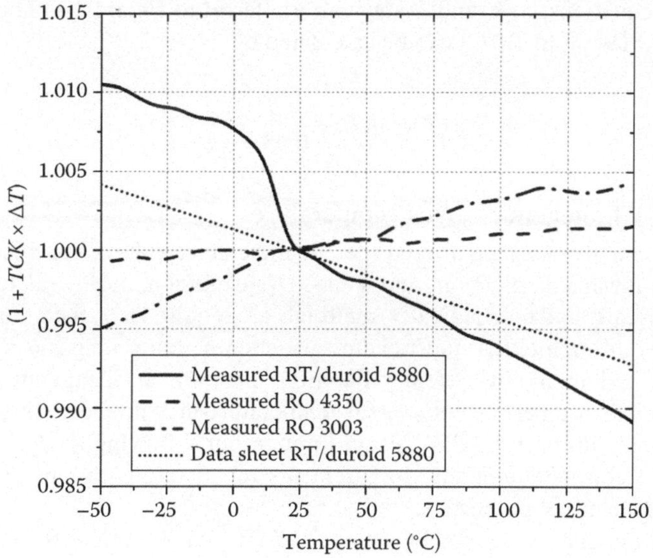

FIGURE 1.29 Measured (1 + TCK × ΔT) curves of different commercial dielectric substrates.

The coefficient of thermal expansion (CTE) means the fractional change in length of a material for a given unit change of temperature [29]. It is another important factor when the variation in temperature is unavoidable.

PTFE material is recognized for providing outstanding electrical properties at high frequencies. However, PTFE-based laminates have a relatively high CTE. For example, their typical *z*-direction CTE values are on the order of 200 ppm/°C [30]. Ideally, the CTE along the *z*-axis should be matched to the bonded metal to avoid failures of plated through-holes. Copper has a CTE, CTE(c), of approximately 17 ppm/°C, which is significantly lower than CTEs of PTFE laminates. Therefore, there will be a reliability problem with the metalized vias of SIW circuits manufactured in PTFE laminates, when exposed to large temperature variation. As such, the ceramic-PTFE composite laminate is a better choice [28].

To simplify the discussion, the CTE in the *z*-axis of a copper-plated through-hole is viewed to be the same as that of laminate. The CTEs of substrate A in different axes are 31 ppm/°C for the *x*-axis, 48 ppm/°C for the *y*-axis, and 237 ppm/°C for the *z*-axis according to the data sheet [26]. The axis orientation is the same as that in Figure 1.1. Let substrates B and C have the same CTE values along the *xyz*-axis. Thus, the temperature change in SIW dimensions can be expressed as

$$a(T + \Delta T) = a(T)(1 + CTE(y) \times \Delta T) \tag{1.43}$$

$$s(T + \Delta T) = s(T)(1 + CTE(x) \times \Delta T) \tag{1.44}$$

$$h(T + \Delta T) = h(T)(1 + CTE(z) \times \Delta T) \tag{1.45}$$

$$d(T + \Delta T) = d(T)(1 + CTE(c) \times \Delta T) \tag{1.46}$$

In these equations, a, s, and d are with a referral to Figure 1.1. The equivalent width of the SIW, w, at T (°C) can be calculated by

$$w(T) = a(T) - \frac{d^2(T)}{0.95 \times s(T)} \tag{1.47}$$

1.1.4.1 SIW Impedance versus Temperature

Circuits are usually designed around specific impedance values. Variation in impedance causes unwanted reflections and reduced transmission. In general, temperature-stable (electrically and mechanically) materials are required to satisfy the application requirements of circuits that must endure wide temperature ranges (such as –50 to 150°C) and still maintain system reliability. However, an ideal laminate having unchanged dielectric permittivity versus temperature may not be the best choice for the temperature-insensitive SIW design through our following analysis. There are some special considerations and requirements for materials due to SIW's unique physical and electrical properties.

The SIW impedance, Z_0, for TE_{10} mode at T (°C) can be calculated as

$$Z_0(T) = \frac{h(T)}{w(T)} \frac{120\pi}{\sqrt{\varepsilon_r(T) - \left(\frac{\lambda_0}{2w(T)}\right)^2}} \tag{1.48}$$

With the change in temperature, the SIW's actual parameters deviate from the designed values. Changes in these parameters contribute to the variation in Z_0. The separated temperature effect of each variable on Z_0 is studied at first, namely, the impact of temperature dependence of each SIW design variable on Z_0 taken one at a time with the remaining variables held constant. It can be found that the temperature effect of a, h, or ε_r on the SIW impedance is far greater than that of s or d. Thus, the temperature property of Z_0 is related to CTE(y), CTE(z), and TCK, and nearly independent of CTE(x) and CTE(c).

Next, the comprehensive temperature effect of different variables is discussed. A larger h contributes to a larger Z_0, a smaller a contributes to a larger Z_0, and a smaller ε_r contributes to a larger Z_0. As a general rule, solids expand as the temperature increases and shrink when they are cooled. Thus, the values of h and a always increase with temperature. But, TCK values may be positive or negative depending on different dielectric materials, as shown in Figure 1.29. Therefore, such complex temperature effects on Z_0 may be heightened or weakened.

Now, a thorough study focuses on the issue of temperature self-compensating, which aims to get the required TCK with a known CTE. Here, a uniform temperature distribution across the entire SIW structure is assumed.

According to Equation (1.48), the impedance of the SIW at $T + \Delta T$ (°C) is $Z_0(T + \Delta T)$. If

$$Z_0(T) = Z_0(T + \Delta T) \tag{1.49}$$

such an SIW has temperature-independent impedance. Substitution of Equation (1.48) into Equation (1.49) yields:

$$\frac{\varepsilon_r(T+\Delta T)}{\varepsilon_r(T)} = C_1\left(\frac{h(T+\Delta T)}{h(T)} \times \frac{w(T)}{w(T+\Delta T)}\right)^2 + C_2\left(\frac{w(T)}{w(T+\Delta T)}\right)^2 \tag{1.50}$$

In Equation (1.50),

$$C_1 = 1 - \frac{\lambda_0^2}{4w^2(T)\varepsilon_r(T)} \qquad C_2 = \frac{\lambda_0^2}{4w^2(T)\varepsilon_r(T)} \tag{1.51}$$

C_1 and C_2 take values between 0 and 1, and their sum is equal to 1.

1. If the effect of CTE in the *y*-axis (varied *w*) is far greater than that in the *z*-axis (varied *h*), Equation (1.50) can be simplified to be

$$\frac{\varepsilon_r(T+\Delta T)}{\varepsilon_r(T)} \approx \left(\frac{w(T)}{w(T+\Delta T)}\right)^2 < 1 \tag{1.52}$$

   Therefore, a negative TCK mitigates the effect of temperature variation.
2. If the effect of CTE in the *y*-axis is much smaller than that in the *z*-axis, Equation (1.50) can be simplified to be

$$\frac{\varepsilon_r(T+\Delta T)}{\varepsilon_r(T)} \approx C_1\left(\frac{h(T+\Delta T)}{h(T)}\right)^2 + C_2 > C_1 + C_2 = 1 \tag{1.53}$$

   Therefore, a positive TCK mitigates the effect of temperature variation.

For these investigated substrates A to C, the CTE(h) is far greater than CTE(w). Thus, the positive TCK is preferred. The temperature-dependent SIW impedances on substrates A to C were calculated considering all temperature effects. As shown in Figures 1.30 and 1.31, substrate B has the best temperature stability, which coincides with the above inference.

Now, we conduct a further study to realize an SIW having the temperature-independent impedance. As an example, for the 30 GHz SIW with an invariable impedance of 37.9 Ω, the required ε_r can be calculated by Equations (1.48) and (1.49)

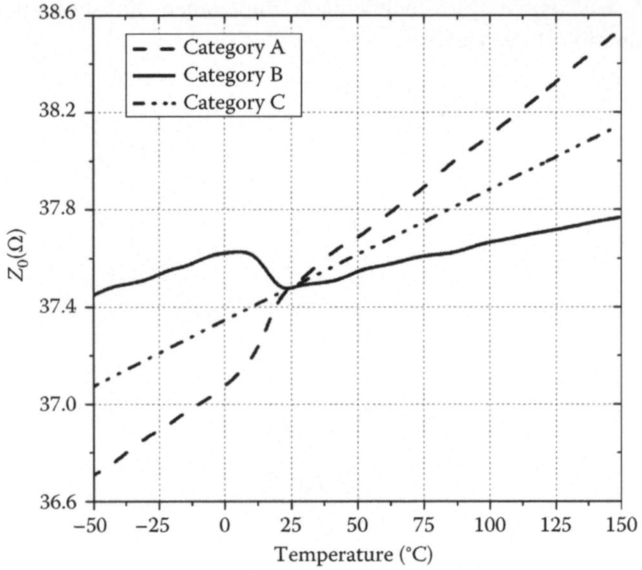

FIGURE 1.30 SIW impedance versus temperature at 10 GHz after considering all temperature effects (a = 15 mm, h = 1.575 mm, s = 0.9 mm, d = 0.5 mm, and ε_r = 2.2 at 23°C).

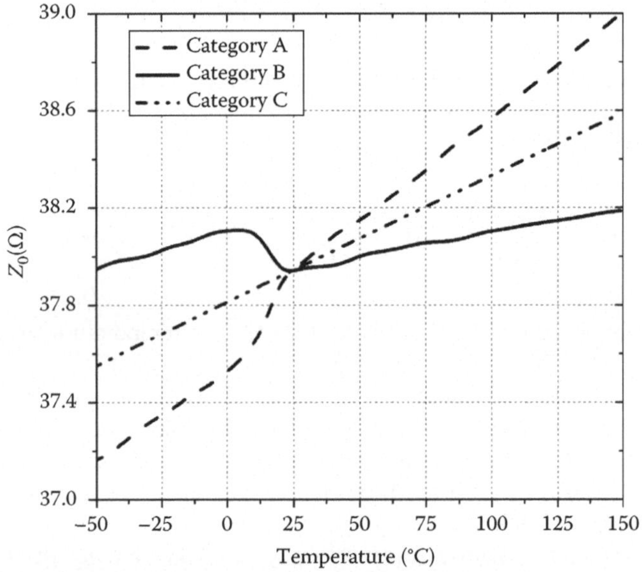

FIGURE 1.31 SIW impedance versus temperature at 30 GHz after considering all temperature effects (a = 5 mm, h = 0.508 mm, s = 0.8 mm, d = 0.4 mm, and ε_r = 2.2 at 23°C).

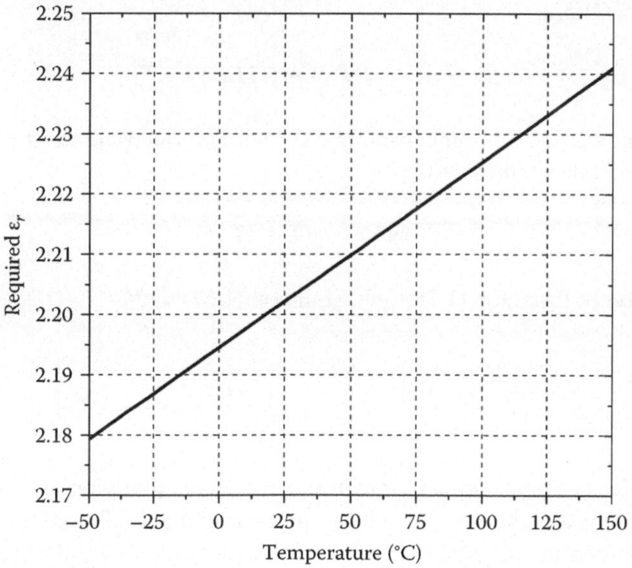

FIGURE 1.32 Required ε_r to realize an SIW with temperature-independent impedance ($a = 5$ mm, $h = 0.508$ mm, $s = 0.8$ mm, $d = 0.4$ mm, and $\varepsilon_r = 2.2$ at 23°C).

and depicted in Figure 1.32 to achieve the complete compensation, meaning an average TCK of 147 ppm/°C. Maybe it is difficult to find or develop an appropriate laminate to meet the requirement of the temperature compensation, but the design principle on how to mitigate the effect of temperature variation should be remembered. Choosing a material with proper TCK and CTE is likely to eliminate the temperature effect as much as possible.

1.1.4.2 SIW Electrical Length versus Temperature

The electrical length is very important for an SIW as well. Variation in electrical length results in serious problems for phase-sensitive applications. If a dielectric material with low tan δ is used, the SIW phase constant β at T (°C) can be calculated as

$$\beta(T) = \sqrt{\left(\frac{2\pi\sqrt{\varepsilon_r(T)}}{\lambda_0}\right)^2 - \left(\frac{\pi}{w(T)}\right)^2} \tag{1.54}$$

With the change in temperature, the SIW actual parameters are changed. As such, β deviates from the expected value. The investigation of the respective temperature effect of each variable indicates that the temperature property of β is related to CTE(y) and TCK and nearly independent of CTE(x) and CTE(c). Although CTE(x) may have minimal impact on β, it has direct impact on the physical length of an SIW, l. Hence, the electrical length, $\varphi(T)$, at T (°C) of an SIW can be calculated by

$$\varphi(T) = \beta(T)l(T) \tag{1.55}$$

In Equation (1.55),

$$l(T + \Delta T) = l(T)(1 + CTE(x) \times \Delta T) \tag{1.56}$$

In a similar way, a self-compensating SIW with an electrical length insensitive to temperature can be synthesized by

$$\varphi(T) = \varphi(T + \Delta T) \tag{1.57}$$

Substitution of Equation (1.55) into Equation (1.57) yields

$$\frac{\varepsilon_r(T+\Delta T)}{\varepsilon_r(T)} = C_1\left(\frac{l(T)}{l(T+\Delta T)}\right)^2 + C_2\left(\frac{w(T)}{w(T+\Delta T)}\right)^2 \tag{1.58}$$

In Equation (1.58), C_1 and C_2 are the same as in Equation (1.51). In general, Equation (1.58) is less than 1 if $\Delta T > 0$ and greater than 1 if $\Delta T < 0$. Hence, a negative TCK will compensate thermal expansion. The results after considering all temperature effects shown in Figures 1.33 and 1.34 verify our above inference. This conclusion is opposite the previous one for the SIW impedance. Thus, there exist different selection criteria for different temperature stability considerations.

Furthermore, an SIW having a temperature-independent electrical length can also be synthesized by Equations (1.55) to (1.57). As an example, the required average

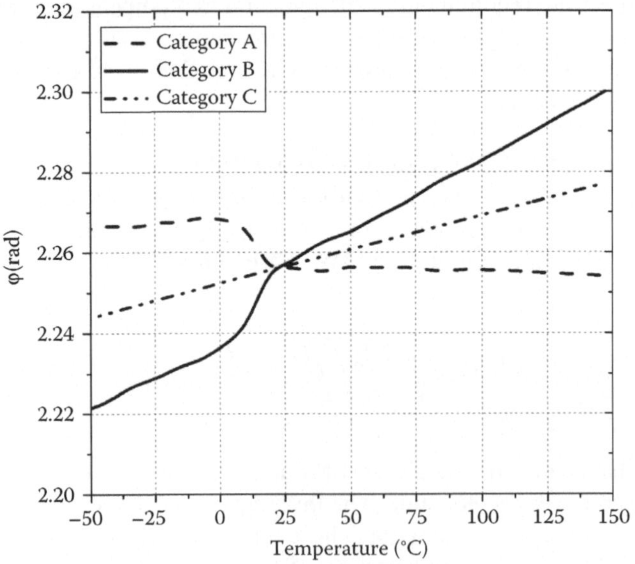

FIGURE 1.33 SIW electrical length versus temperature at 10 GHz after considering all temperature effects (a = 15 mm, s = 0.9 mm, d = 0.5 mm, l = 10 mm, and ε_r = 2.2 at 23°C).

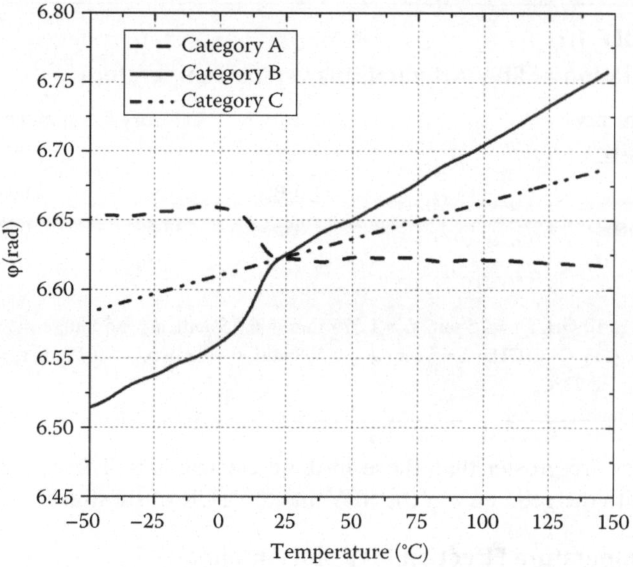

FIGURE 1.34 SIW electrical length versus temperature at 30 GHz after considering all temperature effects ($a = 5$ mm, $s = 0.8$ mm, $d = 0.4$ mm, and $\varepsilon_r = 2.2$ at 23°C).

TCK is −81 ppm/°C for the SIW operating at 30 GHz with an unchanged electrical length of 6.62 rad.

1.1.4.3 SIW Attenuation versus Temperature

Analytical equations for the conductor and dielectric losses for the TE_{10} mode in a rectangular waveguide at T (°C) can be rewritten as

$$\alpha_c(T) = \frac{R_m}{\eta(T)h(T)\sqrt{1-\left(\dfrac{\lambda_0}{2w(T)\sqrt{\varepsilon_r(T)}}\right)^2}}\left(1+\frac{\lambda_0^2 h(T)}{2w^3(T)\varepsilon_r(T)}\right) \tag{1.59}$$

$$\alpha_d(T) = \frac{\pi}{\lambda_0}\sqrt{\varepsilon_r(T)}\frac{1}{\sqrt{1-\left(\dfrac{\lambda_0}{2w(T)\sqrt{\varepsilon_r(T)}}\right)^2}}\tan\delta \tag{1.60}$$

Generally, tan δ is changed by temperature, but its accurate temperature properties should be measured in a specialized laboratory. Thus, we suppose tan δ keeps unchanged at different temperatures in the following investigation.

As shown in Table 1.1, both conductor and dielectric losses of SIWs on three substrates drop with the increase of environment temperature, but these changes in

TABLE 1.1
Variation of SIW Attenuations versus Temperature

Frequency		Category A	Category B	Category C
10 GHz	$\alpha_c\vert_{T=-50°C}/\alpha_c\vert_{T=150°C}$	1.0553	1.0570	1.0560
	$\alpha_d\vert_{T=-50°C}/\alpha_d\vert_{T=150°C}$	1.0100	1.0077	1.0088
30 GHz	$\alpha_c\vert_{T=-50°C}/\alpha_c\vert_{T=150°C}$	1.0553	1.0591	1.0572
	$\alpha_d\vert_{T=-50°C}/\alpha_d\vert_{T=150°C}$	1.0101	1.0196	1.0199

Note: 10 GHz: a = 15 mm, h = 1.575 mm, s = 0.9 mm, d = 0.5 mm, ε_r = 2.2 at 23°C; 30 GHz: a = 5 mm, h = 0.508 mm, s = 0.8 mm, d = 0.4 mm, ε_r = 2.2 at 23°C.

conductor loss are greater than those in the dielectric loss. The changing extent is associated with the geometry of the SIW and the TCK of substrate.

1.1.4.4 Temperature Effect on SIW Slot Antenna

If antennas have to work in some extreme environments like outer space with large temperature variation, electrical properties of antennas will suffer from unwanted change [31]. An SIW-based slot antenna exposed to large temperature variation is investigated here. A uniform temperature distribution across the entire SIW antenna structure is assumed as well.

For an SIW slot antenna, changes in SIW dimension (a, d, s) and substrate parameters (ε_r, h) caused by the variation in temperature affect the performance of the antenna. Besides, the length (sl), width (sw), and offset (so, se) of each slot must be selected accurately to reach expected electric field intensity. Thus, these slot dimensions varying with temperature also affect the performance of the antenna.

The investigated SIW antenna with a single longitudinal slot is designed to operate at 30 GHz at 23°C. The configuration of the proposed antenna is shown in Figure 1.35. The corresponding design parameters are a = 5 mm, s = 0.8 mm, d = 0.4 mm, ε_r = 2.2 mm, h = 0.508 mm, sl = 4.3072 mm, sw = 0.2 mm, so = 0.3965 mm, and se = 2.8412 mm. From 29.471 to 30.479 GHz, S_{11} values are simulated to be below −10dB at 23°C. The center frequency is 30.002 GHz. Simulated E- and H-plane patterns at 30 GHz at 23°C are shown in Figure 1.36.

The radiation characteristics of such an SIW single-slot antenna are almost the same over the large change in ambient temperature. However, the resonant frequency of the slot (defined as the minimal reflection frequency) is offset. CTE(sl) and CTE(sw) are related to the CTE of copper. As temperature increases, parameters sl and sw should be smaller with a negative CTE. Table 1.2 lists the simulated isolated temperature effects of a, h, ε_r, s, d, sl, sw, so, and se on S_{11} of the SIW slot antenna. As shown in this table, varied a, ε_r of substrate B or h with temperature lowers the center frequency, while varied ε_r of substrate A or sl or so with temperature raises the center frequency. The effects of s, d, sw, or se are relatively small. Here, the metalized vias within the SIW structure were also supposed to have the same thermal

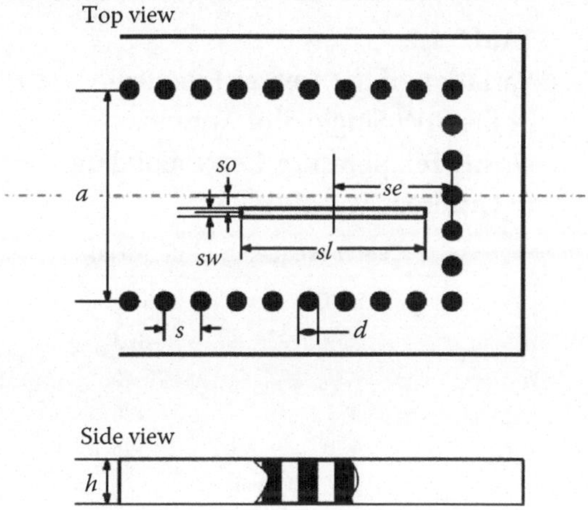

FIGURE 1.35 Configuration of the SIW single-slot antenna.

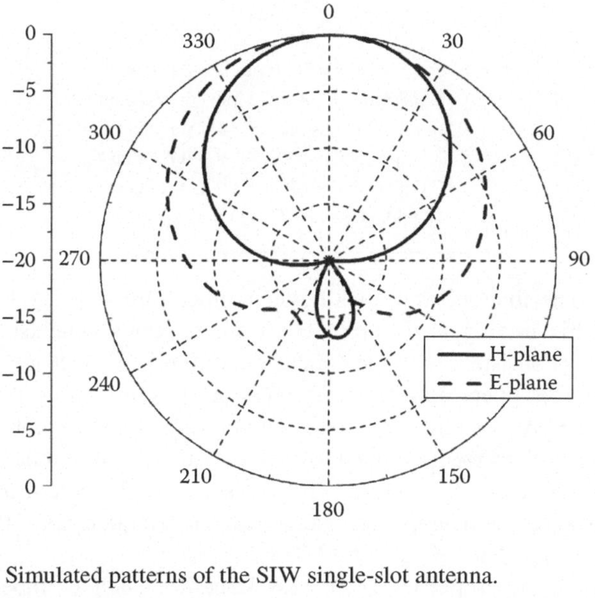

FIGURE 1.36 Simulated patterns of the SIW single-slot antenna.

expansion value with the laminate in the vertical axis. For substrate A, the CTE and TCK have a contrary trend. Thus, the temperature effect of the SIW slot antenna on such a substrate may be the smallest one.

Using the same values of CTE and TCK as described before, the S_{11} values of slot antennas were full-wave simulated and recorded after considering all temperature effects. As shown in Figures 1.37 to 1.39, the slot antenna on substrate A induces a frequency shift of 69 MHz, on substrate B induces a frequency shift of 411 MHz, and on substrate C induces a frequency shift of 170 MHz when temperature varies from

TABLE 1.2
Variation of the Center Frequency of the SIW Single-Slot Antenna versus Temperature Corresponding to Different Variables

	Center Frequency Shift (MHz)		
	–50°C		150°C
a	59		–103
h	38		–56
ε_r (A)	–122		126
ε_r (B)	130		–125
s		Small	
d		Small	
sl	–16		34
sw		Small	
so	–6		20
se		Small	

Note: a = 5 mm, s = 0.8 mm, d = 0.4 mm, h = 0.508 mm, sl = 4.3072 mm, sw = 0.2 mm, so = 0.3965 mm, se = 2.8412 mm, and ε_r = 2.2 at 23°C; CTE(sl) = –17 ppm/°C, CTE(sw) = –17 ppm/°C, CTE(so) = 48 ppm/°C, and CTE(se) = 31 ppm/°C.

–50 to 150°C. The first substrate has better temperature stability for slot antenna applications. This finding is different from the previous conclusion of the SIW impedance but is consistent with the conclusion of the SIW electrical length.

Furthermore, designers may focus on reducing or eliminating the performance variation of the SIW slot antenna versus temperature. The concept of temperature compensating is used as well. In this design, the required ε_r is simulated to be 2.2138 and 2.1802 for –50 and 150°C, respectively. If so, the reflection of the slot antenna will be quasi-insensitive to temperature and the center frequencies of the slot antenna will remain unchanged at –50, 23, and 150°C.

In this section, the properties of an SIW structure, such as impedance, electrical length, and attenuation, were investigated versus temperature. Through mathematic analysis, several useful conclusions were summarized to evaluate the effect of temperature variation and mitigate such an effect. The TCK influence of the used material could be compensated by the thermal expansion influence of the SIW configuration. As an example, an SIW with a temperature-insensitive electric length can be achieved based on a substrate with negative TCK. But the conclusion is opposite that of an SIW with a temperature-insensitive impedance. Besides, the substrate with negative TCK is preferred to synthesize the temperature-insensitive SIW slot antenna.

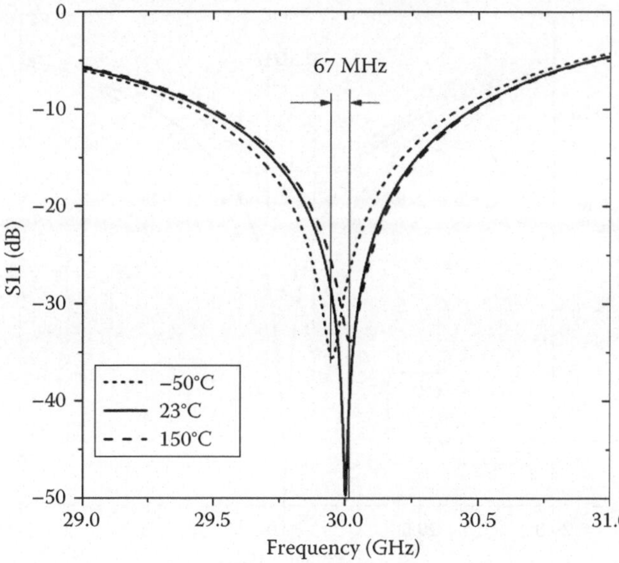

FIGURE 1.37 S_{11} of the SIW single-slot antenna on substrate A versus temperature after considering all temperature effects (a = 5 mm, s = 0.8 mm, d = 0.4 mm, h = 0.508 mm, sl = 4.3072 mm, sw = 0.2 mm, so = 0.3965 mm, se = 2.8412 mm, and ε_r = 2.2 at 23°C; CTE(sl) = −17 ppm/°C, CTE(sw) = −17 ppm/°C, CTE(so) = 48 ppm/°C, and CTE(se) = 31 ppm/°C).

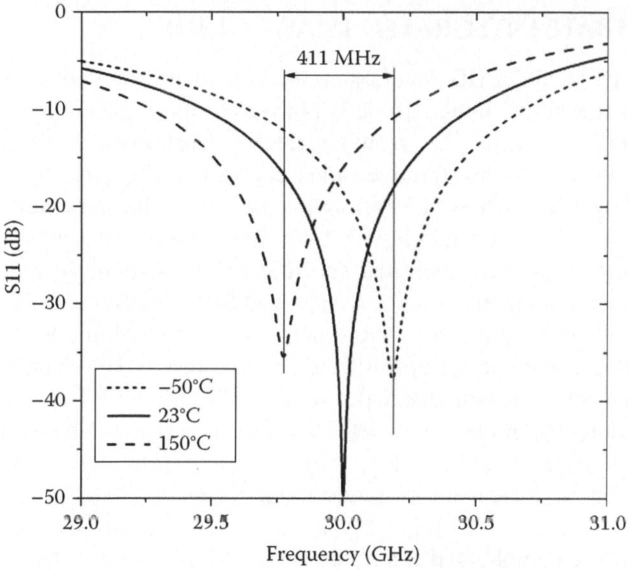

FIGURE 1.38 S_{11} of the SIW single-slot antenna on substrate B versus temperature after considering all temperature effects (a = 5 mm, s = 0.8 mm, d = 0.4 mm, h = 0.508 mm, sl = 4.3072 mm, sw = 0.2 mm, so = 0.3965 mm, se = 2.8412 mm, and ε_r = 2.2 at 23°C; CTE(sl) = −17 ppm/°C, CTE(sw) = −17 ppm/°C, CTE(so) = 48 ppm/°C, and CTE(se) = 31 ppm/°C).

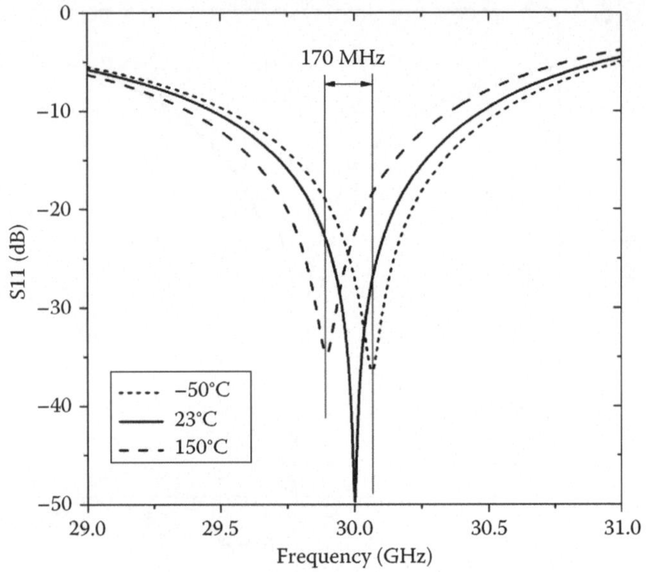

FIGURE 1.39 S_{11} of the SIW single-slot antenna on substrate C versus temperature after considering all temperature effects (a = 5 mm, s = 0.8 mm, d = 0.4 mm, h = 0.508 mm, sl = 4.3072 mm, sw = 0.2 mm, so = 0.3965 mm, se = 2.8412 mm, and ε_r = 2.2 at 23°C; CTE(sl) = −17 ppm/°C, CTE(sw) = −17 ppm/°C, CTE(so) = 48 ppm/°C, and CTE(se) = 31 ppm/°C).

1.2 SUBSTRATE INTEGRATED IMAGE GUIDE

As another type of SIC, SIIG was first studied and demonstrated in [32].A conventional three-dimensional image guide is converted into a planar configuration and synthesized on the metallic plate from a dielectric substrate of the proper thickness. The field is confined by a refractive index contrast in the horizontal plane of the structure [33], which is achieved by removing part of the dielectric materials close to the guiding channel, as shown in Figure 1.40. The permittivity is effectively lowered by this airhole perforation of dielectric substrate at both sides of the guiding channel.

Compared with microstrip line, strip line, and SIW structures, the SIIG in a high-permittivity substrate seems to be the transmission line with the lowest loss. Thus, it is a promising candidate for up-millimeter-wave design. However, this semiopen structure is subject to unavoidable leakage, when discontinuities exist, such as bends and corners, along the guiding channel. Therefore, this type of SIC is suitable for the development of antenna feeders, along which the discontinuities can be controlled.

Although the equivalent image guide model of an SIIG can be used, the theoretical model of an SIIG is frequently involved as its boundary condition. Mode behaviors become complicated for simulations and parameter extractions. A lot of computer memory and computation time are wasted in simulation. Besides, because of the periodic geometry, the SIIG is also subject to bandgap or stopband restriction similar to the SIW technology over certain higher-frequency ranges. Some complicated design methods should be used to avoid such a bandgap effect. Moreover,

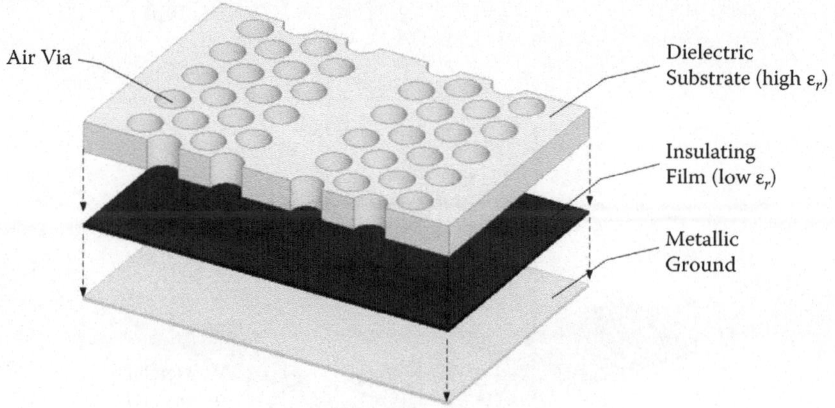

FIGURE 1.40 Configuration of a typical SIIG.

the potential air gap is difficult to avoid between the metallic ground plane and the dielectric strip, which may lead to serious problems in millimeter-wave frequency bands. Thermal bonding of the insulating film to the substrate and ground plane is an immature technology, which makes obvious deviations between expected and measured results [33]. Due to these inherent drawbacks, the conventional SIIG has not been well developed.

As described in [34], an SIIG can be fabricated through the LTCC technology as illustrated in Figure 1.41, which provides a feasible solution to those above problems. The LTCC technology is easier to realize blind and buried vias among different multilayered technologies. It is especially able to provide continuous rectangular air grooves to form a standard dielectric strip. The center guiding channel of an SIIG is realized by removing the adjacent dielectric material. In this case, the classical analysis and design method for a rectangular image guide [35, 36] can be employed directly in our design without any equivalent transformation. Besides, the bandgap effect is avoided because of this structure without periodic geometry. Meanwhile, the LTCC-based SIIG also has an advantage over the conventional SIIG in conquering those above assembly problems, including the inaccurate position fix among different layers and the unavoidable air gap between the ground plane and the dielectric strip.

In the standard LTCC fabrication technology, it is not allowed that the air cavity directly reaches the metallic plate. Thus, an isolating film is inserted between the ground plane and the dielectric strip. Such a film is realized by a single dielectric layer and integrated with the dielectric strip as a unified whole. Insertion of an insulating dielectric film with low permittivity between the substrate and the metallic ground plane forms an insulated image guide, which can further reduce the conductor loss. However, the isolating film in our design has the same substrate properties as the dielectric strip. As such, it only separates the dielectric strip from the metallic ground, but cannot reduce the loss as usual.

The influence of strip dimensions on the SIIG operating frequency range is demonstrated in Figures 1.42 and 1.43 by means of full-wave simulations. Its dielectric

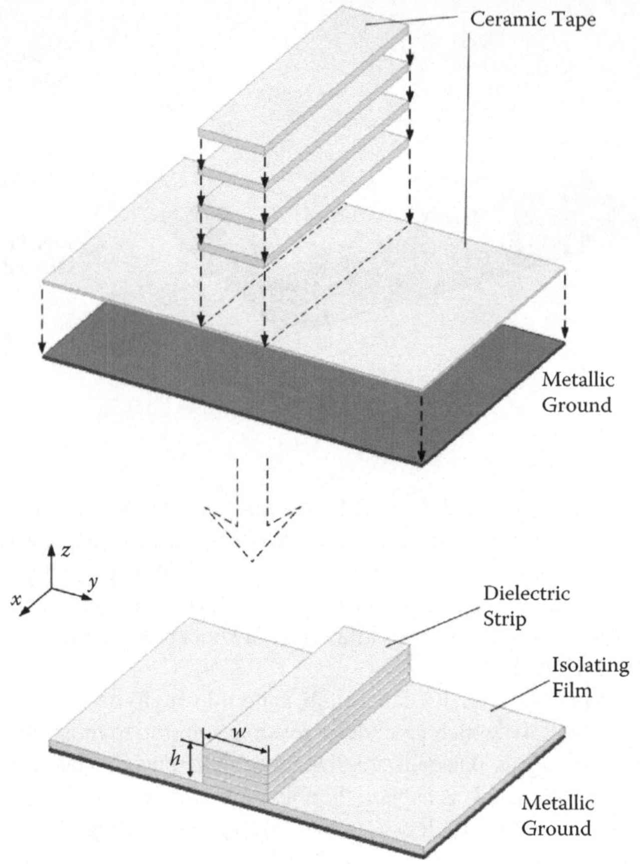

FIGURE 1.41 Configuration of an LTCC-based SIIG.

strip consists of several stacked ceramic tapes. The used material is Ferro A6-M with a relative permittivity of 5.9 and a loss tangent of 0.002 at 60 GHz. As shown, the strip width, *w*, has a smaller influence on the dominant mode E_{11}^{z}, but plays a major role in controlling the single-mode bandwidth. Furthermore, the upper bandwidth limit is shifted to a little lower frequency when the isolating film is used.

Then, the propagation characteristics of LTCC-based SIIGs are investigated. Figures 1.44 and 1.45 present the normalized phase constants versus different dielectric strip widths and thicknesses at different frequency bands. The larger strip width and thickness lead to a larger phase constant. After that, the normalized phase constants of two example SIIGs are depicted in Figures 1.46 and 1.47, respectively. Both of them operate at the dominant mode E_{11}^{z}.

Figure 1.48 presents the loss comparison among a 50 Ω microstrip line, a 50 Ω strip line, an SIW, and an SIIG with the isolating film. The SIW operates at the dominant mode TE_{10}, while the SIIG operates at the dominant mode E_{11}^{z}. The traditional planar transmission lines, such as microstrip, have significant conductor losses, because the existent electromagnetic field singularities at the conductor strip

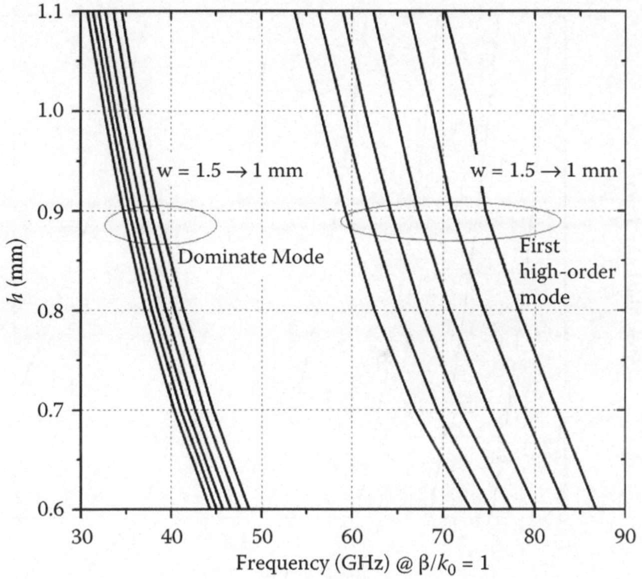

FIGURE 1.42 Normalized phase constants of the LTCC-based SIIG without the isolating film versus strip height and width.

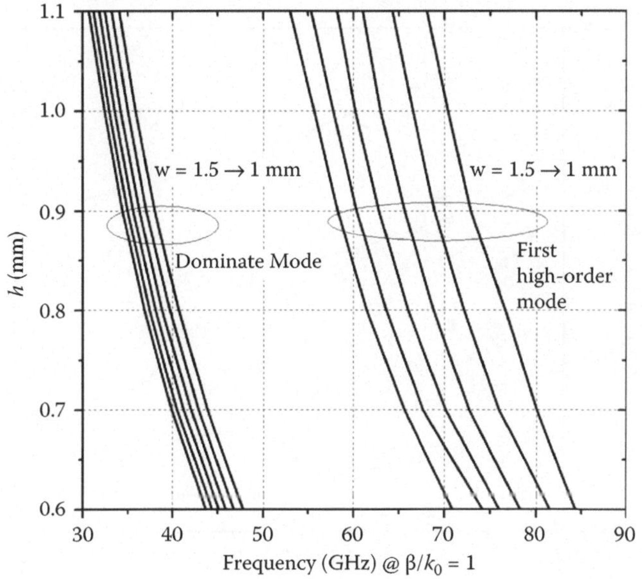

FIGURE 1.43 Normalized phase constants of the LTCC-based SIIG with the isolating film versus strip height and width.

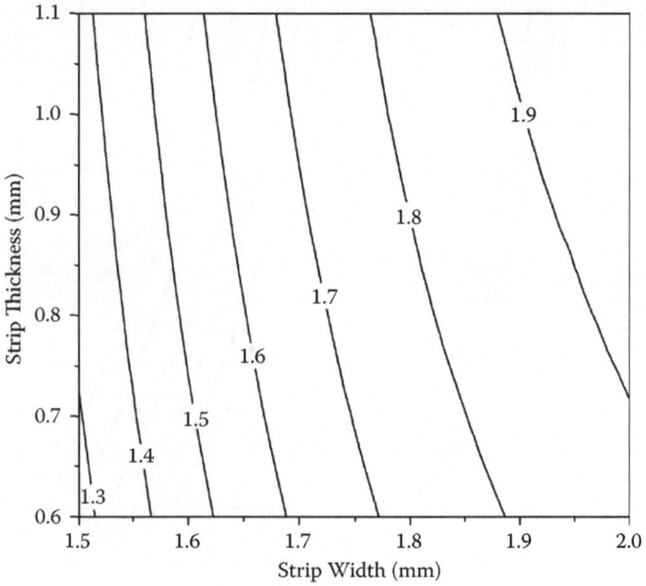

FIGURE 1.44 Normalized phase constant of the LTCC-based SIIG versus strip width and thickness at 60 GHz.

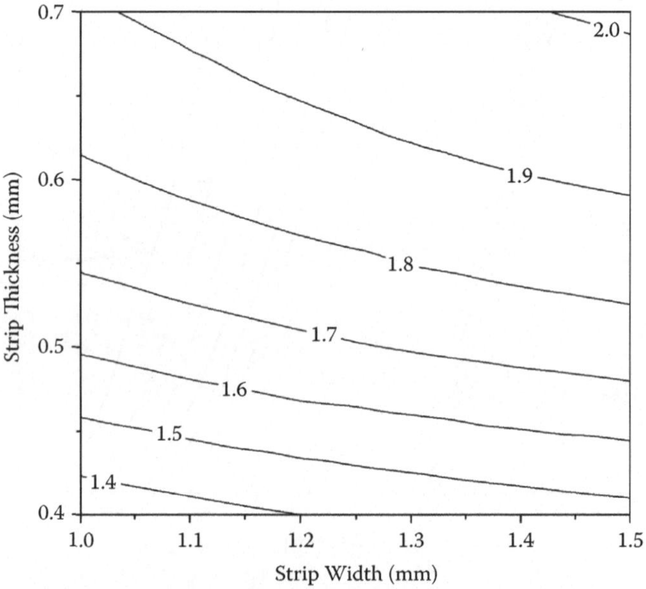

FIGURE 1.45 Normalized phase constant of the LTCC-based SIIG versus strip width and thickness at 94 GHz.

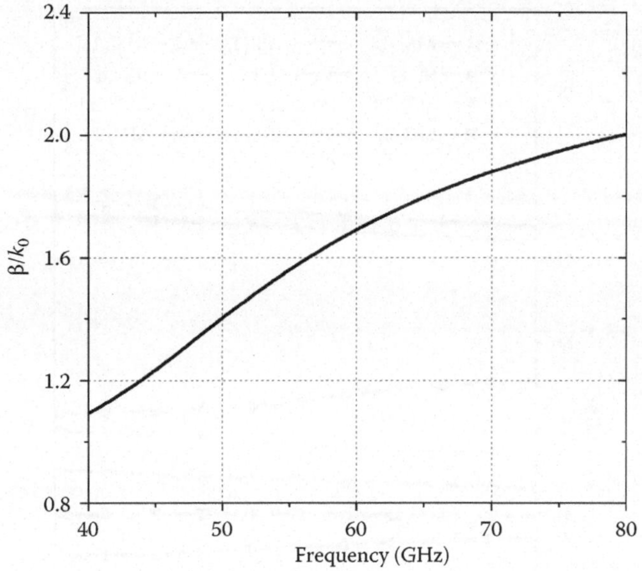

FIGURE 1.46 Normalized phase constant of the LTCC-based SIIG versus frequency (dielectric strip 1.8 mm, dielectric thickness 0.7 mm, isolating film thickness 0.1 mm).

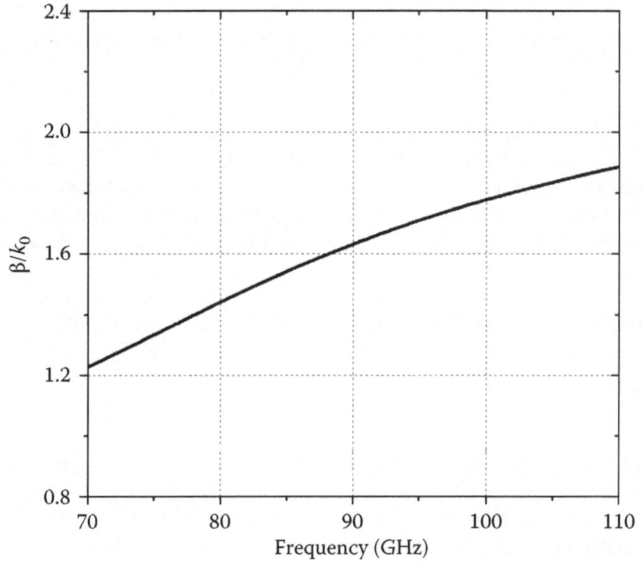

FIGURE 1.47 Normalized phase constant of the LTCC-based SIIG versus frequency (dielectric strip 1.25 mm, dielectric thickness 0.4 mm, isolating film thickness 0.1 mm).

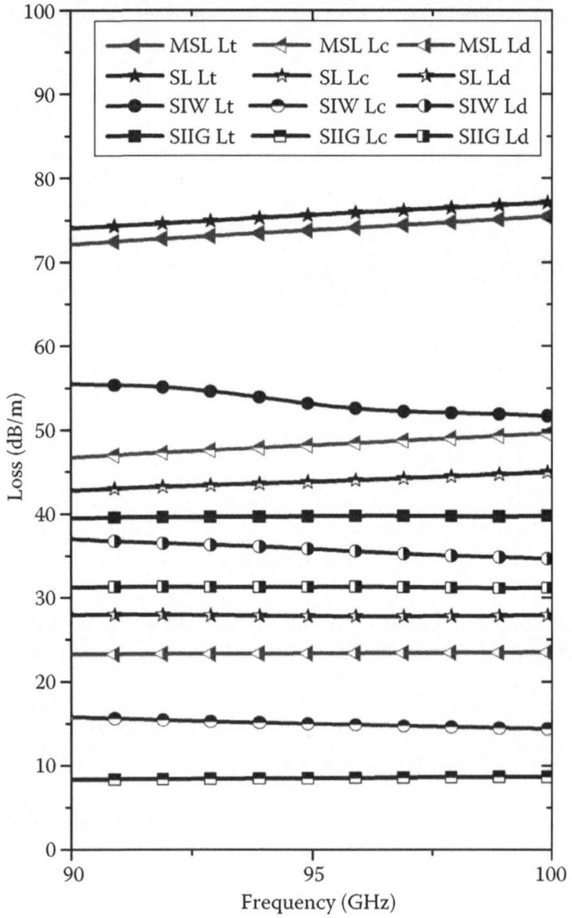

FIGURE 1.48 Loss comparison among different planar transmission lines: microstrip line (MLS, substrate thickness 0.1 mm); strip line (SL, substrate thickness 0.4 mm); SIW (substrate thickness 0.6 mm, SIW width 1.1 mm, distance between metalized vias 0.25 mm, diameter of the metalized via 0.1 mm); SIIG (strip thickness 0.4 mm, strip width 1 mm, film height 0.1 mm).

edges deteriorate the loss problem. As shown, the SIIG has better total loss performance than other guided-wave structures fabricated on the same dielectric substrate. Even their dielectric losses are comparable.

1.3 SUBSTRATE INTEGRATED NONRADIATIVE DIELECTRIC GUIDE

A rectangular dielectric slab sandwiched between two metallic plates can be used to synthesize the NRD guide. Its operating frequency range is determined by the spacing between the metallic plates. The slab width is also important to determine the operating bandwidth.

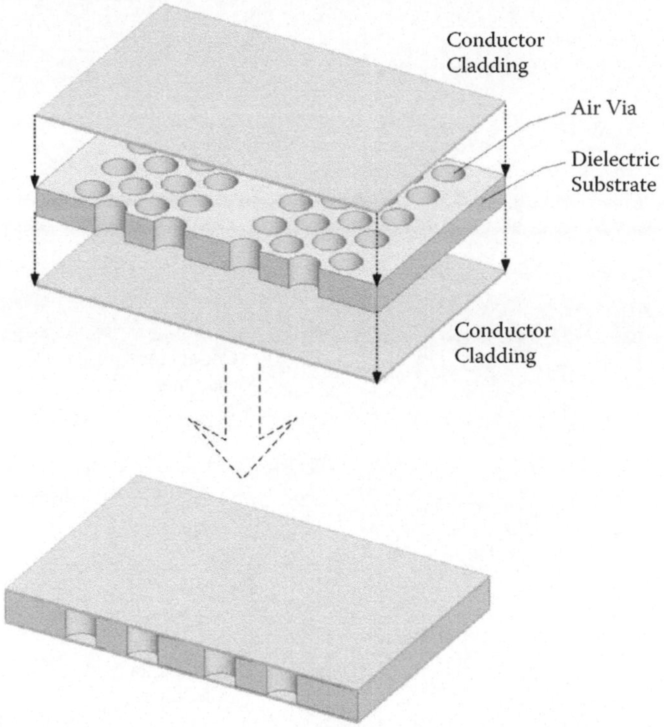

FIGURE 1.49 Configuration of a typical SINRD guide.

As shown in Figure 1.49, the SINRD guide [37] is an integrated and planar structure derived from the conventional NRD guide. Similar to the SIIG case, a material of the lower permittivity fills bilateral extents of the guiding channel. The original way of lowering permittivity in bilateral sides of the center guiding channel is also the airhole perforation. Absolutely, this can be made for broadband operations [38].

Besides, an SINRD guide directly through the conventional PCB was proposed in [39], which can be realized without employing mechanical assembly as usual. It has the configuration as shown in Figure 1.50. However, airholes or slots left on the conductor cladding may cause an energy leakage and deteriorate the guided modes. To deal with the problem, the strip width of the SINRD guide should be increased to minimize the surface currents on the regions of via holes.

Within the SINRD guide, two fundamental modes (LSE (longitudinal section electric) mode and LSM (longitudinal section magnetic) mode) are identified supporting low loss. The SINRD guides can be effectively used to construct antenna feeders or leaky-wave antennas. Different excitation schemes of the mode of SINRD guide using a rectangular waveguide or a microstrip line are presented in [37, 40, 41].

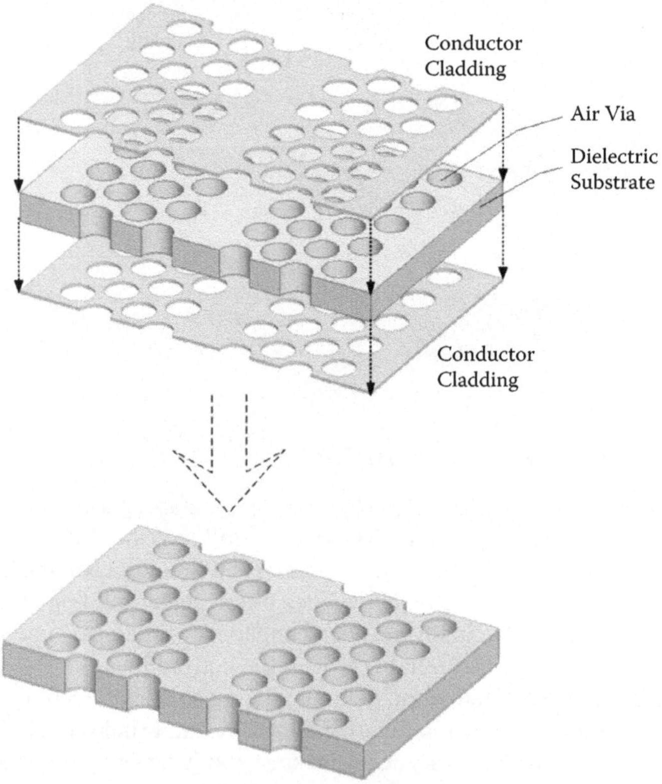

FIGURE 1.50 Configuration of a modified SINRD guide made of a normal PCB process.

2 Substrate Integrated Feeding Network

Substrate integrated feeding technology plays an important role in the construction of substrate integrated antenna arrays. Its performance directly affects the quality of the whole antenna array. The total radiation efficiency of an antenna array, which is the most widely used definition, is defined as the ratio of the realized gain and the directivity. The realized gain contains the conductor loss, the dielectric loss, and the mismatch factor of the feeding network.

In this chapter, an overview of popular and representative substrate integrated feeding technology is introduced, including the substrate integrated waveguide (SIW) divider, SIW phase shifter, SIW directional coupler, and transition between the substrate integrated circuit (SIC) and other transmission lines.

2.1 SUBSTRATE INTEGRATED DIVIDER

2.1.1 SIW Parallel Feeding Divider

A parallel feeding divider has a symmetric configuration, which guarantees all transmission coefficients are equal within a wide band. Here, two configurations are discussed: the T-type divider and Y-type divider.

The configuration of a typical SIW T-type divider is presented in Figure 2.1. This kind of divider is descended from a standard rectangular waveguide divider [42, 43]. In order to achieve low return losses at the input port, an inductive metalized via is employed in the T-junction of the divider [44]. Parameters of the power divider are optimized for attaining two goals: to minimize $|S_{11}|$ and to keep $|S_{21}|$ and $|S_{31}|$ close to –3 dB. Both the diameter, *r*, *d*, and the offset position, *o*, of the inductive matching via can greatly affect the return loss and the resonant frequency at the input port.

As shown in Figure 2.2, a Y-junction straight structure is adopted in the design of the SIW power divider [44, 45]. A bifurcated waveguide junction is used in the power division section. The width of the input SIW and the distance between the two discontinuities, *l*, can be optimized to achieve the required matching and power division. The compact topology of the Y-type divider is convenient for feeding an antenna array.

As an example, a W-band 16-way SIW parallel feeding divider was designed with a configuration shown in Figure 2.3. The classical T-type SIW divider requires at least an inductive via, as shown in Figure 2.1. Considering the limitation of the normal printed circuit board (PCB) process, the matching metalized via is usually relatively large for the W-band SIW, which narrows the bandwidth of the T-type divider. In this case, a four-order divider with good performance is impossible to synthesize. Therefore, a modified Y-type divider is used. Its performance is much better

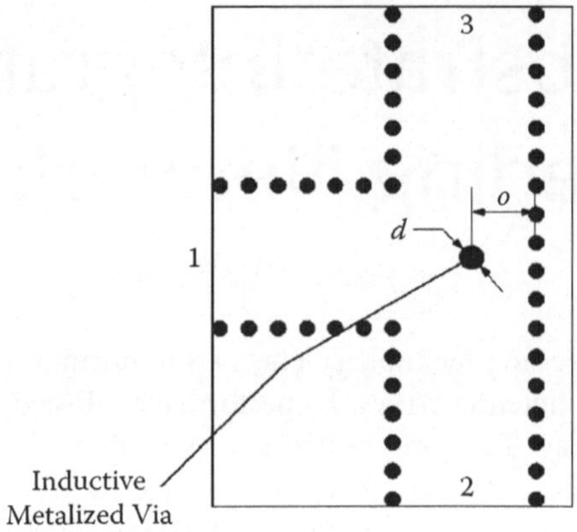

FIGURE 2.1 Configuration of the SIW T-type divider.

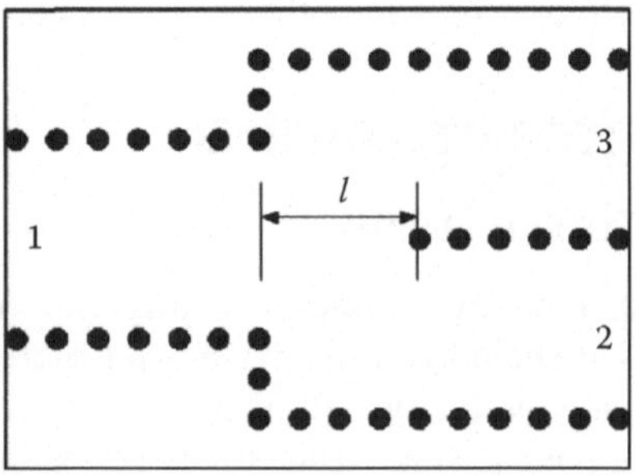

FIGURE 2.2 Configuration of the SIW Y-type divider.

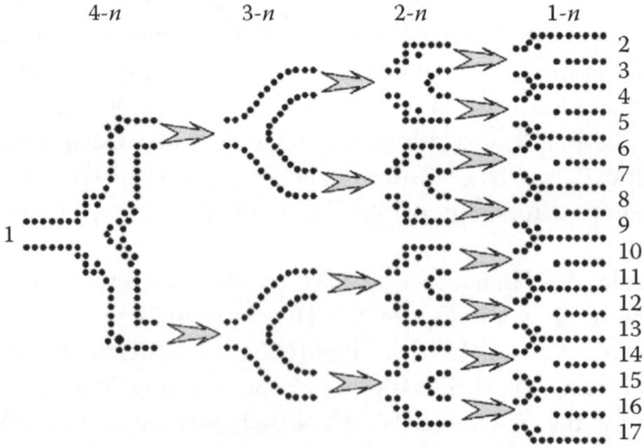

FIGURE 2.3 Configuration of the W-band 16-way SIW parallel feeding divider.

than that of the T-type SIW divider. Several stage two-way dividers are grouped in order to realize a 2^n-way large-scale divider. A two-way divider (1-*n*) is designed at first. In the second step, 1- and 2-*n* should be combined and modeled together, including two 1-*n* dividers and a 2-*n* divider. Only the parameters and dimensions of the 2-*n* divider are adjusted to optimize the whole four-way divider. In a similar way, 3- and 4-*n* dividers are designed, respectively. The symmetric configuration guarantees all transmission coefficients are equal. Thus, the divider is optimized under one goal, i.e., to minimize $|S_{11}|$. For the bandwidth over which return loss is better than 15 dB, a 2-way divider presents more than 90–98 GHz, a 4-way divider grouped by two-stage 2-way dividers is 91.4–97.5 GHz, an 8-way divider grouped by three-stage 2-way dividers is 91.8–97 GHz, and a 16-way divider grouped by four-stage 2-way dividers is 92.2–97.7 GHz.

The feeding network based on the treelike divider topology, as illustrated in Figure 2.3, becomes very large with the increase of the array scale. For the purpose of miniaturization, we hope that the used feeding network can be placed underneath the radiating elements. That means the total size of the integrated array antenna is equal to that of the radiating aperture. To achieve this goal, many T-type dividers are placed back-to-back, as shown in Figure 2.4, to feed the array elements in phase with the broadband characteristic, but also to keep a compact configuration. This type of divider can be called the H-type.

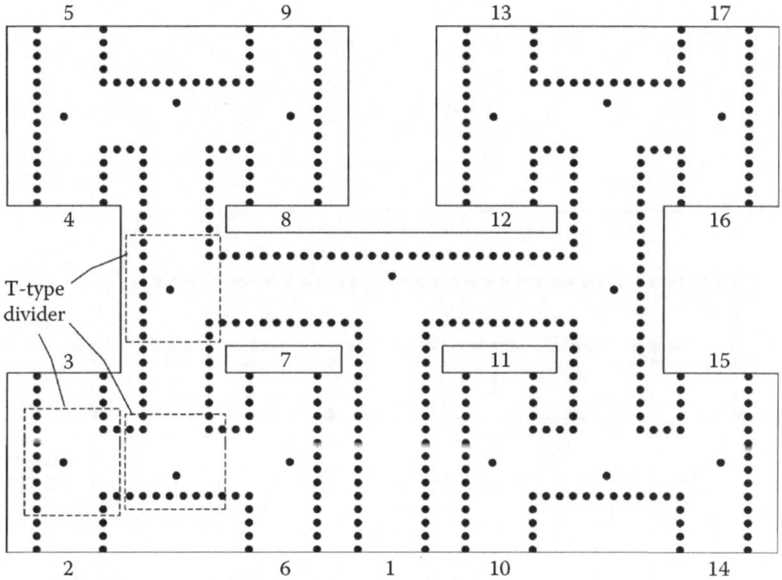

FIGURE 2.4 Configuration of the 16-way SIW H-type divider.

2.1.2 SIW Series Feeding Divider

Based on the symmetry of its structure, it is easy to realize broad bandwidth, good amplitude, and phase balance with a parallel feeding divider. However, it has equal amplitude output, large area, and high insertion loss. Besides, it is not easy to achieve unequal amplitudes while keeping equal phase.

An alternating phase multiple-way power divider was presented in [46] with the configuration shown in Figure 2.5. It can be considered a series of feeding dividers, which are able to achieve unequal amplitude output easily while keeping a small size. The type of divider consists of several modified T-junctions. This structure is transversely symmetrical. When the symmetrical plane is considered an equivalent magnetic wall, half of the divider is taken into account. A center metalized via is used for reflection suppression. In order to satisfy the design requirements, coupling window dimensions of each T-junction and dimensions of each tuning metalized via at each T-junction may be different. Thus, the design process is a little complicated. The input SIW width, the feeding SIW width, and the output SIW width can be different, thus increasing design flexibility.

In principle, the width of the feeding SIW, w_f, should be selected to make the distance between adjacent outputs, d_o, equal to half of the guided wavelength of the feeding SIW for alternating phases. That is

$$d_o = \frac{\lambda_g}{2} = \frac{\lambda}{2\times\sqrt{1-\left(\frac{\lambda}{2\times w_f}\right)^2}} \tag{2.1}$$

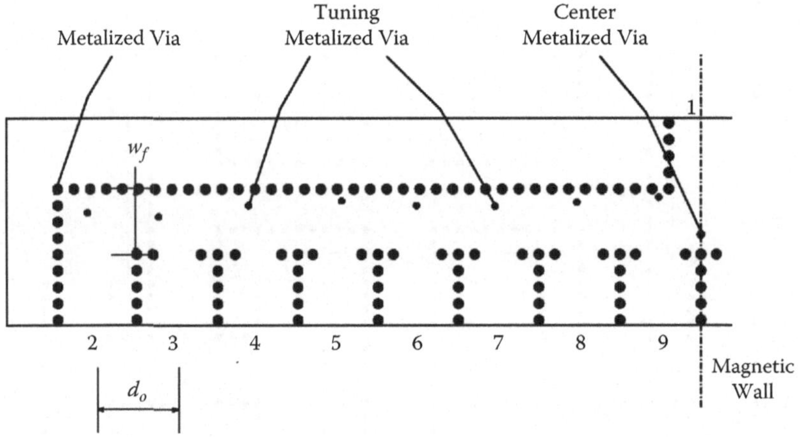

FIGURE 2.5 Configuration of a half of the 16-way SIW series feeding divider.

Equation (2.1) can be rewritten as

$$w_f = \frac{\lambda}{2\times\sqrt{1-\left(\frac{\lambda}{2\times d_O}\right)^2}} \tag{2.2}$$

The power ratio of the divider is mainly determined by the width of the coupling window, while the phase shift can be tuned slightly by modifying the offset of the coupling window. That means the required amplitude and phase distribution can be obtained by adjusting the coupling window parameters. Furthermore, reflection can be improved by optimizing diameters and positions of the tuning vias properly.

If the width of the SIW is a, the circuit size of the series feeding divider can be estimated by

$$S_s = N \times a^2 \tag{2.3}$$

In Equation (2.3), N is the number of output ports. The circuit size of the parallel feeding divider can be calculated roughly by

$$S_p = N \times \log_2 N \times a^2 \tag{2.4}$$

Thus,

$$\frac{S_s}{S_p} = \log_2 N \tag{2.5}$$

Table 2.1 lists the circuit size comparison of two types of SIW dividers. Compared with the parallel feeding divider, the series one is superior in circuit size. However, its bandwidth is relatively narrow considering its inherent cascading characteristic.

TABLE 2.1
Size Comparison of Two Types of SIW Dividers

N	S_s	S_p	S_p/S_s
2	$2a^2$	$2a^2$	1
4	$4a^2$	$8a^2$	2
8	$8a^2$	$24a^2$	3
16	$16a^2$	$64a^2$	4
32	$32a^2$	$160a^2$	5

2.2 SUBSTRATE INTEGRATED PHASE SHIFTER

Many phase shifters have been implemented in the form of different guided-wave structures in order to be easily integrated into the whole circuits. They focus on performances such as broadband and amplitude balance. The Schiffman differential phase shifter is one of the well-known broadband phase shifters. Its original work was based on the stripline structure [47], and then this type of phase shifter was designed mainly in the microstrip line form [48–50]. Another class of a broadband phase shifter consisting of several E-plane stubs was realized in the waveguide technology [51, 52]. A dual dielectric-slab-filled waveguide phase shifter is introduced in [53] and is also able to provide a wide operation bandwidth.

There are some developed SIW phase shifters, as shown in Figure 2.6. Among them, delay lines [45], slow-wave structures [54], and inserted metallic posts [55] have very poor bandwidth-related performances and are sensitive to frequency. Another popular type was realized by the use of different SIW widths [56–58]. A fixed phase difference can be obtained easily between equal-length structures over about 10% relative bandwidth. This is because the dispersion curves of the phase constant versus different SIW widths behave almost in parallel with each other.

To achieve a better performance, a broadband self-compensating SIW phase shifter was introduced in [59]. This technique was proposed based on the fact that

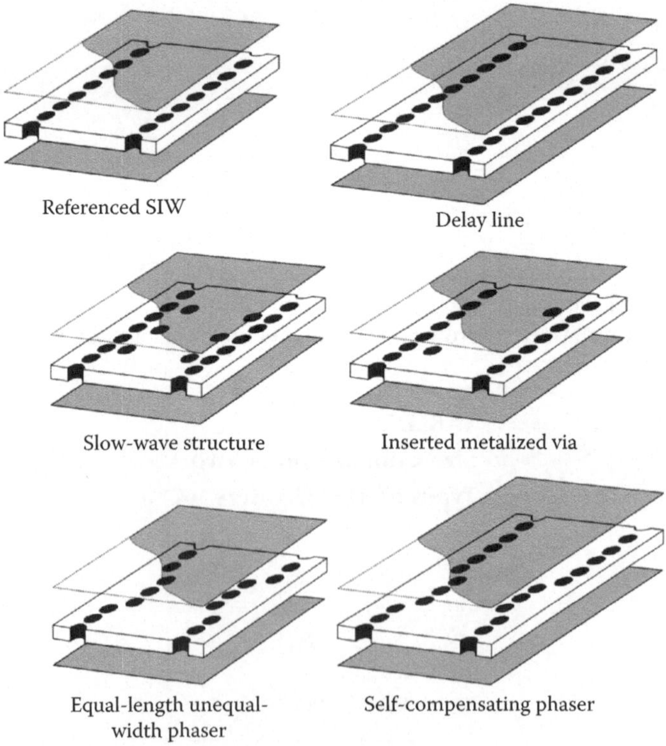

FIGURE 2.6 Configurations of different types of phase shifters.

the changes in the trend of phase shift obtained by the delay line and the equal-length unequal-width structure versus frequency are opposite. Thus, a concept of phase compensation can be implemented. The total phase shift is accomplished by different phase velocities of waves and different electrical lengths at the same time. The operation mechanism of the self-compensating phase shift is unique not only for the SIW, but also for the conventional waveguide.

2.2.1 Phase Self-Compensating Theory

Now, three types of phase shifter are investigated. They are the delay line, the equal-length unequal-width phase shifter, and the self-compensating phase shifter. To simplify the following analyses, all of them are supposed to be actualized in equivalent dielectric-filled rectangular waveguides with solid wall, as shown in Figure 2.7.

As described before, the dispersion characteristic of a TE_{10}-like mode in the SIW is almost identical to that of the TE_{10} mode in the dielectric-filled rectangular waveguide with an equivalent width. In general, the phase shift $\varphi(f)$ varies with frequency. At a given operating frequency f, there is

$$\varphi(f) = \beta(f)l \tag{2.6}$$

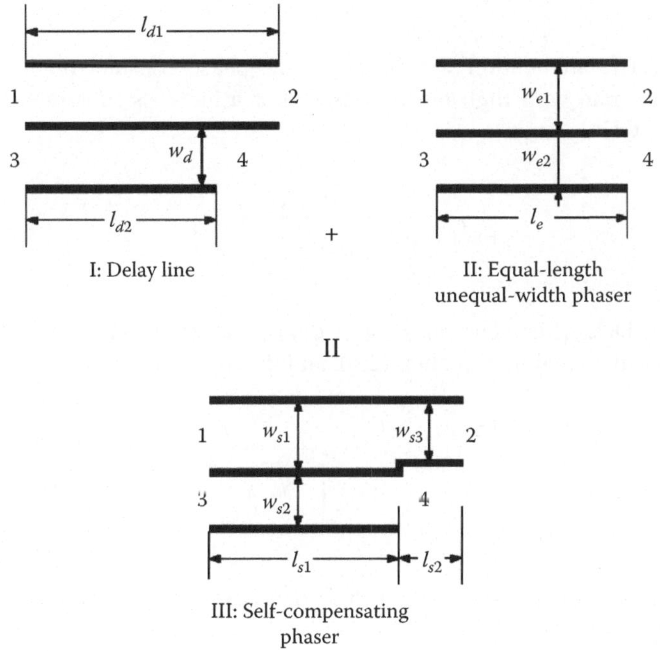

FIGURE 2.7 Sketches of three different types of phase shifter.

In Equation (2.6), $\beta(f)$ is the phase constant and l is the length of the transmission line. The used dielectric substrate is lossy, and it usually has the complex permittivity [60]

$$\varepsilon = \varepsilon'(1 - j\tan\delta) \tag{2.7}$$

In Equation (2.7), ε' is the material's relative permittivity ε_r, and tan δ is the dielectric loss tangent. There are

$$\gamma = \sqrt{k_c^2 - k^2} = \alpha + j\beta \tag{2.8}$$

In Equation (2.8), k_c is the cutoff wave number. There are

$$\begin{aligned} \alpha^2 - \beta^2 &= k_c^2 - \omega^2\mu\varepsilon' \\ 2\alpha\beta &= \omega^2\mu\varepsilon'\tan\delta \end{aligned} \tag{2.9}$$

Thus,

$$\beta = \sqrt{\frac{\sqrt{\left[\omega^2\mu\varepsilon' - k_c^2\right]^2 + \left[\omega^2\mu\varepsilon'\tan\delta\right]^2} + \omega^2\mu\varepsilon' - k_c^2}{2}} \tag{2.10}$$

The dielectric attenuation will influence the phase constant obviously when a dielectric substrate with high tan δ is used. For a low-loss substrate, $\beta(f)$ can be simplified (f, GHz; w, mm):

$$\beta(f) = \sqrt{\left(\frac{2\pi\sqrt{\varepsilon_r}f}{300}\right)^2 - \left(\frac{\pi}{w}\right)^2} \tag{2.11}$$

1. Type I: Delay line. The phase shift $\varphi(f)$ (phase(S21–S43)) of a delay line can be calculated by Equation (2.6), and the first-order derivative of $\varphi(f)$ is

$$\frac{d\varphi(f)}{df} = \frac{4\pi\varepsilon_r(l_{d2} - l_{d1})}{300^2} \frac{f}{\sqrt{\left(\frac{2\sqrt{\varepsilon_r}f}{300}\right)^2 - \left(\frac{1}{w_d}\right)^2}} < 0 \tag{2.12}$$

   This equation suggests that $\varphi(f)$ decreases almost linearly with frequency. As an example, a –90° phase shifter was synthesized on a substrate with $\varepsilon_r = 2.2$. The dimensions are $l_{d1} - l_{d2} = 2.28$ mm and $w_d = 5$ mm. The calculated phase shift versus frequency is presented in Figure 2.8.

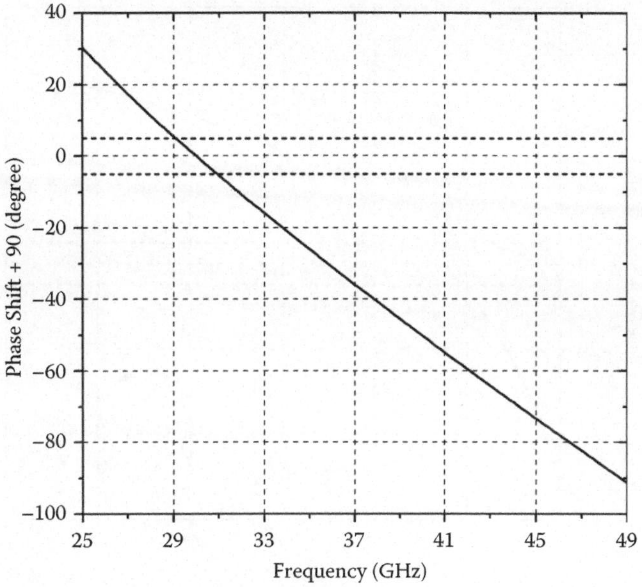

FIGURE 2.8 Phase shift of a –90° delay line.

2. Type II: Equal-length unequal-width phase shifter. The equal-length unequal-width phase shifter has two branches with the same length and different equivalent widths w_{e1} and w_{e2} ($w_{e1} > w_{e2}$). The phase shift $\varphi(f)$ (phase(S21–S43)) of this phase shifter can be calculated by

$$\varphi(f) = l_e\left[\sqrt{\left(\frac{2\pi\sqrt{\varepsilon_r}f}{300}\right)^2 - \left(\frac{\pi}{w_{e2}}\right)^2} - \sqrt{\left(\frac{2\pi\sqrt{\varepsilon_r}f}{300}\right)^2 - \left(\frac{\pi}{w_{e1}}\right)^2}\right] \tag{2.13}$$

In this case, the first-order derivative of $\varphi(f)$ is

$$\frac{d\varphi(f)}{df} = \frac{4\pi\varepsilon_r l_e}{300^2} \times \left(\frac{f}{\sqrt{\left(\frac{2\sqrt{\varepsilon_r}f}{300}\right)^2 - \left(\frac{1}{w_{e2}}\right)^2}} - \frac{f}{\sqrt{\left(\frac{2\sqrt{\varepsilon_r}f}{300}\right)^2 - \left(\frac{1}{w_{e1}}\right)^2}}\right) > 0 \tag{2.14}$$

This expression means that $\varphi(f)$ increases with frequency. Based on this principle, a –90° phase shifter was also synthesized on a substrate with $\varepsilon_r = 2.2$. The dimensions are $l_e = 13.2$ mm, $w_{e1} = 5.5$ mm, and $w_{e2} = 4.5$ mm. The calculated phase shift versus frequency is presented in Figure 2.9.

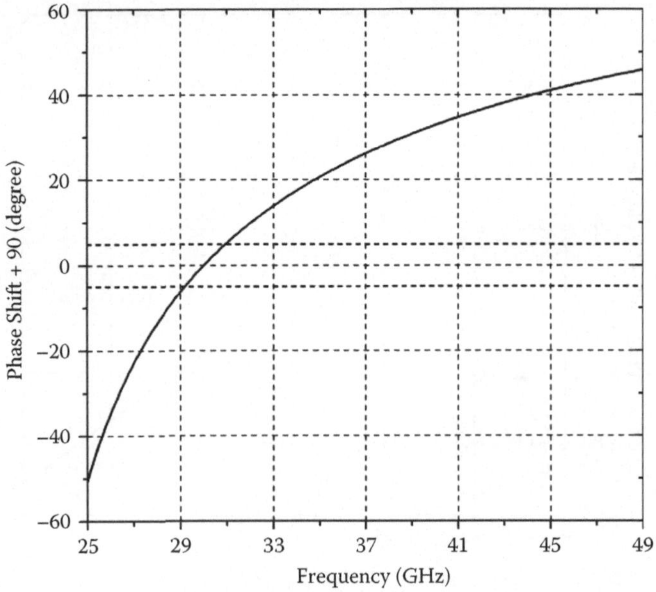

FIGURE 2.9 Phase shifts of a −90° equal-length unequal-width phase shifter.

3. Type III: Self-compensating phase shifter. The two above-mentioned structures can be combined together to construct a self-compensating phase shifter. There exist three different phase constants existing in the new structure: branch 21 *I* $\beta_1(f)$ (length l_{s1}, width w_{s1}), branch 21 *II* $\beta_2(f)$ (length l_{s2}, width w_{s3}), and branch 43 $\beta_3(f)$ (length l_{s1}, width w_{s2}).

The phase shift can be calculated at the frequency f_0 by

$$\varphi(f_0) = l_{s1}[\beta_3(f_0) - \beta_1(f_0)] - l_{s2}\beta_2(f_0) \tag{2.15}$$

The increment of frequency is Δf. It is called the *forward type* when $\Delta f > 0$. The phase shift at frequency $f_0 + \Delta f$ is

$$\varphi(f_0 + \Delta f) = l_{s1}[\beta_3(f_0 + \Delta f) - \beta_1(f_0 + \Delta f)] - l_{s2}\beta_2(f_0 + \Delta f) \tag{2.16}$$

If

$$\varphi(f_0) = \varphi(f_0 + \Delta f) \tag{2.17}$$

is satisfied, the dispersion of the phase coefficient will be suppressed significantly. A similar process can be used to analyze a $\Delta f < 0$ case, called the *backward type*.

TABLE 2.2
Parameters of Different Self-Compensating Phase Shifters with Different Δf (unit: mm)

Δf (GHz)	l_{s1}	l_{s2}
1	6.30	1.07
3	6.68	1.01
5	7.00	0.96
7	7.28	0.92
–1	5.89	1.14
–3	5.40	1.23
–5	4.81	1.39
–7	3.80	1.73

Note: The remaining parameters are w_{s1} = 5.5 mm, w_{s2} = 4.5 mm, and w_{s3} = 5 mm.

Through the above analyses, several self-compensating phase shifters, including the forward and backward cases with different Δf, were designed to achieve a –90° phase shift. If the phase shift at the center frequency of 30 GHz is not equal to –90°, a wider bandwidth will be realized. Table 2.2 lists the dimensions of these phase shifters synthesized on a substrate with $\varepsilon_r = 2.2$. Figures 2.10 and 2.11 show the calculated phase shift of different phase shifters versus frequency.

As shown in Figures 2.8 to 2.11, from 29.2 to 30.9 GHz, the type I phase shifter covers a 5.6% relative bandwidth over which the phase error is less than 5°. From 29.1 to 30.9 GHz, the type II phase shifter covers a 6% relative bandwidth over which the phase error is less than 5°. For the forward case of type III ($\Delta f = 7$ GHz), the relative bandwidth of 90° ± 5° can be made up close to 70%. The backward case also supports a wide band over the entire Ka-band. The relative bandwidth of 90° ± 5° can be designed up close to 52% for $\Delta f = -1$ GHz. The performance of the self-compensating phase shifter is improved greatly compared with that of other counterparts.

The forward and backward type phase shifters offer wide bandwidth, but it is still possible to further improve the phase coefficient near the center frequency. This type of phase shifter can be called the *bidirectional type*, which is implemented by compensating the forward and backward self-compensating phase shifters again. As an example, a –90° bidirectional type phase shifter was designed. The used forward self-compensating phase shifter has $\Delta f = 2$ GHz, while the backward self-compensating phase shifter has $\Delta f = -2$ GHz. That means each one realizes half of the required phase shift, i.e., –45°. After combining them together, the phase difference of –90.25° ± 0.25° can be obtained within 28.4–31.7 GHz, as shown in Figure 2.12. Within the same frequency band, the relative error reduces by half.

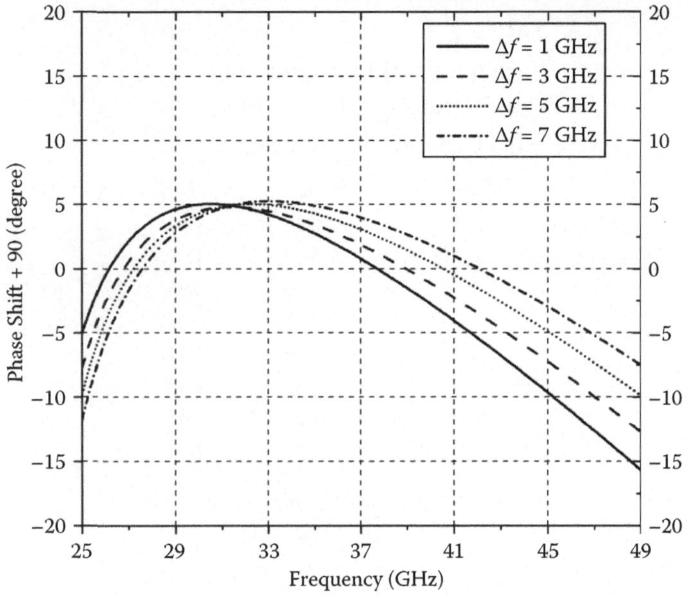

FIGURE 2.10 Phase shifts of several –90° forward self-compensating phase shifters.

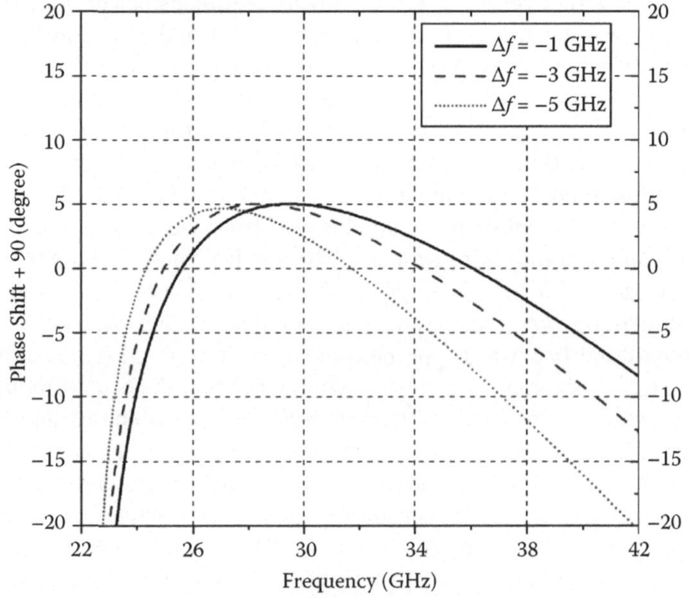

FIGURE 2.11 Phase shifts of several –90° backward self-compensating phase shifters.

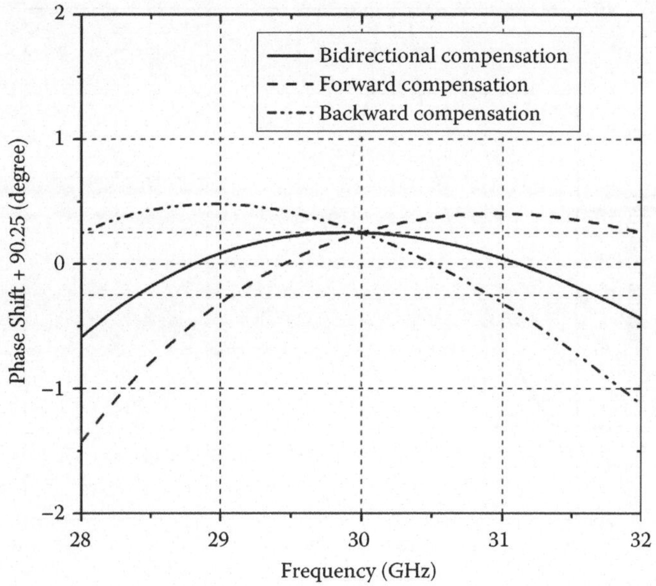

FIGURE 2.12 Phase shifts of different types of self-compensating phase shifters.

Now, the design process of the self-compensating phase shifter can be concluded as follows:

1. Determine the operating frequency band, the required phase shift, the used dielectric substrate (i.e., ε_r, and h), and the parameters of the rest SIW circuits (i.e., a, s, and d).

   In this design, the center frequency is set to be 30 GHz. The required phase shifts are –22.5°, –45°, –67.5°, and –90°, respectively. Besides, $\varepsilon_r = 2.2$, $h = 0.508$ mm, $a = 6$ mm, $s = 0.8$ mm, $d = 0.4$ mm, and $\Delta f = 10$ GHz. The equivalent width of rest SIW, w, can be calculated to be 5.79 mm.
2. Choose the widths of different SIW branches. For a self-compensating phase shifter, there are three equivalent widths: w_{s1}, w_{s2}, and w_{s3}. The last one, w_{s3}, can be selected with the same value of the rest SIW (i.e., 5.79 mm). The optional range of w_{s1} and w_{s2} can be determined as follows.

   First, w_{si} should be properly selected to guarantee that the SIW operates in the dominant mode zone. Here, it is in the range of 3.37–6.74 mm.

   Next, the mismatching owing to the change in SIW widths should be taken into consideration. The impedance variations caused by the SIW discontinuities are expressed in the return loss, RL, which can be calculated by

$$RL = -20\log_{10}\frac{|\Delta Z_0|}{2Z_0 - |\Delta Z_0|} \tag{2.18}$$

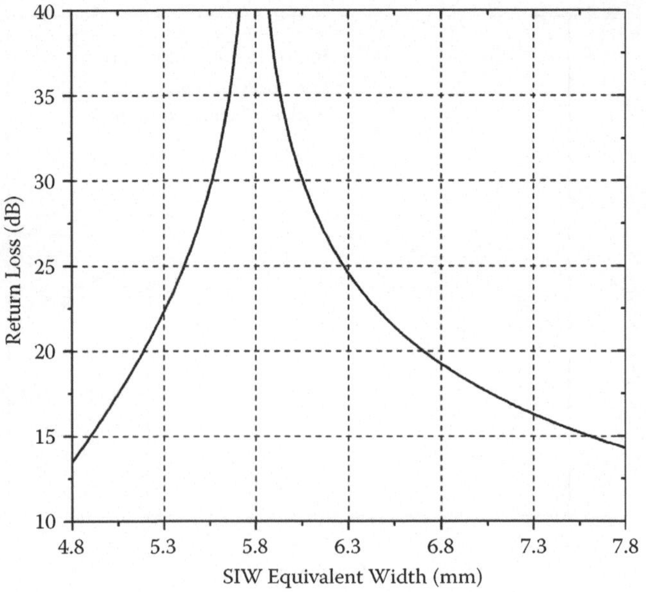

FIGURE 2.13 Calculated return loss versus variable SIW widths.

The values of RL were calculated with different SIW widths considering the equivalent width of the rest SIW is 5.79 mm. As shown in Figure 2.13, w_{si} within 4.90–7.59 mm guarantees return loss better than 15 dB.

Combined with the above-mentioned analysis, the value of w_{si} should be in the range of 4.90–6.74 mm. In this design, w_{s1} = 5.79 mm, w_{s2} = 5.11 mm, and w_{s3} = 5.79 mm.

3. Calculate l_{s1} and l_{s2}. l_{s1} and l_{s2} can be calculated by using Equations (2.15) to (2.17), and then simulated and optimized. The final calculated and optimized parameters are (a) –22.5°, l_{s1} = 3.19 mm (calculated) and 2.80 mm (optimized), l_{s2} = 0.26 mm (calculated) and 0.32 mm (optimized), and w_{s2} = 5.11 mm (calculated) and 5.24 mm (optimized); (b) –45°, l_{s1} = 6.39 mm (calculated) and 6.96 mm (optimized), l_{s2} = 0.52 mm (calculated) and 0.55 mm (optimized), and w_{s2} = 5.11 mm (calculated) and 5.24 mm (optimized); (c) –67.5°, l_{s1} = 9.58 mm (calculated) and 10.44 mm (optimized), l_{s2} = 0.78 mm (calculated) and 0.80 mm (optimized), and w_{s2} = 5.11 mm (calculated) and 5.21 mm (optimized); and (d) –90°, l_{s1} = 12.77 mm (calculated) and 12.80 mm (optimized), l_{s2} = 1.04 mm (calculated) and 1.13 mm (optimized), and w_{s2} = 5.11 mm (calculated) and 5.19 mm (optimized). The remaining parameters are w_{s1} = 5.79 mm and w_{s3} = 5.79 mm.

The corresponding calculated and simulated results are depicted in Figure 2.14. The first self-compensating phase shifter can provide a phase shift of –22.5° ± 0.5° covering 30.4% of relative bandwidth from 25.8 to 34.9 GHz. The second one is able to provide a phase shift of –45° ± 1°

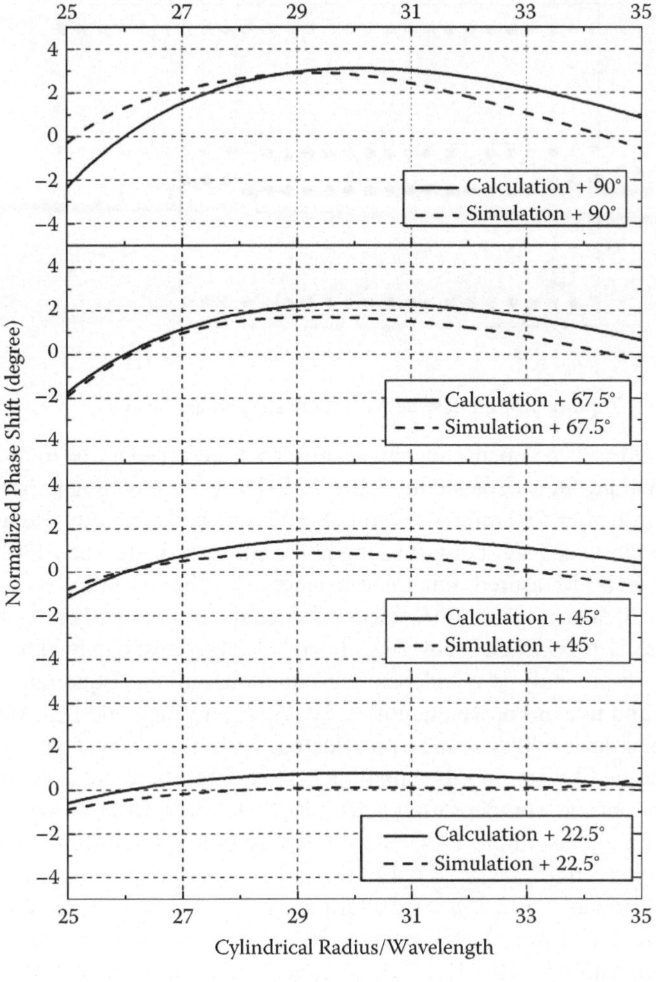

FIGURE 2.14 Calculated and simulated results of four self-compensating phase shifters with desired different phase shifts.

covering 42% of relative bandwidth from 24.8 to 37.4 GHz. The third one can provide a phase shift of −67.5° ± 1.8° covering 46.4% of relative bandwidth from 25.2 to 39.1 GHz. The fourth one is able to provide a phase shift of −90° ± 3° covering 44% of relative bandwidth from 24.2 to 37.4 GHz, respectively.

4. Transform the dielectric-filled waveguide based phaser into the SIW-based phaser. Until now, the developed phase shifters have been designed in the equivalent dielectric-filled waveguide with solid wall. Using Equation (1.1) and considering the fabrication requirement, solid walls are replaced with several rows of metalized vias. The waveguide-based self-compensating phase shifter is transformed into the SIW form.

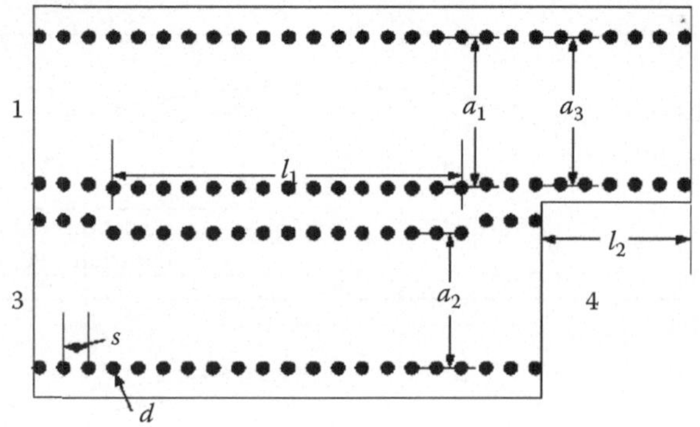

FIGURE 2.15 Topology of an SIW self-compensating phase shifter.

A –90° self-compensating phase shifter was designed as the first showcase. The final layout is depicted in Figure 2.15. The corresponding parameters are $a_1 = 6$ mm, $a_2 = 5.32$ mm, $a_3 = 6$ mm, $l_1 = 12$ mm, and $l_2 = 1$ mm. Return loss of such a fabricated self-compensating phase shifter is better than 13 dB within 27–37 GHz. Measured amplitude balances of different branches are within 0.2 dB. Measured phase differences between the two branches are –90° ± 2.5° from 25.11 to 39.75 GHz covering about 49% of relative bandwidth.

There are only two independent equations among Equations (2.15) to (2.17) and five unknown quantities, i.e., l_{s1}, l_{s2}, w_{s1}, w_{s2}, and w_{s3}. Any combination of three of them can be preselected. Then, the two rest parameters can be obtained by solving the equations. Therefore, there are several ways to design this self-compensating phase shifter to meet different requirements.

As another example, a –45° self-compensating phase shifter was designed with fixed length, i.e., $l_{s1} = 6.4$ mm and $l_{s2} = 0.6$ mm. The parameters were preselected as $\varepsilon_r = 2.2$, $h = 0.508$ mm, $a = 5$ mm, $s = 0.8$ mm, $d = 0.4$ mm, and $w_{s3} = 4.79$ mm. It was proposed to operate at the center frequency of 33 GHz with $\Delta f = 10$ GHz. If these equations are inconsistent, the approximate values can be found by use of the least squares method. The values of w_{s1} and w_{s2} were calculated to be 4.445 and 4.789 mm, respectively. Through full-wave simulation and optimization, parameters of the –45° self-compensating phase shifter were updated to $a_1 = 5$ mm, $a_2 = 4.69$ mm, $a_3 = 5$ mm, $l_1 = 6.4$ mm, $l_2 = 0.6$ mm, $ml = 7.5$ mm, and $mw = 2$ mm. The definitions of these symbols are also shown in Figure 2.15. Measured return loss is better than 10.8 dB within 27–40 GHz. Amplitude balance of different branches is also good, and the measured value between the two paths is within 0.28 dB. The measured phase difference between the two branches is –45° ± 3.5° from 24 to 40 GHz.

2.2.2 Tolerance Analysis for Self-Compensating Phase Shifter

The performance of the SIW self-compensating phase shifter is primarily determined by the SIW actual widths a_1, a_2, and a_3, the distance between neighboring metalized vias s, the diameter of the metalized via d, the substrate relative permittivity ε_r, and the lengths of transmission line l_1, l_2. The errors of all these parameters lead to the deviation of phase shift φ. Therefore, a tolerance analysis based on the sensitivity approach becomes necessary.

As described in Section 1.1.2, the largest value of $\Delta\varphi$ is related to the worst case, which can be predicted by the first-order effects of parametric variations.

$$\frac{|\Delta\varphi|_{\max}}{\varphi} = \left|\frac{\Delta a_1}{a_1} S_{a_1}^{\varphi}\right| + \left|\frac{\Delta a_2}{a_2} S_{a_2}^{\varphi}\right| + \left|\frac{\Delta a_3}{a_3} S_{a_3}^{\varphi}\right| + \left|\frac{\Delta \varepsilon_r}{\varepsilon_r} S_{\varepsilon_r}^{\varphi}\right| + \left|\frac{\Delta s}{s} S_{s}^{\varphi}\right| + \left|\frac{\Delta d}{d} S_{d}^{\varphi}\right| + \left|\frac{\Delta l_1}{l_1} S_{l_1}^{\varphi}\right| + \left|\frac{\Delta l_2}{l_2} S_{l_2}^{\varphi}\right| \tag{2.19}$$

In Equation (2.19), Δa_1, Δa_2, Δa_3, $\Delta\varepsilon_r$, Δs, Δd, Δl_1, and Δl_2 denote the maximum errors of a_1, a_2, a_3, ε_r, s, d, l_1, and l_2, respectively.

On the other hand, to determine the required fabrication accuracy of l_2, there is

$$\Delta l_2 = \frac{l_2}{|S_{l_2}^{\varphi}|}\left[\begin{array}{l} \frac{|\Delta\varphi|_{\max}}{\varphi} - \left|\frac{\Delta a_1}{a_1} S_{a_1}^{\varphi}\right| - \left|\frac{\Delta a_2}{a_2} S_{a_2}^{\varphi}\right| - \left|\frac{\Delta a_3}{a_3} S_{a_3}^{\varphi}\right| \\ - \left|\frac{\Delta \varepsilon_r}{\varepsilon_r} S_{\varepsilon_r}^{\varphi}\right| - \left|\frac{\Delta s}{s} S_{s}^{\varphi}\right| - \left|\frac{\Delta d}{d} S_{d}^{\varphi}\right| - \left|\frac{\Delta l_1}{l_1} S_{l_1}^{\varphi}\right| \end{array}\right] \tag{2.20}$$

As an example, the required phase shift φ equals –90°, and the values of a_1, a_2, a_3, ε_r, s, d, l_1, and l_2 are 6 mm, 5.32 mm, 6 mm, 2.2, 0.8 mm, 0.4 mm, 12.7741 mm, and 1.0354 mm. The values of Δa_1, Δa_2, Δa_3, $\Delta\varepsilon_r$, Δs, Δd, and Δl_2 are assumed to be ±0.02 mm, ±0.02 mm, ±0.02 mm, ±0.02, ±0.02 mm, ±0.02 mm, and ±0.01 mm, respectively. The required fabrication accuracy of Δl_1 can be calculated to be ±0.0167 mm by Equation (2.20), allowing a ±5° error of $\Delta\varphi$.

2.3 SUBSTRATE INTEGRATED COUPLER

Directional couplers are a four-port device and used widely in feeding structures for arrays. As is well known, the conventional 90° 3 dB directional coupler has three important characteristics: (1) equal power distribution between two output ports excited at each input port, (2) 90° phase difference between two output ports excited at each input port, and (3) two input ports are isolated. The ideal coupler is also lossless and matched at all ports.

There are two different types of SIW directional couplers introduced in this part. One is the H-plane coupler, and the other is the E-plane coupler. The coupling section of these SIW-based couplers is a continuous aperture to achieve a strong coupling.

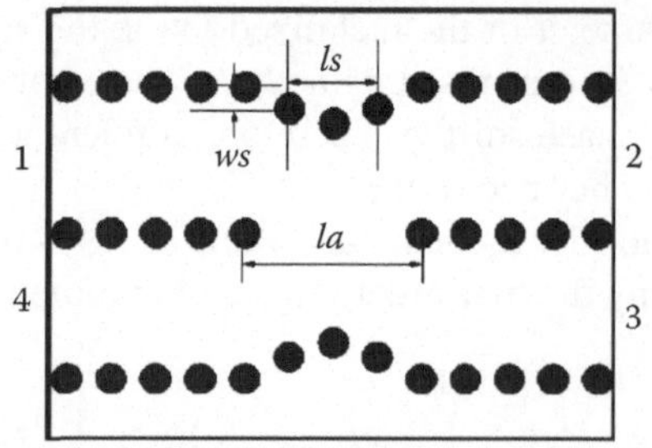

FIGURE 2.16 Configuration of the SIW H-plane coupler.

2.3.1 SIW H-Plane Coupler

The configuration of an SIW H-plane coupler is shown in Figure 2.16. Waveguide steps are used to achieve the good matching. The even- and odd-mode theory can be employed in the design. There are two modes propagated within the coupling section, i.e., the TE_{10} and TE_{20} modes in the SIW section. Their phase constants are β_1 and β_2, respectively. Thus, Equation (2.21) can be used to calculate the required length of the coupling section, *la*.

$$(\beta_1 - \beta_2) \times la = \frac{\pi}{2} \tag{2.21}$$

The coupler is optimized under the following two goals: to minimize $|S_{11}|$ and $|S_{41}|$ and to keep $|S_{21}|$ and $|S_{31}|$ close to –3 dB. First, the step width, *ws*, is optimized to obtain equal $|S_{21}|$ and $|S_{31}|$. Second, the aperture length, *la*, and step length, *ls*, are optimized to obtain the minimum of $|S_{11}|$ and $|S_{41}|$. Because the value of *ls* and *la* also affects the power distribution, it is necessary to repeat these two steps to achieve an optimal result. If feasible, the SIW width can also be optimized to achieve a better performance.

An example was designed and optimized at the V-band. The corresponding parameters are $la = 2.4$ mm, $ls = 1.2$ mm, and $ws = 0.5$ mm. The reflection and coupling coefficients are almost below –20 dB within the frequency band of 55–65 GHz. It also has well-balanced outputs. The phase difference between output ports is 90.5° ± 1° within the frequency band of 57–65 GHz.

2.3.2 SIW E-Plane Coupler

A double-slot E-plane coupler is introduced in SIW technology, which has the configuration shown in Figure 2.17. Ports 1 and 2 are located in one layer, while ports 3 and 4 are in the other layer. A pair of long slots is cut on the middle conductor cladding of the SIW broadside near the metalized via wall. In the coupling region, there are two kinds of modes: the TE_{10} mode in the SIW and the transverse electromagnetic (TEM) mode in the equivalent rectangular coaxial line. The length of the coupling region, *sl*, can be determined on the basis of the even- and odd-mode theory as well.

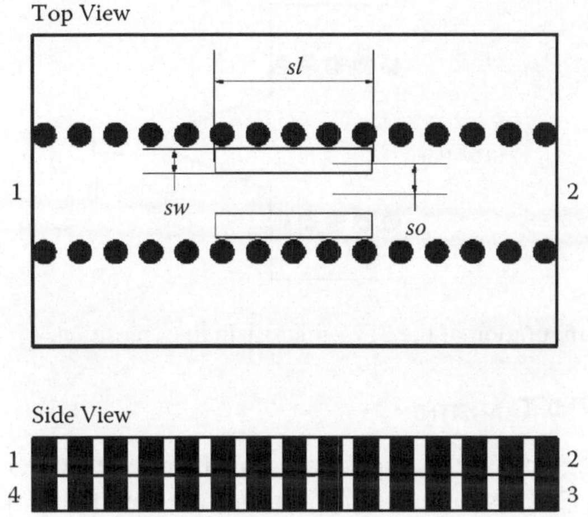

FIGURE 2.17 Configuration of the SIW E-plane coupler.

$$(\beta_1 - \beta_2) \times sl = \frac{\pi}{2} \tag{2.22}$$

In Equation (2.22), β_1 and β_2 are the phase constants of the TE_{10} mode and the TEM mode, respectively. They can be determined by simulation tools. Three parameters, *sl*, the slot width, *sw*, and the slot offset, *so*, are optimized to minimize $|S_{11}|$ and $|S_{41}|$ and to keep $|S_{21}|$ and $|S_{31}|$ close to –3 dB.

A V-band E-plane SIW coupler was designed and optimized. The related parameters are sl = 2.645 mm, sw = 0.4 mm, and so = 0.55 mm. Both reflection and coupling coefficients are below –28 dB within the frequency band of 55–65 GHz. The phase difference between output ports is 90.25° ± 0.5° within the interested frequency band.

2.4 SUBSTRATE INTEGRATED CIRCUIT–RELATED TRANSITION

2.4.1 SIW–Microstrip Line Transition

Figure 2.18 shows the transition proposed in [61] from a microstrip line to an SIW within the same dielectric substrate. The structure consists of a tapered microstrip line section that connects a 50 Ω microstrip line and the integrated SIW. The taper is used for transforming the quasi-TEM mode in the microstrip line into the TE_{10} mode in the SIW. Because the electric fields of the two dissimilar structures are approximately oriented in the same direction, the microstrip line is well suited to excite the SIW, and they also share the same profile.

The design of such a transition is simple and straightforward. First, the size of the SIW can be determined by considering the dielectric effect and well-established theory as described in Chapter 1. The width and length of the tapered section are then designed and optimized to achieve only one goal: to minimize $|S_{11}|$.

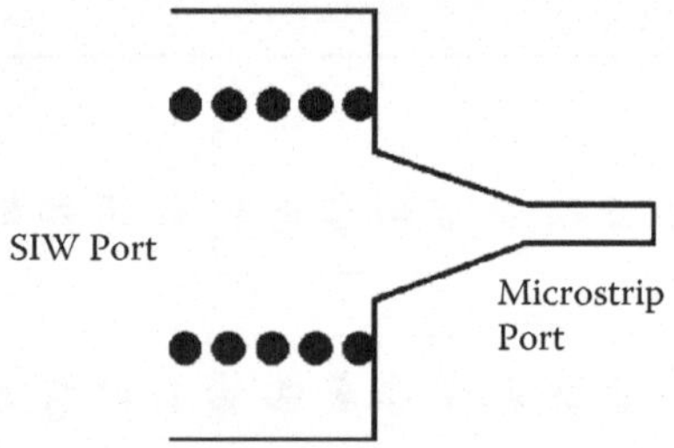

FIGURE 2.18 Configuration of the SIW–microstrip line transition.

2.4.2 SIW–CPW Transition

The coplanar waveguide (CPW) is a very promising microwave and millimeter-wave transmission line. Because its electrical properties are almost independent of the dielectric thickness, it is compatible with thick substrates.

Figure 2.19 shows the transition structure from an SIW to a CPW integrated on the same substrate layer [62]. It consists of a CPW with a 90° bend on each slot. A stub is added on the CPW line to match the transition. There may be a parallel plate mode propagating between the CPW and the back plane before the SIW. Two metallic vias are added to suppress such a mode. The length of each bend slot on the CPW is approximately $\lambda/4$ and ends with a short circuit. The electric fields on these slots match the TE_{10} field distribution very well within the SIW. The CPW is a uniplanar transmission line, but the SIW is a structure with the top and bottom claddings. It is difficult to achieve a good matching when this discontinuity exists. Therefore, a stub has to be added to obtain good insertion loss. The stub length, the stub width, the slot length, and the slot width must be optimized to minimize insertion loss and match the transition over a wide bandwidth.

Another transition between a grounded coplanar waveguide (GCPW) and an SIW was investigated in [63] and its configuration is shown in Figure 2.20. The proposed scheme employs a current probe, not the voltage probe described in [62], to transfer power between these two different transmission lines. By interfacing the SIW with the GCPW other than the microstrip line, substrate thickness can be increased without incurring a penalty because of transmission loss. Thus, it is possible to achieve higher-quality components.

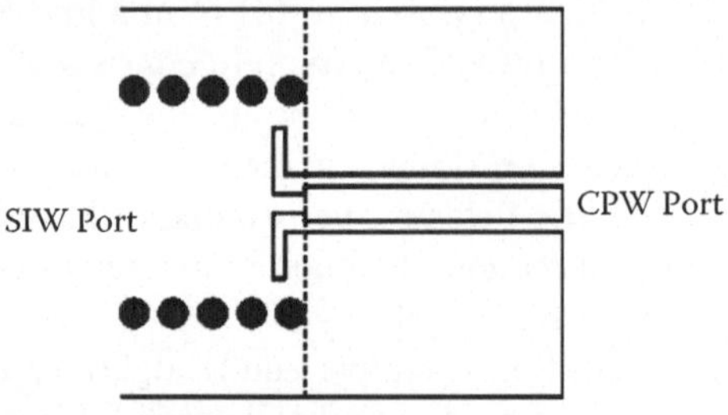

FIGURE 2.19 Configuration of the SIW–CPW transition.

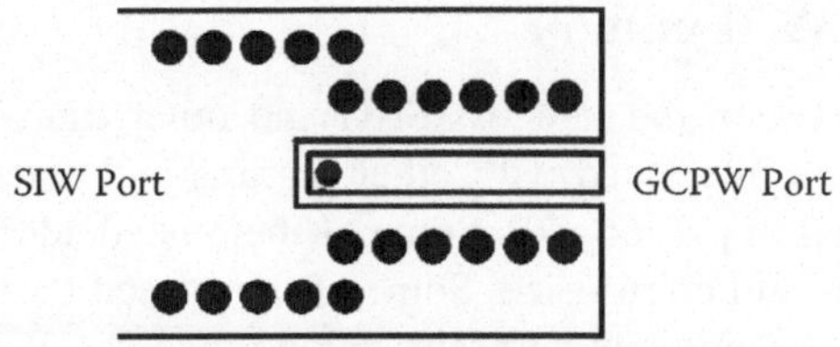

FIGURE 2.20 Configuration of the SIW–GCPW transition.

2.4.3 SIW–Waveguide Transition

As described above, an SIW can be fed by conventional planar transmission lines, namely, the microstrip line, the CPW, etc. However, in antenna applications, the power will be radiated by these open feedlines, which deteriorates the antenna performance. The radiation patterns may present higher SLL, especially in the range of high frequency. The metallic shield can be used to avoid the leakage. Besides, the SIW can be fed by waveguides to remove the direct radiation from the feed discontinuity.

The simplest configuration of the transition between an SIW and an air-filled standard waveguide is illustrated in Figure 2.21. A rectangular coupled aperture is etched on the top conductor cladding. A stepped SIW is used since this structure is less sensitive to mechanical tolerance than one that is not stepped.

Besides, several works focus on how to expand the bandwidth of the SIW–waveguide transition. The first design was introduced in [64], which uses a Y-type divider to combine two parallel SIW branches together. These two SIWs are shorted at one end and connected to a single SIW port on the other end through a Y-type divider. The microwave energy is transferred into an air-filled waveguide through the coupled aperture etched on the top conductor claddings of two SIWs at the same time.

Another example of the SIW–waveguide transition was developed in [65]. There are two slots etched on the top conductor cladding of the SIW. The energy is coupled into the waveguide through these slots. Two metalized vias are placed in the center of slots to improve the performance of the transition.

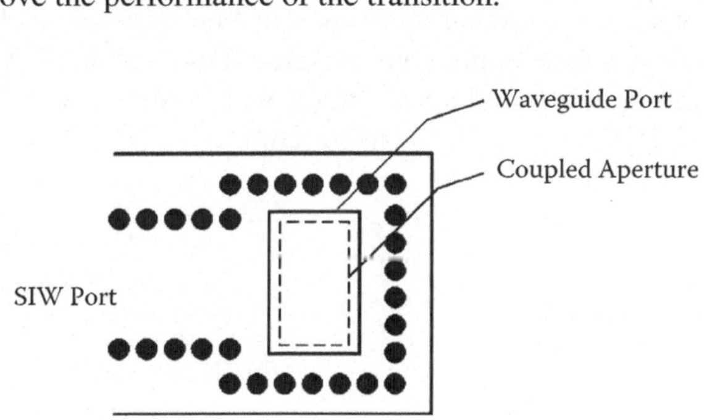

FIGURE 2.21 Configuration of the SIW–waveguide transition.

2.4.4 SIIG–GCPW Transition

A good transition between the low-loss SIIG and other transmission lines is crucial for the SIIG to be integrated with other circuits and antennas. Several works have been investigated in [33, 66–70]. Some of them need additional baluns, which increase the total loss and circuit size. Some of them need tapered dielectric strips, which are inaccurate through low-cost fabrication processes. In some cases, the conductor edge must be very close to the circuit border and cannot be fabricated according to low-temperature cofired ceramic (LTCC) design rules.

To overcome these drawbacks, a transition from a GCPW to an SIIG was proposed, designed, and fabricated using the standard LTCC process [34]. It operates in the V-band and is integrated in a single substrate. The dielectric strip of the example SIIG has seven stacked ceramic tapes, while the isolating film has one additional ceramic tape. Each ceramic tape has a thickness of 0.1 mm. The SIIG strip width is 1.25 mm. The thickness of the 50 Ω GCPW feedline is 0.1 mm.

Such a transition can be divided into two parts. One part is an integrated planar SIW horn to excite the SIIG. The field within the SIW horn is the TE_{10} mode, which matches the dominant mode field in an SIIG, E_{11}^{z} [34]. To match a wide band, the distance between two rows of metalized vias forming the sidewall of the horn is gradually flared. To decrease the radiation leakage owing to the discontinuity, a tapered V-slot is etched on the top conductor cladding. Besides, there exist two rows of reflecting metalized vias at the open end of the horn to further minimize the radiation leakage. The other part is the SIW horn excited by a GCPW. The GCPW line excites the SIW horn through a metalized post with the proper inserting depth. Several metalized vias are placed surrounding the channel to achieve a better field confinement. To improve the reflection, a circular metalized plate is placed at the end of the post.

Compared with other existing SIIG-related transitions, this type of CPW–SIIG transition is able to decrease design difficulty, improve fabrication reliability, and increase design flexibility. Besides, it is able to resolve the contradiction of the substrate thickness between the SIIG and the CPW. In the V-band applications, the CPW needs a substrate thickness of 0.1 mm compatible with integrated circuits, while the SIIG needs a substrate thickness approximately more than 0.6 mm. It usually leads to incompatibility with truly planar fabrication technologies and a challenge to find the solution to this problem in previous designs. Based on the LTCC process, the SIIG and CPW can be fabricated on substrates with different thicknesses, which is able to meet the requirements of different structures.

3 Substrate Integrated Slot Array Antennas

As an important class of antennas, the slot array antenna has advantages in realizing medium-gain radiation patterns easily. The broad or narrow conductor wall of a waveguide is cut by slots with different shapes and rotations. In this case, the flow of current is discontinued and the power is radiated through slots to free space. The slot array antenna is capable of being implemented by the rectangular waveguide and substrate integrated waveguide (SIW) technologies. There are two basic categories: standing-wave antenna and traveling-wave antenna.

In this chapter, an overview of popular and representative substrate integrated slot array antenna technology is introduced, including standing-wave SIW slot array antenna, traveling-wave SIW slot array antenna, and SIW conformal slot array antenna.

3.1 STANDING-WAVE SIW SLOT ARRAY ANTENNA

For decades, researchers have made many contributions to the development of waveguide longitudinal slot array antennas through analytical theories, numerical techniques, and experimental methods. For the first time, the normalized conductance of a single slot etched on a rectangular waveguide was investigated in [71] based on the transmission line theory and Green function in waveguides. A variational solution was employed to compute the accurate impedance of a radiating longitudinal shunt slot in the broad face of a rectangular waveguide [72, 73]. The impedance properties of a waveguide longitudinal slot antenna with arbitrary length and offset in guide walls of finite thickness were discussed in [74] based on the method of moments.

Elliott's design procedure [75–79] has been well accepted to accurately synthesize a slot array. Through solving nonlinear equations, this iterative procedure is able to determine slot parameters to achieve the desired aperture distribution and keep the excellent input matching. All mutual couplings were taken into consideration in the process. Furthermore, an integral equation and the moment method design procedure were given in [80], which removes the characterizing requirement of radiating elements based on their equivalent circuit representations. The infinite array model developed in [81] is quite accurate for large-scale waveguide arrays. An iterative procedure described in [82] aims to determine the off-center-frequency performance of a waveguide slot array. A strategy was presented in [83] for the full-wave analysis of a longitudinal waveguide slot array. In [84], a waveguide slot array configuration was proposed to allow a simpler realization of a slot array antenna. When the requirement of pattern changes, this type of antenna can be reused. This advantage comes from the slots printed on a copper-clad substrate. A technique for the design of a nonuniformly spaced waveguide slot array was proposed in [85]. These slots are in active resonance, and their positions are adjusted so that the sources of the array will have the desired amplitudes and phases. The conductances of the slots depend on the desired pattern and the efficiency. The concept of subarray was introduced in [86] to provide a wide reflection and gain bandwidth. By using an appropriate objective function as described in [87], simulated annealing was found to prevent the emergence of weakly excited slots in waveguide slot arrays designed to produce required patterns. A series of waveguide antennas were developed in [88–94]. The waveguides and the slots are located on the same plate without separation. This topology may alleviate the necessity of electrical contact between the lower waveguide plate and the upper slot plate. In [95], a single-layer waveguide slot array at 76 GHz has a sidelobe level (SLL) below –24 dB and a high gain of 34.5 dBi.

These above-mentioned techniques are applied to developments of slot array antennas on conventional rectangular waveguides. As described before, properties of an SIW are close to those of a rectangular waveguide. Thus, an SIW-based slot array antenna can be developed with a configuration similar to that of the waveguide counterpart. Easy as it might seem to be, the differences in their heights may lead to distinct features.

3.1.1 Standing-Wave SIW Slot Array Antenna Synthesis

The basis of the slot array synthesis is evaluating the properties of a single slot etched on the broad wall of the SIW, as shown in Figure 3.1. The longitudinal slot on an SIW can also be modeled as a shunt element according to its admittance, Y, on a transmission line with the characteristic conductance, G_0 [76]. These characteristics can be obtained by accurate calculations or measurements. For a slot, its parameters include the width, sw, the length, sl, and the offset, x, from the centerline.

As described in [75], it is revealed that the assumption of the shunt element relies on the symmetry of the field in a slot. This assumption is less appropriate for larger

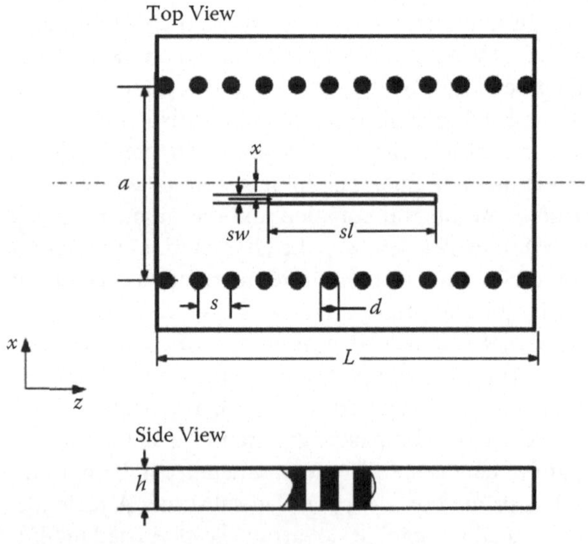

FIGURE 3.1 Configuration of a single slot on the SIW.

offsets, as well as slots on reduced-height waveguides. In [96], a number of simulations were implemented to investigate the transverse electric field in a slot with the resonant length. Although the height is considerably reduced, the asymmetry of the electric field in a slot is small enough to satisfy the shunt element assumption in the range of desired offset values. At large slot offset, the asymmetry of the amplitude and phase appears to be visible and becomes severe. Fortunately, these offsets are not employed in conventional designs.

Simulated S-parameters based on the SIW model are carried out to extract properties of the single slot. To remove the effect of the transmission line, the reference planes at two ports should be shifted inward.

$$S = \begin{pmatrix} e^{j\beta L} & 0 \\ 0 & e^{j\beta L} \end{pmatrix} S^S \begin{pmatrix} e^{j\beta L} & 0 \\ 0 & e^{j\beta L} \end{pmatrix} \tag{3.1}$$

In Equation (3.1), S^S is the scattering parameter matrix, and β is the propagation constant of the TE_{10} mode. Here, the normalized admittance can be calculated by

$$\frac{Y^a}{G_0} = -\frac{2S_{11}}{1+S_{11}} \tag{3.2}$$

Here, the resonant length and G/G_0 can be obtained with an appropriate choice of sl to make $B/G_0 = 0$ at different x. Then, the admittance of the slot versus sl and x can be recorded as a two-dimensional lookup table. The calculated discrete data can be interpolated to generate continuous design curves. Admittances of different x converge at the resonant length and become divergent when sl deviates from the resonant length. The deviations are not obvious for a conventional waveguide design [79], but are noticeable for an SIW design because of its much smaller thickness. For the slot etched on the SIW, there are stronger dependences of admittance properties on parameters of the slot [97].

Several important equations, which are used to design the standing-wave SIW slot array antenna, are listed as follows at first:

$$\frac{Y_n^a}{G_0} = K_1 f_n \frac{V_n^s}{V_n} \tag{3.3}$$

In Equation (3.3),

$$K_1 = \frac{\lambda}{jw}\sqrt{\frac{2\left(\frac{k}{k_0}\right)}{\eta G_0\left(\frac{\beta}{k}\right)k^2 wh}}$$

$$f_n = \frac{\left(\dfrac{\pi}{2kl_n}\right)\cos(\beta l_n)}{\left(\dfrac{\pi}{2kl_n}\right)^2 - \left(\dfrac{\beta}{k}\right)^2}\sin\left(\frac{\pi x_n}{w}\right)$$

w is the SIW equivalent width and h is the SIW height. And,

$$\frac{Y_n^a}{G_0} = \frac{2f_n^2(x_n,l_n)}{\dfrac{2f_n^2(x_n,l_n)}{\dfrac{Y}{G_0}(x_n,l_n)} + j\dfrac{\beta}{k}(k_0 h)\left(\dfrac{w}{\lambda}\right)^3 \displaystyle\sum_{\substack{m=1\\ m\neq n}}^{N} \frac{V_m^s}{V_n^s} g_{mn}(x_m,l_m,x_n,l_n)} \tag{3.4}$$

In Equation (3.4),

$$g_{mn}(x_m,l_m,x_n,l_n) = \int_{-k_0 l_m}^{k_0 l_m} \cos\left(\frac{\lambda_0 z'_m}{4l_m}\right) \left\{ \begin{array}{l} \dfrac{\lambda_0}{4l_n}\left[\dfrac{e^{-jk_0R_1}}{k_0R_1} + \dfrac{e^{-jk_0R_2}}{k_0R_2}\right] + \left[1 - \dfrac{\lambda_0^2}{(4l_n)^2}\right] \\ \times \displaystyle\int_{-k_0 l_n}^{k_0 l_n} \cos\left(\frac{\lambda_0 z'_n}{4l_n}\right)\frac{e^{-jk_0R}}{k_0R}\,dz'_n \end{array} \right\} dz'_m$$

where $\omega = 2\pi f$, $k = \omega\sqrt{\mu_0\varepsilon_0\varepsilon_r}$, $k_0 = \omega\sqrt{\mu_0\varepsilon_0}$, and $\eta = \omega\mu_0/k$. x_n is the offset of the nth slot, and l_n is half the length of the nth slot. V_n^s is the slot voltage in the nth slot, and V_n is the mode voltage. Y_n^a/G_0 is the equivalent admittance of the nth slot considering the mutual coupling, and Y/G_0 is the equivalent admittance of the nth slot ignoring the mutual coupling. $g_{mn}(x_m, l_m, x_n, l_n)$ represents the external mutual coupling between the mth slot and the nth slot. R is the distance between the point (x_m, z'_m) in the mth slot and the point (x_n, z'_n) in the nth slot, R_1 is the distance between the point (x_m, z'_m) in the mth slot and the start point (x_n, l_n) of the nth slot, and R_2 is the distance between the point (x_m, z'_m) in the mth slot and the endpoint $(x_n, -l_n)$ of the nth slot.

To synthesize an N-slot resonant slot SIW array antenna, slots should be alternately spaced $\lambda_g/2$ apart. Each slot voltage can be determined according to the required radiation pattern. Thus, this slot voltage distribution is known. With the generated Y/G_0 curves as a function of sl and x, initial values for lengths and offsets of all slots in the array can be found out. Since the design procedure converges rapidly, it is also sufficient to start with all the slots on the centerline with the resonant length. In this case, the initial value of the following mathematical expression can be calculated:

$$\sum_{\substack{m=1\\ m\neq n}}^{N} \frac{V_m^s}{V_n^s} g_{mn}(x_m,l_m,x_n,l_n) \tag{3.5}$$

Next, a computer search is implemented to find a couplet (x_n, l_n), which will make the denominator on the right-hand side of Equation (3.4) pure real. There is a continuum of couplets (x_n, l_n) satisfying this condition. Similarly, there is a continuum of couplets (x_p, l_p) for the pth slot, which also make the denominator on the right-hand side of Equation (3.4) pure real. According to (3.3),

$$\frac{Y_p^a/G_0}{Y_n^a/G_0} = \frac{f_p}{f_n}\frac{V_p^s}{V_n^s}\frac{V_n}{V_p} \tag{3.6}$$

For a given acceptable couplet (x_n, l_n), there is only one acceptable couplet (x_p, l_p) to satisfy Equation (3.6). The optimal choice is the one that makes the sum of the normalized active admittances, i.e.,

$$\sum_{n=1}^{N} \frac{Y_n^a}{G_0}$$

to be a specific constant since this is the condition for an input matching. For a linear array, the specific constant should be 1.

Such a procedure is able to obtain length and offset values for all slots. Repeat the above steps with new (x_n, l_n) until the last one becomes very close to its predecessor. The whole design process of the SIW slot array antenna is complete. In fact, this procedure is found to converge fast and does not need too many iterative computations.

Generally speaking, there are three different kinds of sources to generate the field in a slot. The first source is the incident field. Another source comes from the coupling among slots through the air, called the external mutual coupling. The last source comes from the internal coupling among slots etched on the same SIW. For the conventional waveguide, this internal mutual coupling is primarily the TE_{10} mode coupling. For the SIW, the internal mutual coupling owing to higher-order modes, especially the TE_{20} mode, cannot be neglected [75, 98], because of its reduced-height configuration. It has been proved that the internal coupling of the TE_{20} mode between two neighboring slots plays a significant role in the SIW. Thus, Equation (3.4) should be modified as described in [77] to obtain a more accurate result after considering the mutual coupling effect.

$$\frac{Y_n^a}{G_0} = \frac{2f_n^2}{\dfrac{2f_n^2}{\dfrac{Y}{G_0}(x_n,l_n)} + MC_n} \tag{3.7}$$

In (3.7),

$$MC_n = j\frac{\beta}{k}(k_0 h)\left(\frac{w}{\lambda}\right)^3 \sum_{\substack{m=1\\ m\neq n}}^{N} \frac{V_m^s}{V_n^s} g_{mn}(x_m,l_m,x_n,l_n)$$
$$+ j\frac{\beta}{\beta_{20}} e^{-\beta_{20} d}\left[b_n b_{n-1}\frac{V_{n-1}^s}{V_n^s} + b_n b_{n+1}\frac{V_{n+1}^s}{V_n^s}\right]$$

$$b_n(x_n, l_n) = 2\frac{\left(\frac{\lambda}{4l_n}\right)\cosh(\beta_{20}l_n)}{\left(\frac{\beta_{20}}{k}\right)^2 + \left(\frac{\lambda}{4l_n}\right)^2}\cos\frac{2\pi x_n}{w}$$

The above iterative method can be used to calculate initial values of the SIW slot array antenna. After that, a further accurate optimization by the full-wave simulation is still needed. Some slight tunings of offset and length values of slots may achieve the desirable radiation pattern better.

3.1.2 Standing-Wave SIW Slot Array Antenna Examples

Radiation pattern synthesis theory and method of the standing-wave SIW slot array antenna have been relatively mature and accurate. Limited by geometrical constraints and fabrication tolerances, the SIW slot array antenna is more appropriate for lower millimeter-wave frequency applications. This type of antenna is able to achieve different gains, shaped beams, and desired polarizations. Because the distance between neighboring slots is related to the half-wavelength, this antenna usually has a narrow frequency band.

As a slot array antenna example, a –30 dB low SLL SIW design was developed in [96]. It is an X-band 16 × 16 SIW slot array antenna with a compact series feeding divider. Root matching sampling of the Taylor continuous distribution was used with $\bar{n} = 6$. To leave enough margin, –35 dB Taylor distributions were used. Several designed linear arrays can be aligned to form a planar array, while the feature of low SLL in the H-plane is nearly unchanged. The SLL in the E-plane of the planar array can be controlled by the power ratio of the feeding divider. However, in the E-plane of the planar array, there exists a big difference between the designed and full-wave simulated results. Compared with the designed –35 dB SLL, a much higher SLL appears due to the E-plane external coupling effects. In order to achieve the SLL suppression goal, the E-plane coupling has to be compensated by slightly tuning the power ratio through the simulation software for the whole antenna. Measured 10 dB return loss frequency band is 9.85–10.12 GHz. Detailed experimental specifications are listed in Table 3.1.

Another design example is a K-band 16 × 32 SIW slot array antenna. Shaped beam, not just low SLL, is essential for the antenna array in various commercial applications. For instance, an array antenna, which is applied in microwave point-to-point communication, should meet the antenna radiation pattern envelope (RPE) (20–24 GHz) of the European Telecommunications Standards Institute (ETSI) standard document [99]. In the ETSI standard document, the antennas operating at different frequencies have different template patterns.

A variety of methods, including the Schelkunoff synthesis, the Chebyshev synthesis, the Taylor synthesis, etc., have been developed to achieve the requirement of low SLL pattern. However, it is difficult to employ these synthesis methods to satisfy the ETSI standard. Let us give a simple explanation for it. Very low SLL is needed in

TABLE 3.1
Measured Performance of the X-Band Low SLL SIW Slot Array Antenna

Frequency (GHz)	9.9	10	10.1
Measured gain (dBi)	23.5	24.5	24.1
3 dB beamwidth in E-plane (°)	10.4	10.2	9.3
3 dB beamwidth in E-plane (°)	9.5	9.6	9.2
SLL in E-plane (dB)	–30.1	–31.3	–30.2
SLL in H-plane (dB)	–32.4	–30.3	–29.9
Efficiency	61.0%	75.9%	60.4%

the shaped-beam design because the Chebyshev synthesis provides an equal-ripple SLL characteristic. On the other hand, the initial SLL in the antenna RPE is not low, which leads to waste of the low SLL pattern synthesized by the Chebyshev method.

As an example, a shaped-beam SIW array antenna has the configuration shown in Figure 3.2. It includes a 16-way series feeding divider and sixteen 32-slot linear arrays. These linear array antennas are placed alternatively to compensate for the out-of-phase caused by the used series feeding divider, except for the center two. Its radiation pattern should satisfy both copolarized and cross-polarized desired levels for the class 3 antenna RFE in 20–24 GHz.

Shaped beam is essential in both the E-plane and the H-plane. It is easy to meet the shaped-beam requirement in the H-plane because it has 32 elements in this plane. Therefore, the main focus is to synthesize the E-plane pattern and satisfy the class 3 antenna RFE requirement. The active element pattern technique [100, 101], which uses the individual element patterns in the array environment to speed up the calculation of the pattern of the fully excited array, can be employed here. Sixteen 32-slot linear array antennas were modeled and full-wave simulated without a divider. Element pattern data of each linear array antenna, including the effects of mutual coupling and element position, were extracted and recorded. Then, the simulated annealing algorithm [102–104] was employed to implement the pattern optimization, which aims at meeting the pattern requirement within the angle range of (–100°, 100°). By using absorbing materials, the E-plane pattern of –180° to –100° and 100° to 180° can be suppressed below the ETSI standard as a result of the surface-wave suppression. After obtaining the desired excited amplitude and phase of each radiating element, the feeding divider can be achieved by using an unequal SIW power divider.

To validate this design, a prototype antenna was fabricated on the Rogers 5880 substrate with a thickness of 1.575 mm. The permittivity and loss tangent of the substrate are 2.2 and 0.0009, respectively. The whole size of the antenna is 255 × 130 mm^2. Its copolarized and cross-polarized patterns were measured in a microwave anechoic chamber. As shown in Figures 3.3 and 3.4, both copolarized and cross-polarized patterns are below the ETSI standard. The radiation patterns at other frequencies within the interested frequency range were measured as well. As shown in Figures 3.5 and 3.6, both copolarized and cross-polarized patterns

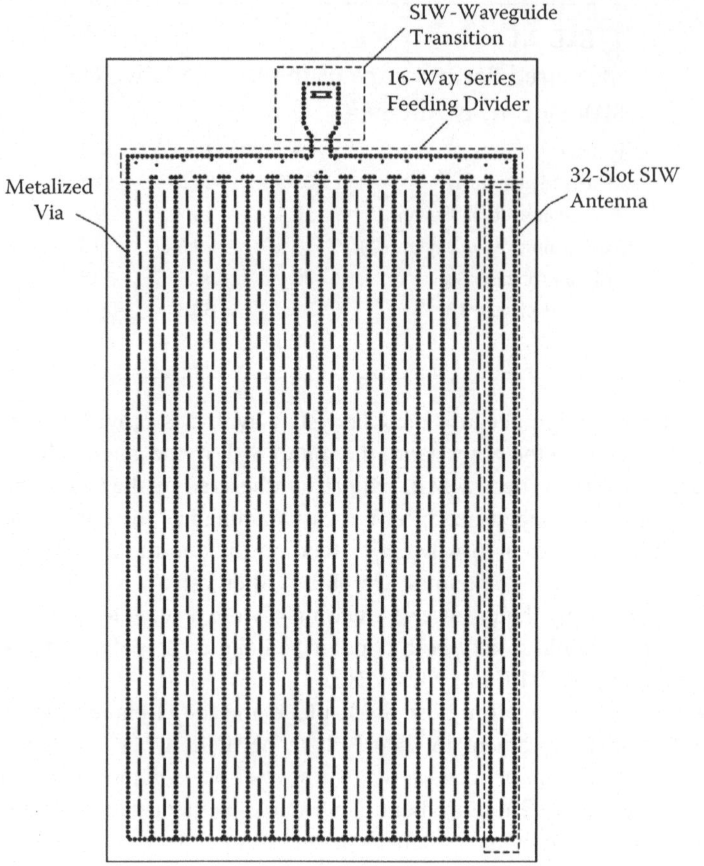

FIGURE 3.2 Configuration of a K-band 16 × 32 SIW slot array shaped-beam antenna.

operating at +75 MHz are below the ETSI standard. As shown in Figures 3.7 and 3.8, both copolarized and cross-polarized patterns operating at –75 MHz are below the ETSI standard.

Another interesting and similar use of a so-called post-wall waveguide technique was first reported for creating an end-feed type of parallel plate slot array at 40 GHz in [105]. The feeding post-wall waveguide with a series of coupling windows is placed at the end of the parallel plate waveguide. A series of slot pairs are etched on the top conductor cladding of the parallel plate waveguide. In this case, the antenna is a kind of series feeding array antenna. Considering the long line effect, the bandwidth of this antenna becomes narrower with the increase of antenna size. To overcome this weakness, the center-feed parallel plate slot array was originally proposed in [106] to expand the bandwidth. The feeding waveguide is placed at the center of the parallel plate waveguide in a single layer, and two subarrays of slot pairs are arranged on both sides of the feeding post-wall waveguide.

A Ka-band 8 × 16 standing-wave slot array antenna with 45° inclined linear polarization (LP) was presented in [107]. This SIW slot array antenna consists of arrays

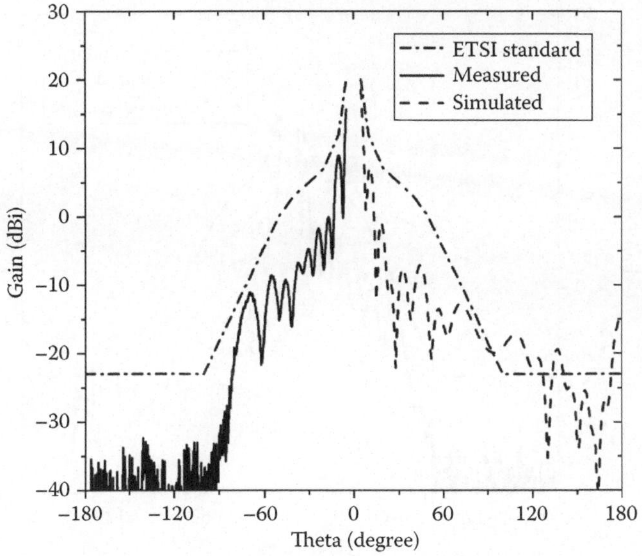

FIGURE 3.3 Measured and simulated copolarized E-plane pattern at 23 GHz of a fabricated K-band 16 × 32 SIW slot array antenna.

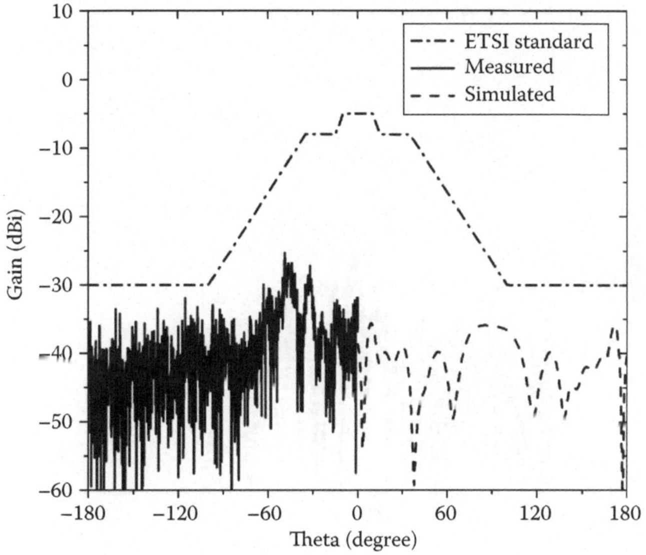

FIGURE 3.4 Measured and simulated cross-polarized E-plane pattern at 23 GHz of a fabricated K-band 16 × 32 SIW slot array antenna.

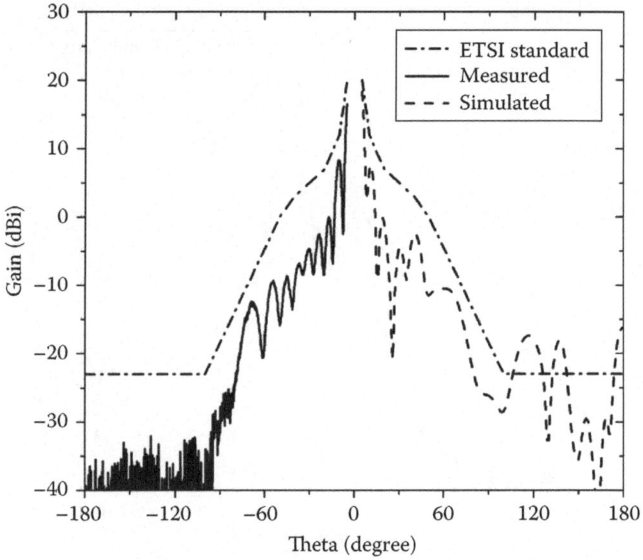

FIGURE 3.5 Measured and simulated copolarized E-plane pattern at 23.075 GHz of a fabricated K-band 16 × 32 SIW slot array antenna.

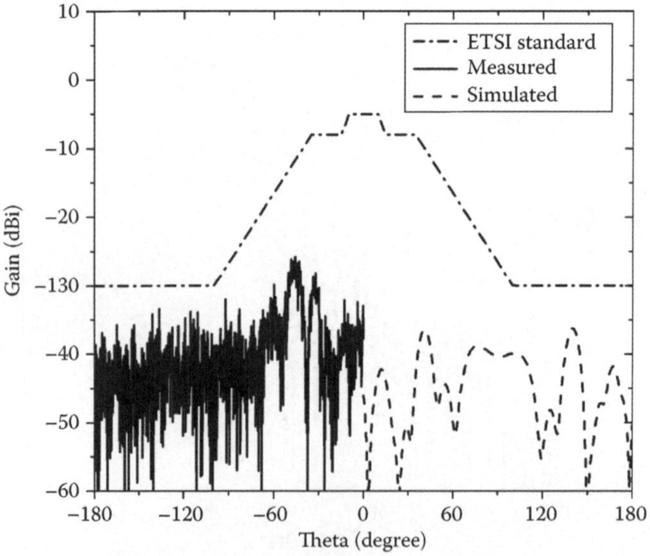

FIGURE 3.6 Measured and simulated cross-polarized E-plane pattern at 23.075 GHz of a fabricated K-band 16 × 32 SIW slot array antenna.

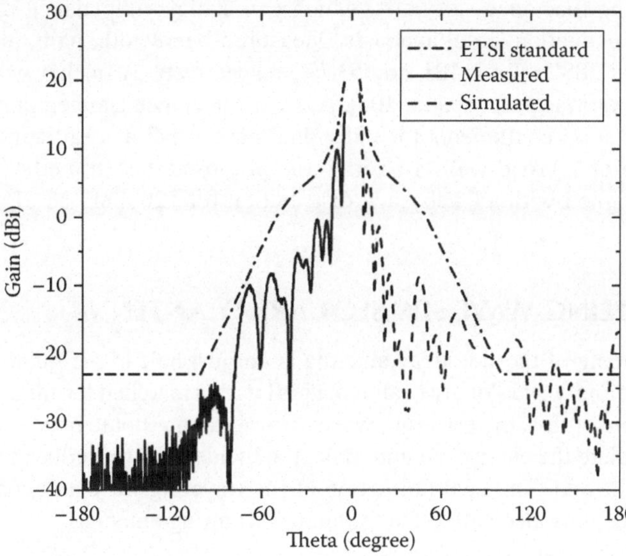

FIGURE 3.7 Measured and simulated copolarized E-plane pattern at 22.925 GHz of a fabricated K-band 16 × 32 SIW slot array antenna.

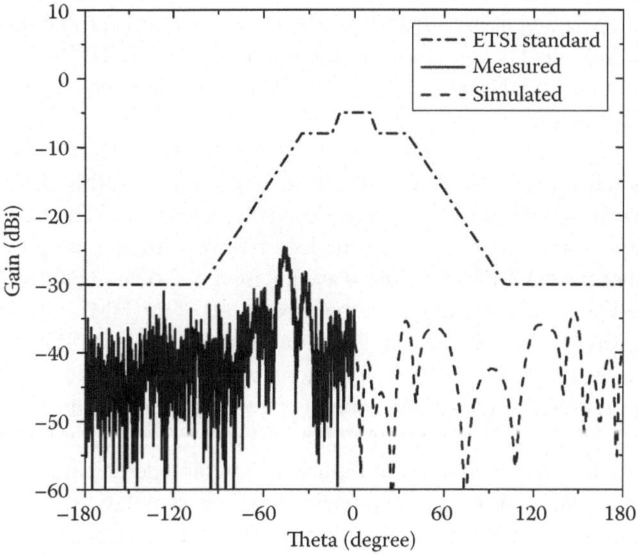

FIGURE 3.8 Measured and simulated cross-polarized E-plane pattern at 22.925 GHz of a fabricated K-band 16 × 32 SIW slot array antenna.

with alternating reactance slot pairs, which can achieve impedance matching and uniform field excitation simultaneously. Measured bandwidth, gain, and efficiency are 990 MHz (2.88%), 24.3 dBi, and 53.7%, respectively. A further work about the SLL suppression was proposed in [108]. Axial displacements are employed to create arbitrary excitation coefficients for individual center-inclined radiating slots along the centerline of a broad wall. To verify the proposed design method, two examples were designed with a Chebyshev distribution for –20 and –26 dB SLLs in the Ka-band.

3.2 TRAVELING-WAVE SIW SLOT ARRAY ANTENNA

If the slot spacing of the slot array antenna is not just half of the guide wavelength, there is a traveling-wave slot array antenna [79] with a matched termination placed at the end. In spite of this, the traveling-wave array antenna should have enough slots to effectively radiate the energy out and make the incident power reduced to an acceptable level at the end. The beam direction of the traveling-wave antenna varies with frequency; thus, it is also called the frequency scanning antenna.

3.2.1 HMSIW Traveling-Wave Antenna with Quadri-Polarization

A Ka-band half-mode substrate integrated waveguide (HMSIW) traveling-wave antenna, which is a planar passive circuit and capable of dynamically changing the state of polarization, was investigated and synthesized in [109]. This antenna is able to operate in either linear or circular polarization (CP). Compared with the SIW structure, the original HMSIW is nearly half of it. However, due to the high current density along the open edge of the HMSIW, it has higher element-to-element parasitic coupling operating at high frequency band. To solve this problem, an additional row of metalized vias is added close to the open edge of the HMSIW. The distance between these vias and the open edge of the HMSIW should be enough to eliminate the perturbation of the propagating field within the HMISW.

This traveling-wave antenna has to be excited at both ends for the polarization diversity. Therefore, a symmetrical structure is preferred in this design. Based on this, a uniformly spaced traveling-wave slot array is selected. The beam is scanned toward the direction of propagation of the leaky wave with operating frequency. The configuration of the HMSIW traveling-wave antenna is shown in Figure 3.9. There are sixteen 45° radiating slots arranged on the broad wall of the HMSIW. Simulated voltage standing wave radio (VSWR) and residual power of the HMSIW traveling-wave antenna are shown in Figure 3.10. It is demonstrated that the residual power is down by operating frequency. This nonresonant array achieves a wide band because of the phase differences between the reflections from different radiating slots [79]. The curve of $1 - |S_{11}|^2 - |S_{21}|^2$ is plotted in Figure 3.11, which demonstrates excellent efficiency of this antenna. Simulated copolarized and cross-polarized E-plane radiation patterns are presented in Figures 3.12 to 3.15 at different frequencies. The beam directions are 26°, 21°, 17°, and 13° corresponding to 34, 35, 36, and 37 GHz, respectively.

Undoubtedly, such a HMSIW slot array antenna with –45° or 45° inclined slot can generate a –45° or 45° LP mode. If two slot antennas with –45° and 45° LP,

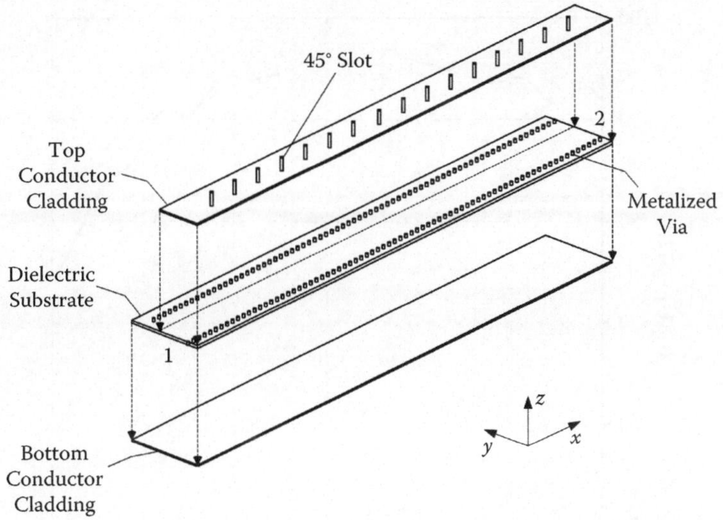

FIGURE 3.9 Configuration of the Ka-band HMSIW traveling-wave antenna.

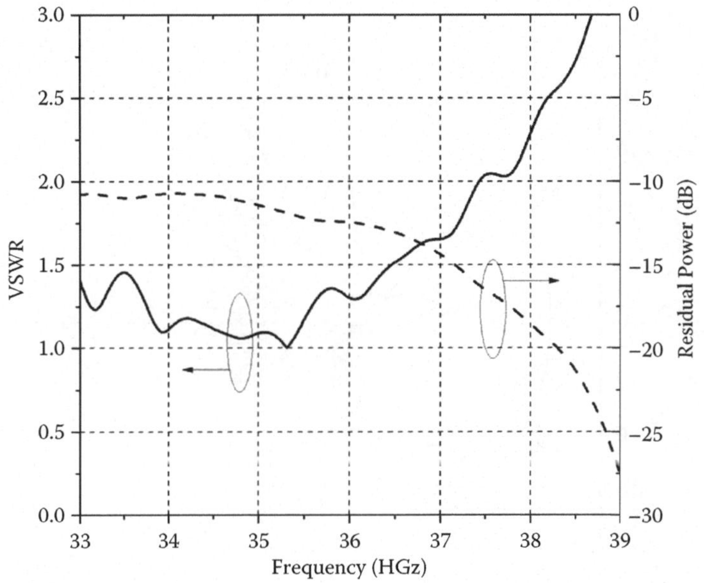

FIGURE 3.10 Simulated VSWR and residual power of the Ka-band HMSIW traveling-wave antenna.

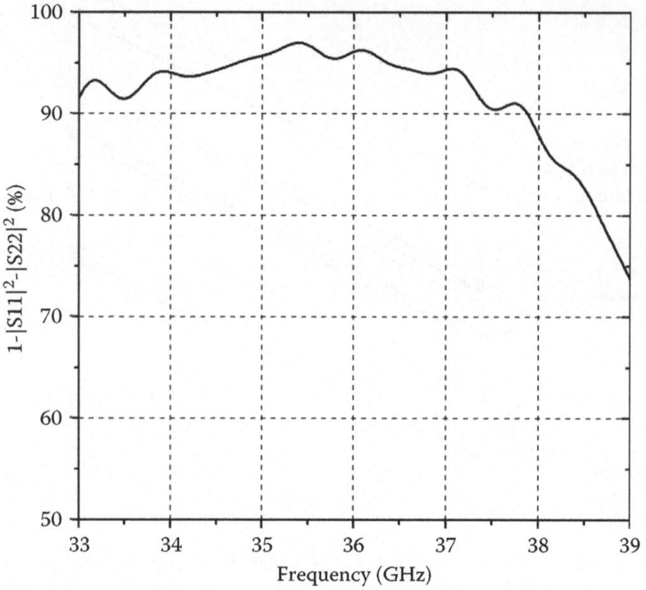

FIGURE 3.11 Simulated efficiency of the Ka-band HMSIW traveling-wave antenna.

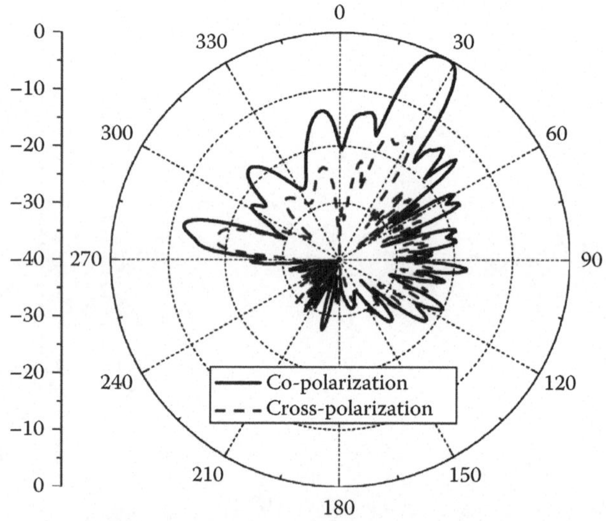

FIGURE 3.12 Simulated E-plane patterns of the HMSIW traveling-wave antenna at 34 GHz.

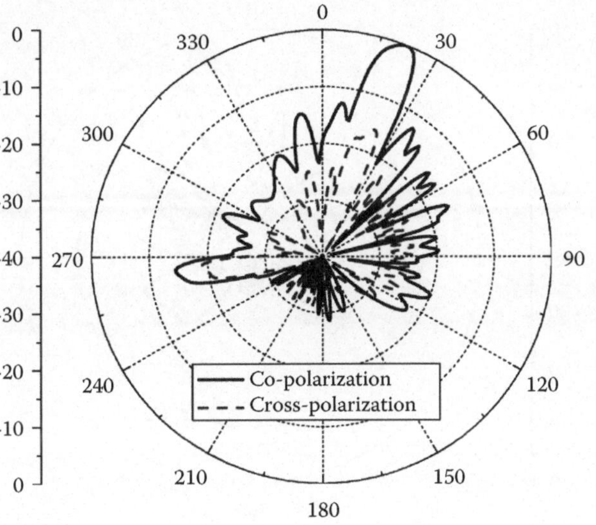

FIGURE 3.13 Simulated E-plane patterns of the HMSIW traveling-wave antenna at 35 GHz.

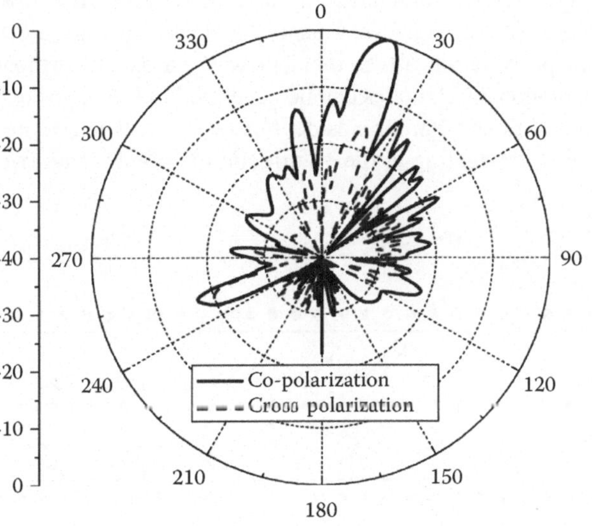

FIGURE 3.14 Simulated E-plane patterns of the HMSIW traveling-wave antenna at 36 GHz.

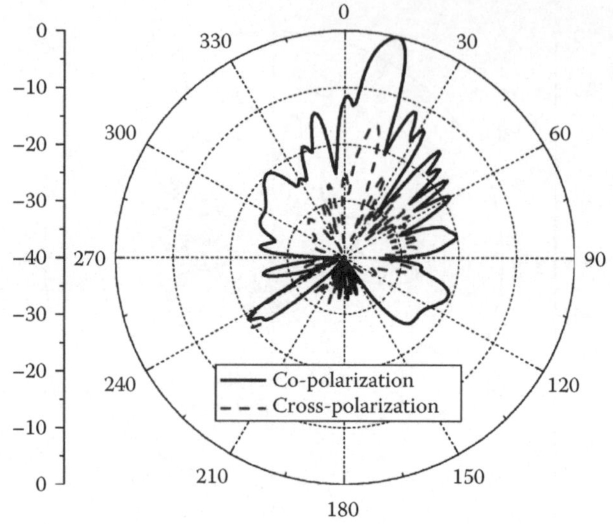

FIGURE 3.15 Simulated E-plane patterns of the HMSIW traveling-wave antenna at 37 GHz.

respectively, are combined symmetrically and excited with 90° relative phase difference, a CP wave will be generated. Therefore, a two-element array can be used to generate left-handed circular polarization (LHCP) and right-handed circular polarization (RHCP) modes. To suppress the mutual coupling between –45° and 45° traveling-wave antennas, an offset distance equal to half the slot distance is required between two traveling-wave antennas.

As shown in Figure 3.16, the multiple-polarization HMSIW antenna array consists of two traveling-wave antennas and a 3 dB directional coupler realizing the 90° phase difference feeding mechanism. As shown in Figure 3.17, by changing the direction of propagation of the incident wave, a pair of orthogonal LP modes and a pair of orthogonal CP modes can be generated in the opposite beam directions with a tilting angle rather than broadside. For instance, when excited at port 1 or 4, the incident wave propagating to the $+x$ direction is divided into two equal portions

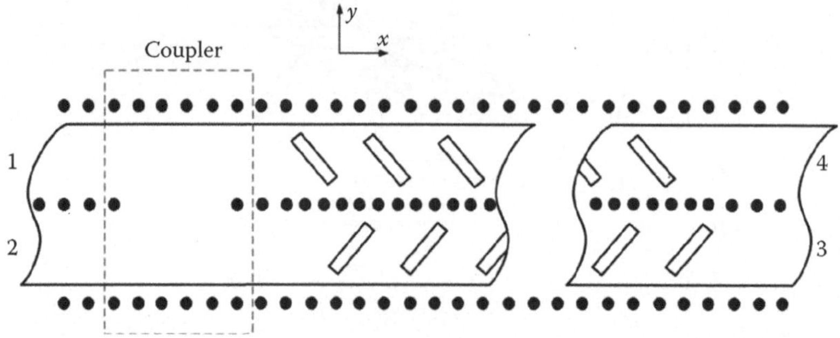

FIGURE 3.16 Configuration of the Ka-band HMSIW quadri-polarization traveling-wave antenna.

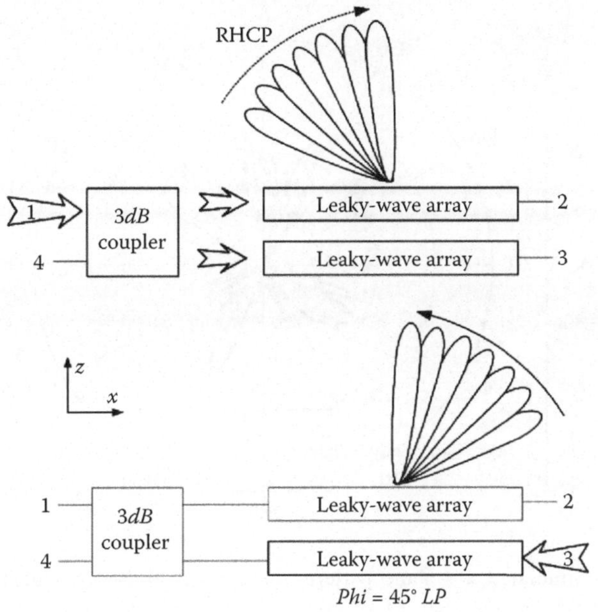

FIGURE 3.17 Operation mechanism of the quadri-polarization traveling-wave antenna.

with a 90° phase difference to generate an ideal RHCP or LHCP beam pointing to $-\theta_0$. When excited at port 2 or 3, a –45° or 45° LP beam can be generated and radiates at the θ_0 angle because the incident wave propagates to the $-x$ direction. Two orthogonal states of polarization, i.e., –45°/45° LP or L/RHCP, can be generated to cover the same angular region.

The performances of those LP beams are similar to the beams generated by a single traveling-wave antenna, as shown in Figures 3.12 to 3.15. Radiation patterns and axial ratios of RHCP beams in the *xoz*-plane are presented in Figures 3.18 to 3.21 at different frequencies. With increasing operating frequency, a wide angular region can be covered by 3 dB beamwidths of continuous scanning LP beams and CP beams. Axial ratios of the RHCP and LHCP modes are excellent at the main beam pointing within the whole frequency band of interest. Isolations between each channel are good as well.

Through the normal printed circuit board (PCB) process, the Ka-band HMSIW quadri-polarization traveling-wave antenna was fabricated on a single-layer substrate on a single Rogers 5880 substrate with a relative permittivity of 2.2 and a thickness of 0.508 mm. Reflection coefficients are below –10 dB within 33–39 GHz. Isolation coefficients are all better than 10 dB within 33–38 GHz. Such a HMSIW quadri-polarization traveling-wave antenna was characterized in an anechoic chamber. A single low-cost method was used to evaluate the CP characteristics. A standard gain LP horn antenna was rotated at a given angle ψ to change the polarization of the transmitter. The measured antenna is rotated from –180° to 180° as a receiver. To estimate the axial ratio, the receiving power level was detected and recorded several times at different angles ψ. Measured results excited at port 1 are listed in Table 3.2.

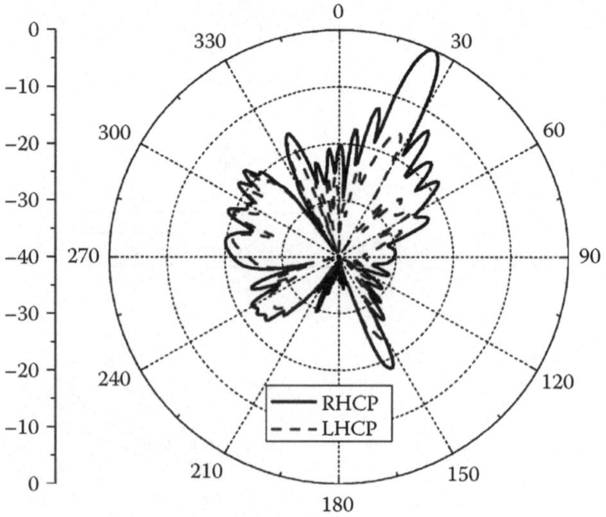

FIGURE 3.18 Simulated *xoz*-plane patterns of the traveling-wave antenna with RHCP mode at 34 GHz.

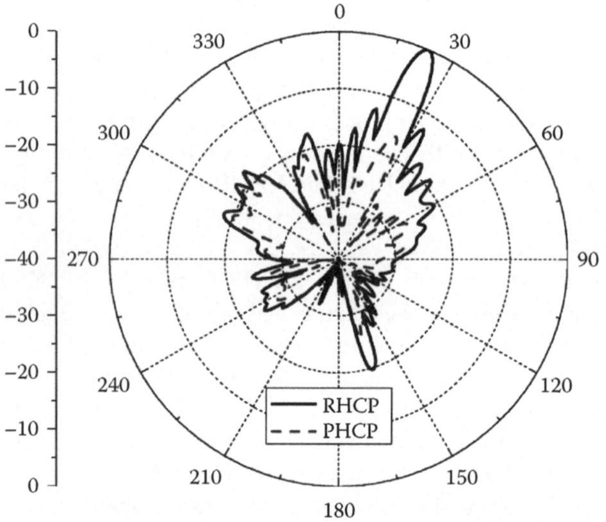

FIGURE 3.19 Simulated *xoz*-plane patterns of the traveling-wave antenna with RHCP mode at 35 GHz.

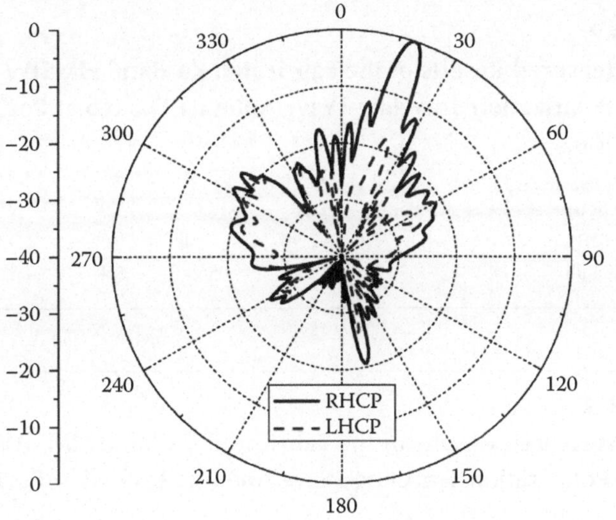

FIGURE 3.20 Simulated *xoz*-plane patterns of the traveling-wave antenna with RHCP mode at 36 GHz.

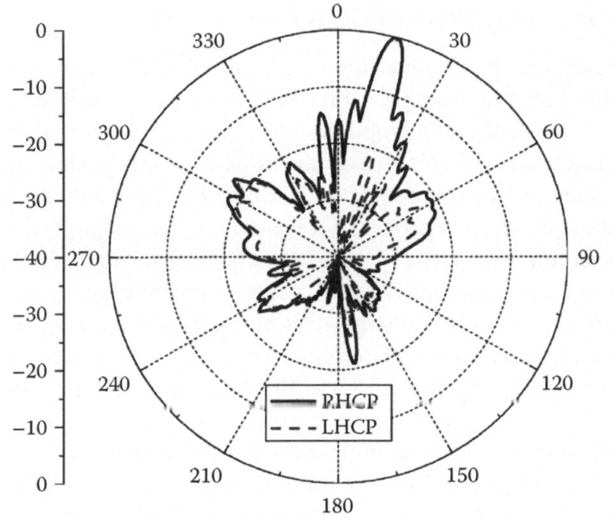

FIGURE 3.21 Simulated *xoz*-plane patterns of the traveling-wave antenna with RHCP mode at 37 GHz.

TABLE 3.2
RHCP Measured Results of the Fabricated Ka-Band HMSIW Quadri-Polarization Traveling-Wave Antenna Excited at Port 1

Frequency (GHz)	33	34	35	36	37	38	39
Gain (dBi)	11.52	12.13	12.88i	13.31	13.55	13.50	11.20
Beam direction	–29.2°	–25.6°	–20.8°	–15.6°	–11.9°	–7.9°	–3.1°
Axial ratio (dB)	4.2	2.71	1.76	1.51	1.66	1.93	2.32
3 dB beamwidth	7.4°	6.1°	6.6°	5.2°	5.4°	4.2°	6.1°

TABLE 3.3
45° LP Measured Results of the Fabricated Ka-Band HMSIW Quadri-Polarization Traveling-Wave Antenna Excited at Port 3

Frequency (GHz)	33	34	35	36	37	38	39
Gain (dBi)	12.92	13.53	14.01	14.78	14.85	14.80	12.50
Beam direction	30.3°	26.4°	21.6°	16.9°	11.9°	8.1°	3.1°
3 dB beamwidth	7.2°	7.2°	6.6°	7.2°	6.4°	6.9°	5.7°

It is able to cover the angular region of (–32.8°, 0.6°) with 3 dB beamwidths. When excited at port 3, the radiation patterns in the *xoz*-plane with 45° LP were also tested and are listed in Table 3.3. It can cover the angular region of (1.3°, 34.8°) with 3 dB beamwidths.

3.2.2 SIW Traveling-Wave CP Antenna

The above CP design is based on a combination of two LP antennas. Certainly, a CP can be generated based on a single-array antenna. Several waveguide slot arrays for generating CP beams have been proposed in [110–112]. Using a normal PCB fabrication process, slots with arbitrary shapes and angles, such as compound slots, can be easily implemented on the top conductor cladding of an SIW with low cost. By use of the X-type slot as shown in Figure 3.22, a Ku-band CP traveling-wave array antenna has been reported for low-profile mobile DBS applications in [113]. Another similar CP traveling-wave array antenna based on the SIW technology was realized in [114]. Both the LHCP wave and RHCP wave can be excited by feeding the two ports of the antenna. At 16 GHz, a 1 × 22 prototype antenna was designed with 16 dB gain and 10% relative bandwidth. From 15.2 to 16.8 GHz, the measured axial ratio in the direction of the main beam is less than 3 dB. All of them show good performance, but they need a large beam tilting angle from broadside, usually 45°.

In [111], a two-compound slot pair cut on the top conductor cladding of waveguides was investigated by the method of moments (MoM) with the configuration, as shown in Figure 3.23. Inherited from this work, a CP traveling-wave linear array antenna was synthesized in [115]. It should be noted that a broadside beam was

FIGURE 3.22 Top view of the X-type CP radiating element.

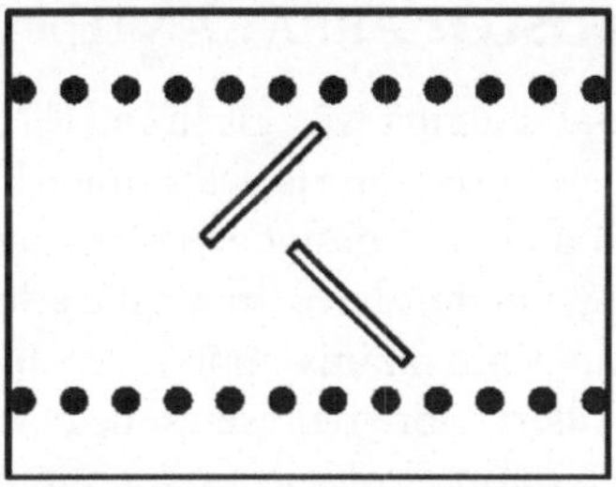

FIGURE 3.23 Top view of the two-compound slot pair CP radiating element.

realized with –20 dB SLL and 0.4 dB axial ratio in simulation. However, its fabrication requires highly accurate machining and high cost. Absolutely, it is possible to transform this rectangular waveguide CP traveling-wave antenna into the SIW version. Unfortunately, the large width-to-height ratio, as well as the nearly zero thickness of the conductor cladding, of an SIW results in deterioration of a conventional two-compound slot pair CP radiating element.

A modified two-compound slot pair CP radiator was introduced in [116]. As illustrated in Figure 3.24, its configuration is similar to the CP radiating element proposed in [115]. However, such a modified CP slot radiator is appropriate for being used in the waveguide with a large width-to-height ratio and thin thickness of the top conductor wall. Thus, it is ideal for SIW applications through the PCB etching process. In [116], the FEM full-wave simulation of a two-element structure was used to evaluate the mutual coupling between two radiators, instead of the MoM method

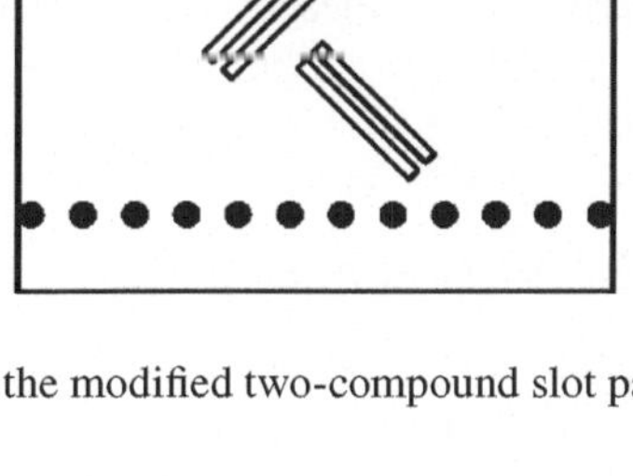

FIGURE 3.24 Top view of the modified two-compound slot pair CP radiating element.

used in [115]. By de-embedding the simulated S-parameters to the center of the two elements, mutual coupling factors between them can be calculated. The parameters of the SIW two-compound slot pair radiator have been adjusted to acquire the desired amplitude distribution and reduce mutual coupling. Finally, a 16-element SIW CP traveling-wave linear array antenna was designed, simulated, and fabricated at the center frequency of 16 GHz. A –23 dB SLL with an axial ratio of 1.95 dB was achieved in measurement. The peak gain is 18.9 dB, while the relative bandwidth is 2.5%. This kind of SIW CP slot radiator shows the potential of being used in high-performance and low-cost planar array applications.

3.3 SIW CONFORMAL SLOT ARRAY ANTENNA

The definition of a conformal antenna was given in [117]. It is an antenna that conforms to a prescribed surface whose shape is determined by considerations other than electromagnetic. Conformal antennas provide several unique advantages compared with the planar counterparts, and thus have drawn the attention of many researchers. At present, conformal antennas are mainly applied in military aircraft and missiles. Besides, they are being considered for civil use as the cost of the processing technology continues to decrease. Examples of successful application include train antennas, car radio antennas, and cellular base station antennas. The conformal antenna is able to save space and make the antenna less visually intrusive when integrating it into existing objects.

A low-profile printed antenna, such as a patch antenna [118–120], is the most popular used form in conformal antennas at present. In order to achieve enough gain or desired pattern, antenna elements should be grouped to be an array. For a conformal patch antenna array, it can be excited by several probe feeders and following coaxial lines. This topology is costly and unpractical, especially in millimeter-wave applications. The conformal patch array can also be fed by a coplanar microstrip divider. In this case, the array radiation pattern and polarization purity will be interfered with by spurious radiation from open edges and discontinuities of these conformal feedlines. SIW has the good conformability and fully closed topology to avoid the unwanted leakage, which is a great impetus for the development of millimeter-wave integrated conformal antenna arrays. Radiating elements and a feed network can be integrated in a single substrate and mounted on a curved surface, as shown in Figure 3.25.

In a few designs of SIW conformal antennas, the synthesis of a planar antenna is directly applied to those conformal antennas and the effects of the structure curvature on the antenna properties are neglected. As discussed in [118], the curved shape of the conformal antenna determines its characteristics to a great extent. Thus, it is impossible to achieve satisfactory performance or special shaped-beam radiation requirements.

3.3.1 Conformal SIW Transmission Line

A conformal SIW transmission line is the basic configuration to form the SIW conformal antenna array or the SIW conformal feeding divider. That is the reason to be

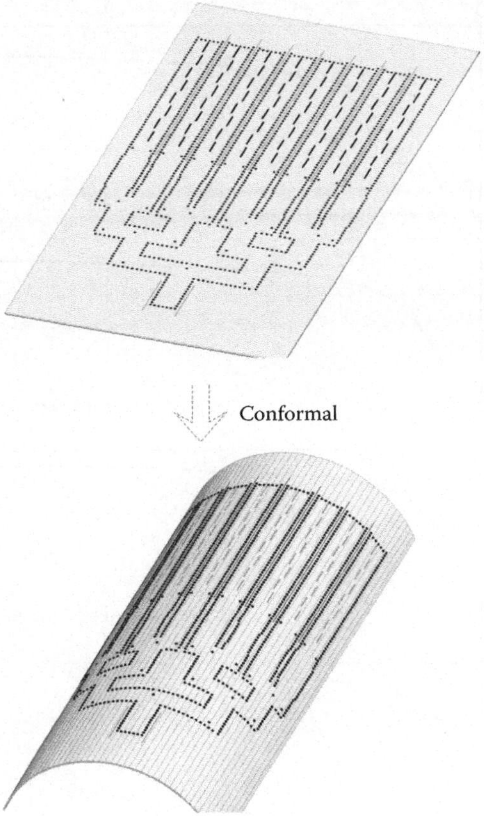

FIGURE 3.25 Configuration of an 8 × 8 SIW conformal antenna array.

precise about the characteristics of a conformal SIW at first. For the SIW conforms to a cylindrical surface, there exist numerous different types, depending on SIW arbitrary orientations. Here, two simple cases, i.e., the axial and the circumferential directions, are investigated to recognize the effects of the structure curvature on SIW propagation characteristics.

As shown in Figures 3.26 and 3.27, the normalized cutoff frequencies and propagation constants of cylindrically conformal SIWs, which are along the z-direction referring to a cylindrical coordinate system, were simulated with different equivalent bottom widths. They conform to cylindrical surfaces with different radii to recognize the effect of the curvature radius on the SIW characteristics. The corresponding planar result is used as a reference for normalization. The used substrate is Rogers 5880 with the relative permittivity of 2.2 and a thickness of 0.254 mm. It is noted that the conformal structure has a lower cutoff frequency and a larger propagation constant than the planar one. A series of conformal SIWs, which are along the φ-direction in a cylindrical coordinate system, is investigated in a similar way. Interestingly, their propagation characteristics almost keep unchanged because their transversal sections remain the same with the planar SIW.

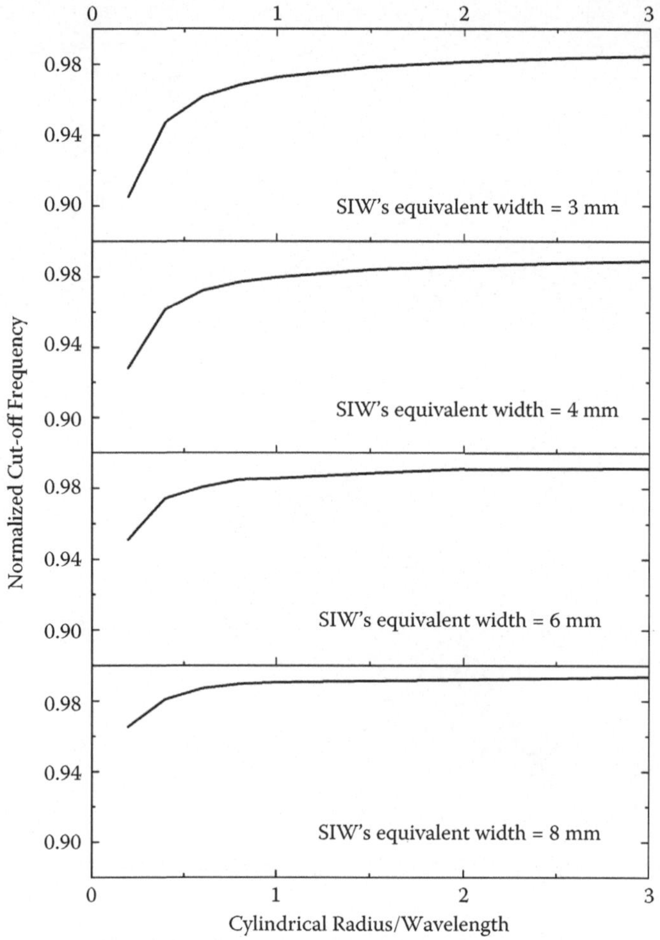

FIGURE 3.26 Normalized cutoff frequencies of cylindrically conformal SIWs with different equivalent widths along the axial direction of cylinders with different radii.

3.3.2 SIW Conformal Slot Array Antenna Design

To design an SIW conformal slot array antenna, the first step is to characterize the resonant response of a single slot. A longitudinal slot etched on a conformal SIW along the axial direction of a cylinder can be modeled as shunt admittance as well. Full-wave simulation is also used to extract the slot parameters, which parallels the method of investigating a slot etched on a planar SIW. With different slot offset values, the slot admittance curves are presented versus the slot length at 35 GHz as shown in Figures 3.28 and 3.29. To compare with the cylindrical result, the planar result is included. There is a one-to-one relationship between them. For example, the offset 0.2° in the cylindrical case corresponds to the offset 0.07 mm in the planar case. Different slot offsets and slot lengths have different resonant characteristics. It

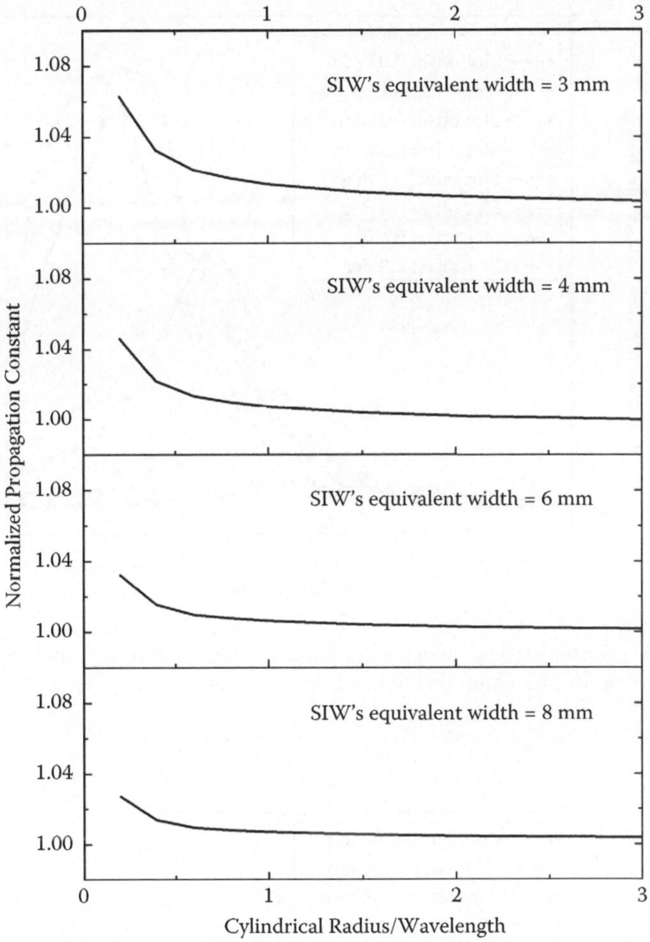

FIGURE 3.27 Normalized propagation constants of cylindrically conformal SIWs with different equivalent widths along the axial direction of cylinders with different radii.

is noted that the conformal slot has a shorter resonant length than its planar counterpart. This can be easily explained by the larger propagation constant of the conformal SIW.

Then, the external mutual coupling between two conformal slots is investigated. The first results in Table 3.4 show the dependence of the mutual coupling on the cylinder radius and the axial distance. The following radii are tested: $R = 5$ mm and $R = 20$ mm. A comparison with the corresponding planar case is also included. When the axial distance is small, the mutual coupling between elements on a curved surface is less than that of the corresponding planar case. With the increase of the radius, the cylindrical result gradually approaches the planar result. Next, the influence of the circumferential direction is analyzed. Different angular interval values, i.e., 1.2° for $R = 20$ mm and 4.8° for $R = 5$ mm, are investigated. Both of them correspond

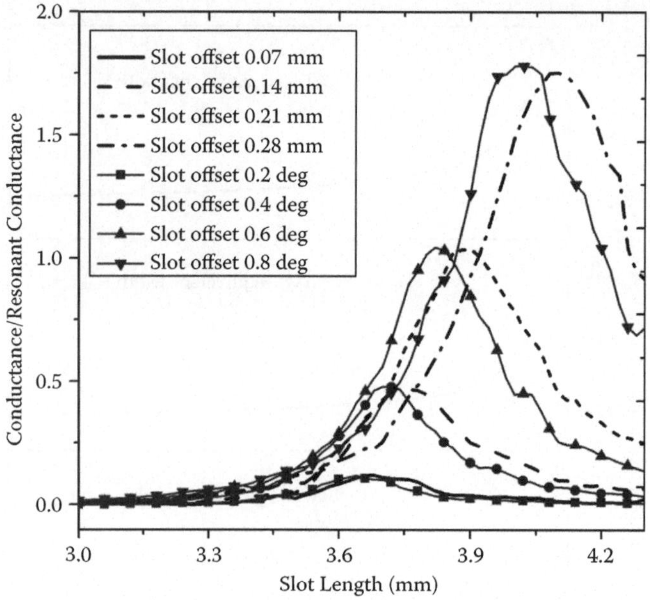

FIGURE 3.28 Normalized conductance of the axial slot (line with symbol, conformed to a cylinder with a radius of 20 mm; line without symbol, planar).

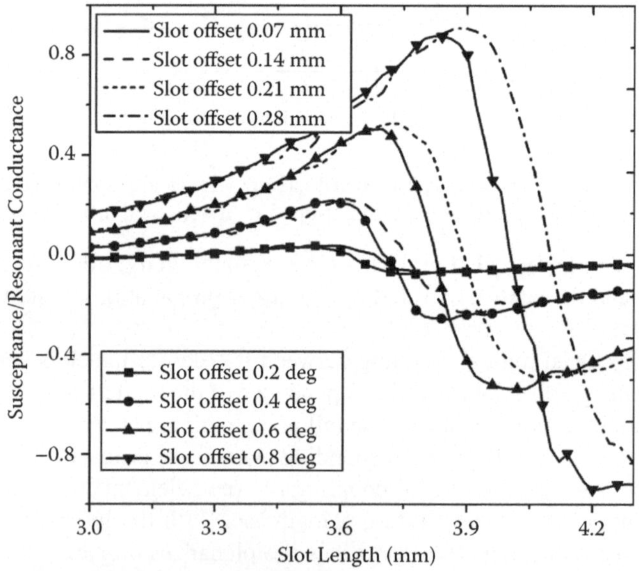

FIGURE 3.29 Normalized susceptance of the axial slot (line with symbol, conformed to a cylinder with a radius of 20 mm; line without symbol, planar).

TABLE 3.4
External Mutual Coupling between Two Slots Etched on a Conformal SIW along the Axial Direction of Cylinders with Different Radii

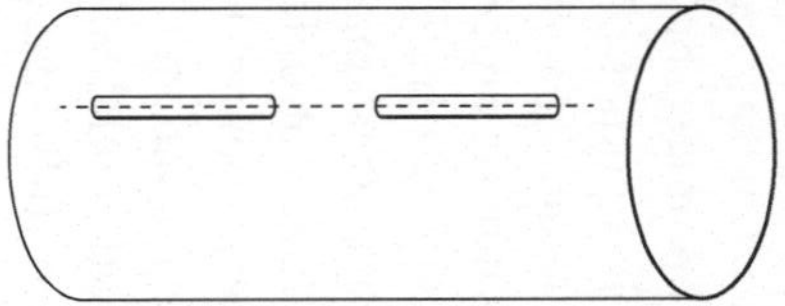

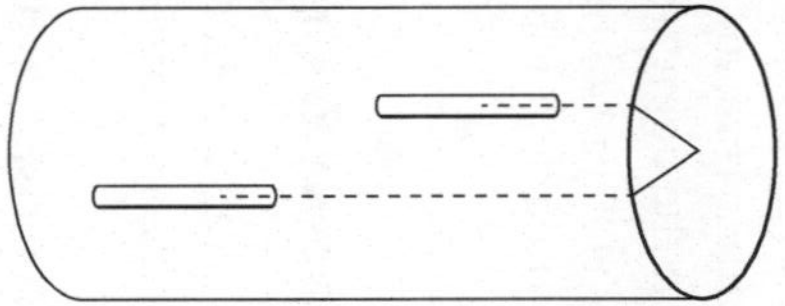

Normalized Axial Distance	Mutual Coupling (dB)					
	Planar	R = 5 mm	R = 20 mm	Planar	R = 5 mm	R = 20 mm
$0.26\lambda_0$	–22.87	–24.59	–23.34	–22.83	–24.95	–24.88
$0.37\lambda_0$	–24.99	–27.58	–25.67	–24.91	–27.87	–27.26
$0.49\lambda_0$	–27.11	–30.58	–28.01	–26.99	–30.80	–29.64
$0.61\lambda_0$	–29.42	–33.57	–30.44	–29.40	–33.73	–31.99
$0.72\lambda_0$	–31.73	–36.56	–32.86	–31.81	–36.65	–34.34
$0.84\lambda_0$	–34.87	–39.24	–35.12	–34.93	–39.32	–36.69
$0.96\lambda_0$	–38.01	–41.91	–37.37	–38.04	–41.99	–39.05
$1.07\lambda_0$	–42.80	–44.52	–39.70	–43.16	–44.65	–41.34
$1.19\lambda_0$	–47.60	–47.12	–42.03	–48.29	–47.30	–43.63

to a distance offset of 0.42 mm in the planar case. As shown in Table 3.4, when the axial distance is small, the magnitude of the mutual coupling between elements on a curved surface is also less than that of the corresponding planar case.

In this design, the SIW is fed by a standard rectangular waveguide, i.e., WR28, to avoid the direct radiation from the feed discontinuity. The aperture transition between an SIW and a waveguide has been described in Chapter 2, but this type of transition is not a good choice in the conformal design. A thin dielectric substrate is preferentially used to construct a conformal SIW, but this aperture transition doesn't perform well in the thin SIW. As such, another transition was introduced in [121]. That is a coupled longitudinal slot etched on the conformal SIW bottom conductor cladding. It can be considered a conformal slot to radiate energy into an air-filled rectangular waveguide. The effect of the cylindrical radius on the performance of the transition is slight.

3.3.3 SIW Conformal Slot Array Antenna Examples

First, an SIW conformal longitudinal slot array antenna is designed with –30 dB SLL. The used substrate is Rogers 5880 with a relative permittivity of 2.2 and a thickness of 0.254 mm.

The conformal SIW propagation characteristic and the conformal slot self-admittance are very important in the synthesis of a conformal linear array. The solid wall structure can be employed to extract the slot resonant parameters, but the low SLL linear slot array should be finally optimized based on the configuration of metalized-via wall. The adjacent radiating slots point to different directions with an angular interval in a conformal slot array antenna. That makes it essential to include the element characteristics in an accurate synthesis procedure. Fortunately, an SIW used in the millimeter-wave band usually has a relatively small cross section. Thus, such an angular interval between neighboring slots is small. The difference of beams generated by them can be neglected. In this case, isotropic radiators are applicable to synthesize an axial conformal array. Another influential factor in the array synthesis is the mutual coupling among radiating slots. As shown in Table 3.4, the planar solution seems to be acceptable for curvatures with a relatively large radius.

An eight-element linear array is modeled, full-wave simulated, and optimized after calculating the initial parameters of each slot through Elliott's iterative procedure. SLL suppression is the major concern in the first optimization stage. It is difficult to achieve the perfect pattern characteristic while keeping good matching meantime because of the mutual coupling between slots caused by the SIW's large width-to-height ratio. This problem becomes more severe for a conformal SIW due to the curved structural requirement, which leads to an extremely large width-to-height ratio. To solve this problem, two matching posts are employed to control the reflection while keeping good radiation performance. The radiation patterns versus frequency are shown in Figures 3.30 to 3.32. As shown, –30 dB SLL can be achieved over 34.2–35.8 GHz.

As described in [121], a 35 GHz SIW slot array antenna conformed to a prescribed curved surface with shaped beam was investigated, which has a low SLL beam in the H-plane and a flattop fan beam in the E-plane to cover a ±35° angular

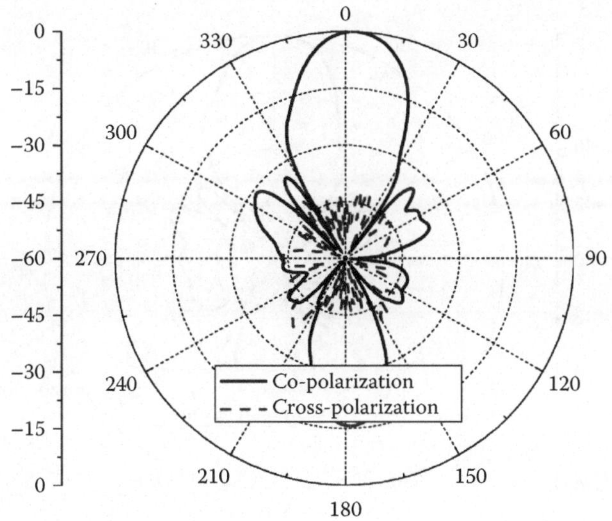

FIGURE 3.30 Radiation patterns of the low SLL SIW conformal slot linear array, including a transition at 34.2 GHz.

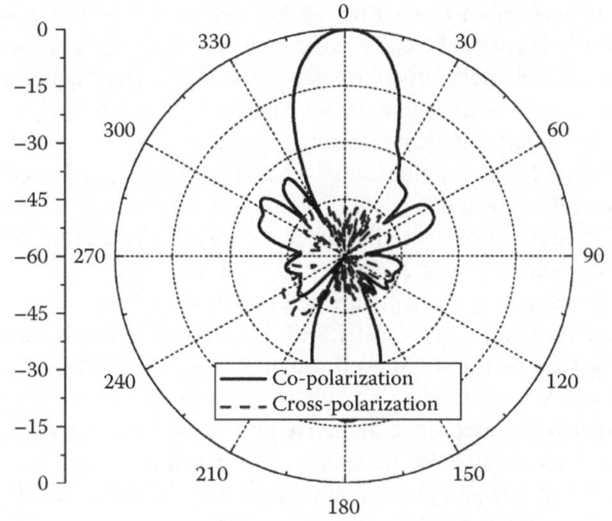

FIGURE 3.31 Radiation patterns of the low SLL SIW conformal slot linear array, including a transition at 35 GHz.

sector, respectively. It consists of eight developed eight-element slot array antennas and conforms to a cylinder with a radius of 20 mm, as shown in Figure 3.25. The angular interval between adjacent elements is 15°. Here, the E-plane orientation circumferentially conforms to a cylinder. Considering array elements pointing to different directions, the E-plane synthesis of the conformal array is difficult. The

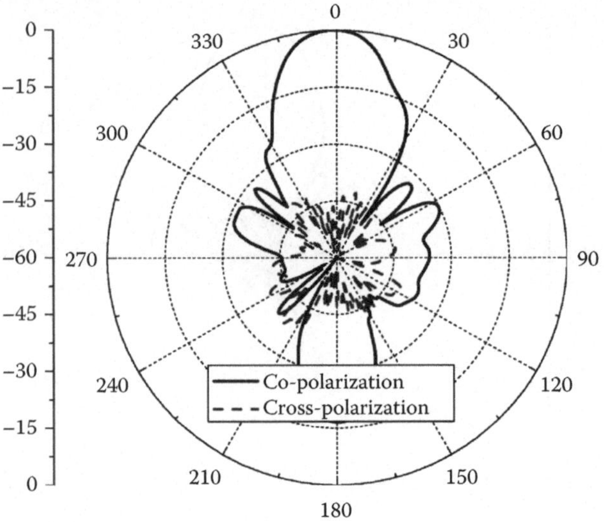

FIGURE 3.32 Radiation patterns of the low SLL SIW conformal slot linear array, including a transition at 35.8 GHz.

total radiation pattern can no longer be synthesized only concentrating on the array factor. As shown in Figure 3.33, each radiating element has its own radiation pattern after considering the element lattice, fixture shape, and mutual coupling. Neglecting these influences causes a considerable deviation between the designed and experimental values, especially in shaped-beam applications. Using the element pattern data, including the effects of mutual coupling and fixture shape, the required shaped-beam pattern can then be synthesized through an iterative method. Meanwhile, the corresponding excitations of the radiating elements can be determined.

Finally, the optimized flattop E-plane pattern of the conformal array is shown in Figure 3.34. The radiating elements are fed in phase and with unequal amplitude. The desired excitation ratio is 0.1:1:3:1:1:3:1:0.1 for numbers 1 to 8 radiating element. The desired excitations are obtained through a conformal planar feeding divider. Several T-type and Y-type dividers, as well as SIW bends, are employed and grouped together. A center-offset metalized inductive post is used at the junction to bring the input power into two output SIWs unequally. The required tapered output magnitude can be achieved, but unequal phase is produced at the same time. Here, the output phase should be compensated by phase shifters. Similar to the conventional SIW, the conformal SIW also has a width-dependent propagation constant. This feature can easily be deployed to realize the required phase shift by adjusting the position of metalized vias to form variable SIW widths. The full-wave simulated radiation pattern, including the divider, is also demonstrated in Figure 3.34, and it is somewhat worse than the calculated one. Besides, the flattop patterns can be synthesized to cover the maximum beamwidth with different ripple levels. As shown in Figure 3.35, the 1 dB ripple can be achieved in the angular range of 78.4°, while the 3 dB ripple can be achieved in the angular range of 92.4°.

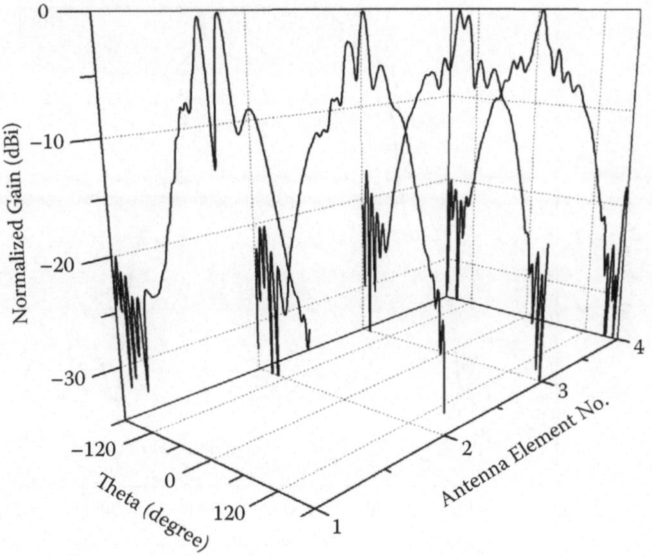

FIGURE 3.33 Isolated radiated pattern for each conformal element, including the influences of mutual coupling and fixture shape.

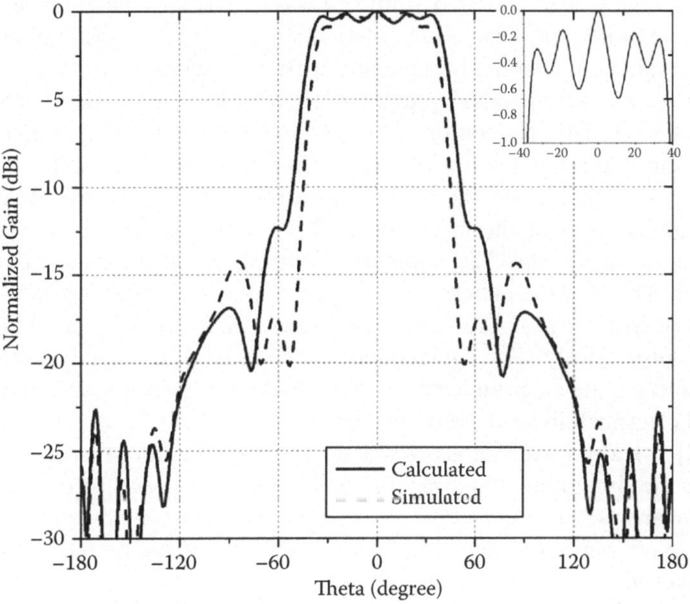

FIGURE 3.34 Calculated and simulated E-plane patterns of the Ka-band SIW conformal antenna array.

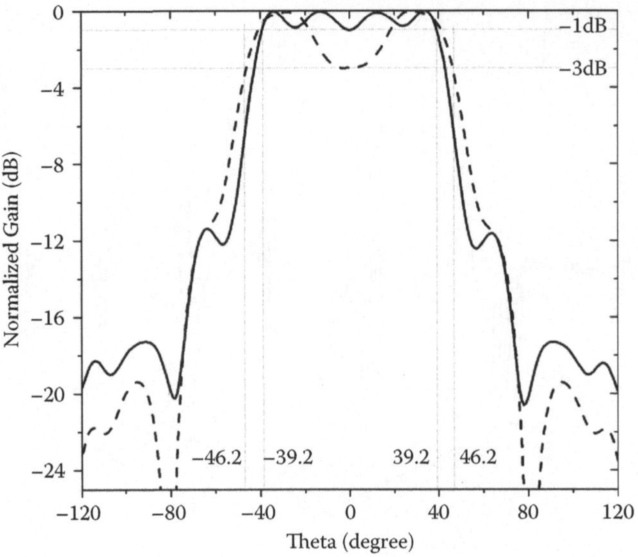

FIGURE 3.35 Optimized E-plane patterns of the Ka-band SIW conformal antenna array covering different angular ranges with different ripple levels.

The SIW conformal array was fabricated on a single-layer Rogers 5880 substrate through the standard PCB process and conforms to a cylindrical surface. Measured reflection is almost below –10 dB within 34.6–35.3 GHz. In the H-plane, its SLL is lower than –27.4 dB, while the cross-polarization is lower than –41.7 dB at the direction of the main beam. In the E-plane, the 3 dB beamwidth is –38 to 37°, while the cross-polarization within such an angular region is less than –37 dB. The E-plane radiating characteristic is deteriorated owing to the assembly and fabrication tolerance, as well as the fixed screw blockage. Other possible reasons for the E-plane pattern degradation include the fabrication error of the conformal phase shifter, the permittivity error of the used Rogers 5880 substrate, and the assembly error or the fixture shape error. The radiation patterns were also measured at different frequencies. The H-plane pattern keeps excellent performance over 34.2–35.8 GHz. The realized SLLs are –24.2, –27.1, –28.5, and –22.5 dB at 34.2, 34.6, 35.4, and 35.8 GHz, respectively. The E-plane has a nice shaped beam only over a narrow band considering the magnitude/phase errors in the feed network and the frequency-dependent electrical distance between elements.

In [122], a leaky-wave slot array was designed on a conformal SIW around the cylinder in the circumferential direction. A 4 × 16 slot array is constructed using four SIW linear arrays. Each conformal leaky-wave antenna has 16 longitudinal slots. When the operating frequency is varied from 24 GHz to 26 GHz, 10° beam steering can be achieved.

4 Substrate Integrated Traveling-Wave Antennas

A traveling-wave antenna uses a traveling wave on a transmission line as the main radiating mechanism. Compared with a resonant antenna, the distinguishing feature of the traveling-wave antenna is that the current travels through the antenna in one direction to generate the electromagnetic waves. The traveling-wave antenna often has a wider bandwidth than the resonant antenna.

This type of nonresonant antenna is very useful in millimeter-wave applications, because it is much easier for the energy to radiate out from the discontinuities of open transmission lines in the high-frequency band. In addition, leakage can be generated by enclosed guided-wave structures with properly designed slots or open apertures. According to different operating principles, traveling-wave antennas fall into two general categories: surface-wave antennas and leaky-wave antennas.

A surface wave radiates only at discontinuities. That means the total radiation pattern of a basic surface-wave antenna is formed by interferences at the beginning and the end of the structure. Thus, it is difficult to obtain high gain or shaped beam. Moderate gain patterns can be achieved with the main beam near the endfire. In this chapter, some substrate integrated circuit (SIC) surface-wave antennas are introduced, including substrate integrated waveguide (SIW)-fed tapered slot antenna (TSA), substrate integrated image guide (SIIG)-fed rod antenna, etc. On the other hand, discontinuities can be placed along the whole surface-wave structure. As representatives, the SIW-fed Yagi-Uda antenna and SIW-fed log-periodic dipole array are introduced as well.

A leaky-wave antenna is basically a waveguide structure, which possesses a mechanism that allows it to radiate waves continuously along its length [120]. The phase constant of the wave controls the beam angle that varies with the frequency. Meanwhile, the attenuation constant of the wave controls the beam shape. In this case, high-gain or low sidelobe level (SLL) beams can be achieved at an arbitrary specified angle. Two types of leaky-wave antennas, i.e., uniform and periodic types, have been developed based on different SIC concepts, including SIW long slot leaky-wave antenna, half-mode substrate integrated waveguide (HMSIW) leaky-wave antenna, SIW periodic leaky-wave antenna, SIIG metal-strip leaky-wave antenna, and SIIG-fed leaky-wave patch antenna. Another closely related antenna structure has been discussed in Chapter 3. That is the SIW slot array leaky-wave antenna.

4.1 SIC SURFACE-WAVE ANTENNA

4.1.1 SIW-Fed Tapered Slot Antenna

TSAs were first introduced in the late 1950s. Their operation is based on a traveling wave propagating along the surface of the antenna taper with a phase velocity less than the speed of light in free space [120]. Endfire radiation happens under this condition. In general, the performance of a wide-frequency bandwidth, moderate gain, and a pair of symmetric beams in both the E- and H-planes can be expected to achieve for a typical TSA. Thus, this structure is a suitable element in microwave, millimeter-wave, and terahertz reflector feeds or stand-alone antennas.

A TSA can be fed by a stripline, finline, or slotline. The current of a typical TSA is planar in nature, etched on a thin, low ε_r dielectric substrate. A balun is often necessary to drive the antenna. As a result, it is important to consider the possibility that the feedline may radiate or couple with the antenna element, which spoils the antenna performance.

Fortunately, the SIW is a balanced structure and can be viewed as a balun. Compared with other feeding technologies, an SIW-based structure is able to avoid the power radiated from the feedline. Figure 4.1 presents the configuration of an SIW-fed TSA. The typical TSA element consists of two gradually flaring conductor claddings on opposite sides of a dielectric substrate. The feeding SIW is designed to only support the TE_{10} mode, and the electric field direction is gradually changed from the vertical to the horizontal, as shown in Figure 4.1.

The conventional TSA usually has high input impedance. The SIW usually has relatively low impedance because of the limitation of the substrate thickness. There exists a mismatching when the SIW and the TSA are integrated directly. To solve this problem, the flaring conductor covers overlap each other [123]. As a result, low input impedance of the TSA can be obtained to achieve good impedance matching with the feeding SIW over a wide band. The bandwidth of the SIW-fed TSA is also determined by the cutoff frequency of the feeding SIW, which determines the lowest operating frequency of the TSA.

A degradation of the radiation pattern has been observed for a narrower TSA width. Thus, corrugation structure was used in [124] to improve the radiation pattern for a narrow-width SIW-fed TSA.

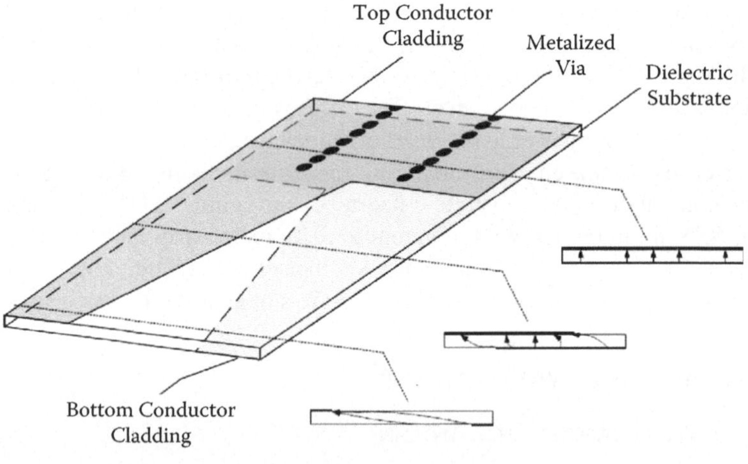

FIGURE 4.1 Configuration of the SIW-fed TSA and the electric field distribution within the antenna.

Furthermore, a silicon-based terahertz SIW-fed TSA has been designed and fabricated on a substrate with a thickness of several microns through the microelectromechanical systems (MEMS) process [125], covering the range of 350–370 GHz. To tackle this difficult process, some special techniques were employed, including substrate removal, micromachining, and membrane support of the conductors.

4.1.2 SIC-Fed Rod Antenna

The dielectric rod antenna, also known as a polyrod, was investigated to increase the radiation efficiency [126, 127]. It belongs to the surface-wave antenna. This type of antenna, a few wavelengths long, offers medium gain around 15 dBi with high radiation efficiency and a wide operating frequency band. Its rod shape should be designed and fabricated accurately, which is crucial for the SLL control and energy radiation.

The rod antenna is commonly fed by a metallic horn. This is highly problematic because the horn is bulky and thus incompatible with planar integrated technologies [128]. An alternative of feeding a planar rod antenna was demonstrated in [129] by using the SIIG technology. By laser, a planar dielectric rod antenna and its SIIG feedline were fabricated into a single sheet of alumina substrate at 94 GHz. Integrated fabrication of the rod antenna and the corresponding feedline on a single layer substrate has become feasible at low cost.

Another wideband planar dielectric rod antenna fed by an SINRD guide was presented in [130]. As shown in Figure 4.2, the rod antenna and its feeding structure are also integrated into a single substrate. Because the SINRD guide is shielded on top and bottom of the dielectric substrate, the SINRD-fed rod antenna is acceptable for multilayer techniques and applications. Measured bandwidth of the antenna is from 93 to 103 GHz, while a relatively flat gain is obtained over the band of interest.

4.1.3 SIC-Fed Yagi-Uda Antenna

The printed Yagi-Uda antenna is another useful surface-wave antenna, depending on its advantages of low profile, low cost, light weight, and easy integration with other planar devices. The feeding technology of the Yagi-Uda antenna mainly determines its performance. In general, baluns should be integrated to drive the antenna.

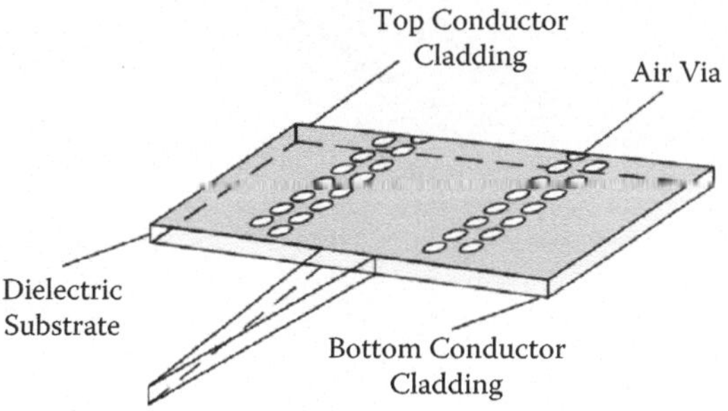

FIGURE 4.2 Configuration of the SINRD-fed rod antenna.

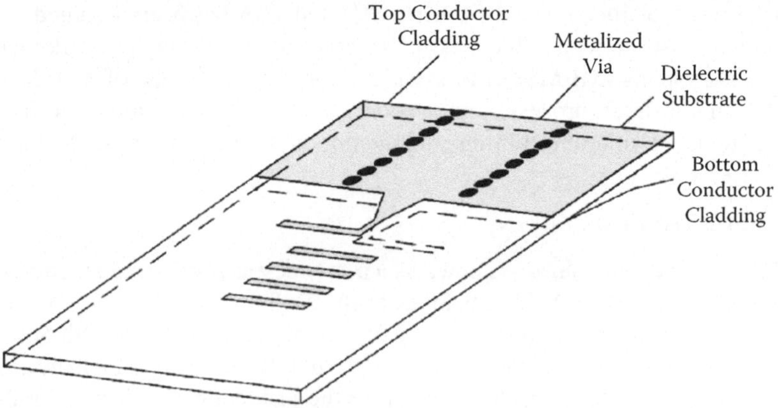

FIGURE 4.3 Configuration of the SIW-fed Yagi-Uda antenna.

As shown in Figure 4.3, a Yagi-Uda antenna fed by SIW technology was proposed in [131] and able to achieve good performance. The conductor claddings on the top and bottom surfaces of the used dielectric substrate serve as reflector elements for the Yagi-Uda antenna. The design example has four director elements. More director elements can increase the antenna gain. Meanwhile, the number of design parameters and the complexity of the design optimization are increased.

The above design needs a tapered coupled line to match the SIW feedline and the printed Yagi-Uda antenna, which complicates the design of the antenna. Besides, the SIW feedline is still a little larger than its microstrip counterparts. To overcome these weaknesses, Figure 4.4 presents an HMSIW-fed Yagi-Uda antenna as described in [132]. In this topology, the HMSIW feedline is directly connected to the Yagi-Uda antenna through a coupled line without tapering. The first arm of the dipole antenna is printed on the top conductor cladding, and the second arm is printed in the opposite direction on the bottom conductor cladding. Besides, there are several director elements printed on the top conductor cladding.

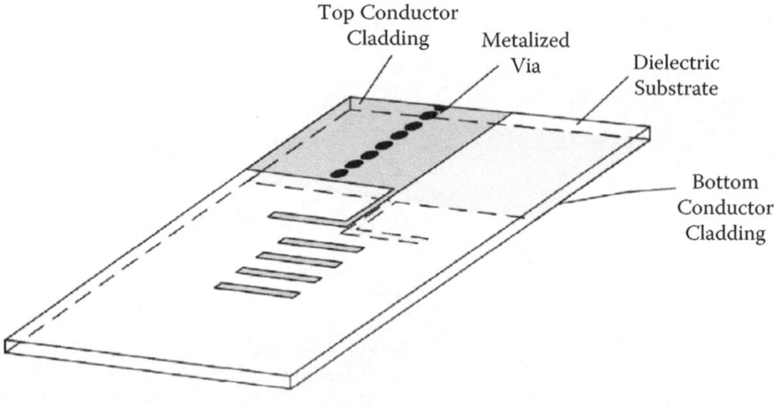

FIGURE 4.4 Configuration of the HMSIW-fed Yagi-Uda antenna.

Another X-band SIW-fed Yagi-Uda antenna was proposed in [133]. The director element is no longer a single microstrip line on one surface of the substrate. It is divided into two parts on the upper and lower surfaces of the substrate, respectively, which could improve the gain of the antenna.

4.1.4 SIC-Fed Log-Periodic Dipole Array Antenna

The printed log-periodic dipole array antenna is a development of the printed dipole antenna. The impedance bandwidth of the log-periodic dipole array antenna increases with the number of the dipole element. When a large number of dipole elements are used, this antenna will achieve very wide impedance bandwidth.

The printed log-periodic dipole array antenna fed by an SIW or an HMSIW was developed in [134]. The configuration of the SIW-fed version is shown in Figure 4.5. The SIW-fed log-periodic dipole array antenna is able to cover almost the whole Ka-band, while the HMSIW-fed version is able to cover both K- and Ka-bands.

4.2 SIC Uniform Leaky-Wave Antenna

As described before, there are two different types of leaky-wave antennas. The categorized standard is whether the geometry of the guided-wave structure is uniform or periodically modulated along its length [120]. Two such types are actually similar to each other in principle, but their design processes and performances are different in several ways.

An important difference between the uniform and periodic leaky-wave antennas is the scan range. The beam scans in the forward quadrant only for the uniform leaky-wave antenna. It cannot approach broadside. However, for the periodic leaky-wave antenna, it can scan from the backward quadrant into the forward quadrant.

This section pays attention to the uniform leaky-wave antenna. Of course, with a small taper along its length, the SLL of the uniform leaky-wave antenna can be controlled and improved.

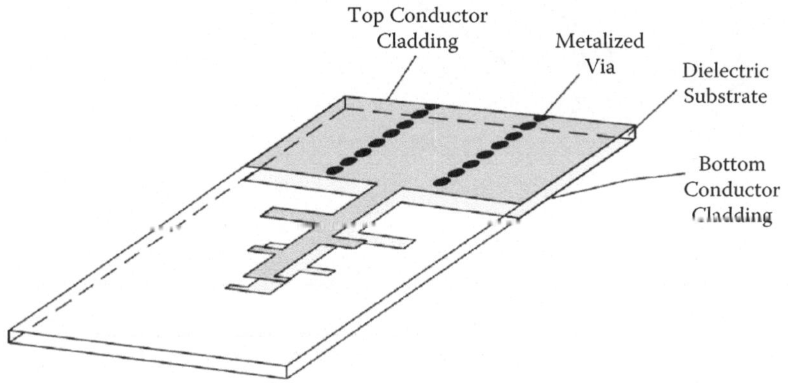

FIGURE 4.5 Configuration of the SIW-fed log-periodic dipole array antenna.

4.2.1 SIC Long Slot Leaky-Wave Antenna

Long slot leaky-wave antennas exhibit many interesting features. Among various types, the longitudinal straight long slot antenna is the simplest one judging by the design and fabrication process. However, it is impossible to control its SLL. To solve this problem, the long slot is meandered from the waveguide centerline toward the sidewall and then back to the centerline to properly control the aperture illumination and achieve an optimum far-field pattern as described in [135–137].

For designs based on a conventional rectangular waveguide, the long slot on the broad wall of the waveguide has to be precisely machined to form a desired meandering shape. Besides, the standard waveguide cannot be directly used and has to be modified. Fortunately, the SIW technology can break the above-mentioned bottleneck.

In [138], a low SLL SIW meandering long slot leaky-wave antenna was developed with the configuration shown in Figure 4.6. The value of offset, x_0, from the centerline determines the leakage rate of the slot.

As discussed in [135], a symmetrical slot can provide better cross-polarization than the asymmetry slot. With the same equivalent length, the cross-polarization level of a straight radiating slot is obviously lower than that of a tapered slot. Thus, we hope to use a long straight slot instead of the meandering one. To achieve low SLL as before, the shapes of the SIW lateral wall are meandered properly, as shown in Figure 4.7. The value of wall offset, x_0, also determines the leakage rate of the slot. The SIW lateral walls are incurved with the same shape, and the distance between them is unchanged. Because the SIW sidewalls are constructed by rows of metalized vias, it is easy to precisely position these vias along the desired meandering shape through the normal PCB process.

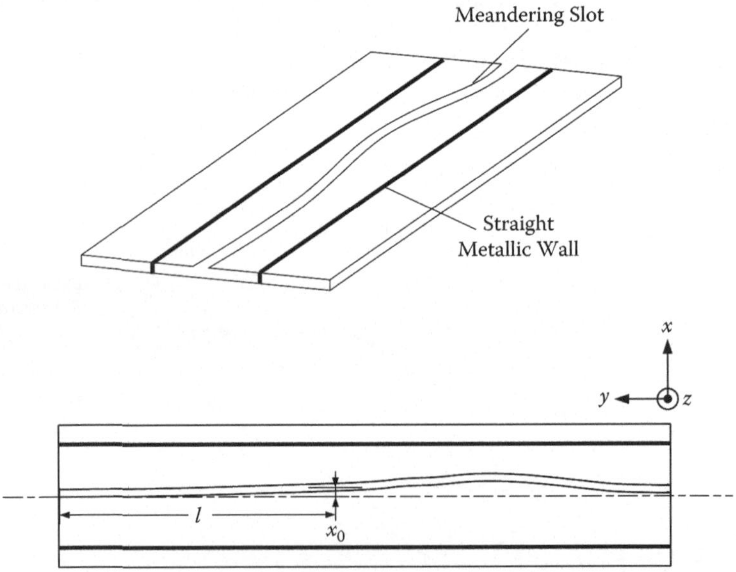

FIGURE 4.6 Configurations of the straight SIW meandering long slot leaky-wave antenna.

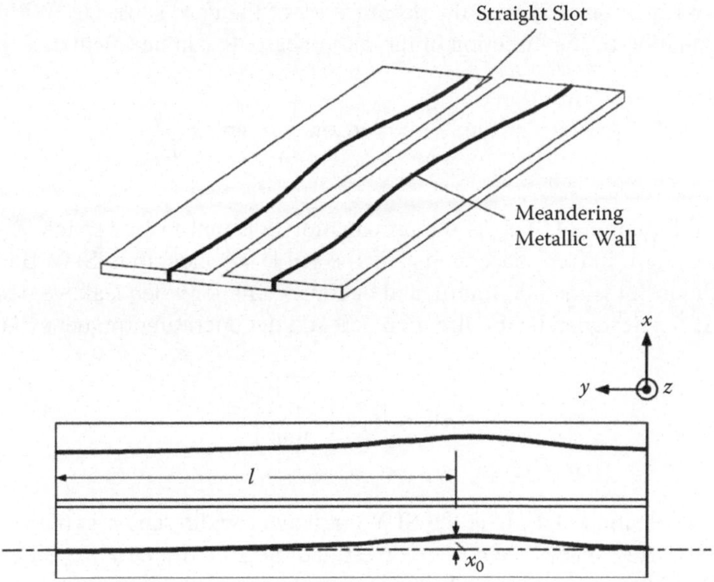

FIGURE 4.7 Configurations of the meandering SIW long straight slot leaky-wave antenna.

Another long slot leaky-wave antenna was presented in [139] based on the SINRD guide. In this design, the fundamental LSM_{11} mode of the SINRD guide is selected. Placing an offset long slot on the top conductor cladding of the SINRD guide, this fundamental mode can be converted to a leaky mode and the power can be effectively radiated out. The bandwidth of the antenna can be controlled by the periodicity of the airhole array. Besides, by appropriately choosing the periodicity of the airhole array, a fast or a slow scanning is possible without changing parameters of the dielectric substrate.

In the following sections, the SIW long slot leaky-wave antenna is introduced in detail.

4.2.1.1 Design Procedure

1. Determination of SIW dimensions. There is a relationship between the propagation constant in SIW, k, and the phase constant in SIW, β:

$$\sin\theta' = \frac{\beta}{k} \tag{4.1}$$

To simplify the analysis, β is regarded as the phase constant without leakage. At the dielectric-air interface, according to Snell's law of refraction, there is

$$\frac{\sin\theta'}{\sin\theta} = \frac{1}{\sqrt{\varepsilon_r}} \tag{4.2}$$

In Equation (4.2), ε_r is the permittivity of the used substrate. With small attenuation α, the direction of the main beam, θ, can be calculated by

$$\theta = \arcsin\frac{\sqrt{\varepsilon_r}\beta}{k} = \arcsin\frac{\beta}{k_0} = \arcsin\frac{\lambda_0}{\lambda_g} \tag{4.3}$$

In Equation (4.3), k_0 is the propagation constant in free space, λ_0 is the wavelength in free space, and λ_g is the guide wavelength in SIW. β in SIW without slot is the maximum, and becomes smaller when leakage occurs.

If the designed beam direction is θ and the operating frequency is f,

$$\sin\theta = \sqrt{\varepsilon_r - \left(\frac{c}{2wf}\right)^2} \tag{4.4}$$

In Equation (4.4), w is the SIW equivalent width, and c is the speed of light in free space. Then, the corresponding SIW dimensions, such as the SIW width a, the distance between neighboring vias s, and the diameter of the metalized via d, can be determined by Equation (1.1).

As an example, the center frequency is set to be 35 GHz. The used substrate has a relative permittivity of 2.2 and a thickness of 1.575 mm, respectively. The SIW equivalent width is selected to be 3.5 mm. Then, the corresponding SIW dimensions are a = 3.83 mm, s = 0.8 mm, and d = 0.5 mm, respectively.

2. Determine the radiation per unit length, α, versus the offset x_0, and β/k_0 versus x_0. The radiation properties of a long slot etched on the SIW can be obtained by the commercial full-wave simulation software. A family of SIWs is modeled and the long straight slot with different offset, x_0, is cut on the top broad wall of each SIW. To obtain more accurate results, the SIW is extended on both sides with the equivalent solid sidewalls and excited by two waveports. The simulated results can be used to recognize radiation per unit length, α, and normalized phase propagation per unit length, β/k_0, according to different x_0.

   Here, x_0 is varied from 1.75 to 2.95 mm at equidistant sequences with a spacing of 0.01 mm. The slot width w_{slot} is set to be 0.45 mm. All conductor material, including metalized vias and conductor claddings, is set to be the perfect electric conductor. Besides, the dielectric material is set with tan δ = 0. In this case, the attenuation value represents the radiated power. As shown in Figure 4.8, α monotonously increases with x_0 and has a considerable dynamic range. Furthermore, the increase of α with x_0 is larger for a wider slot. For example, when w_{slot} = 0.2 mm, the increase of α is only 0.23 dB/mm with Δx_0 = 1 mm. When w_{slot} = 1.4 mm, the increase of α is 0.83 dB/mm with Δx_0 = 1 mm. That means more power is radiated at the beginning of the long slot with the increase of w_{slot}. It is difficult to be radiated over the whole slot to achieve the required aperture taper. Therefore,

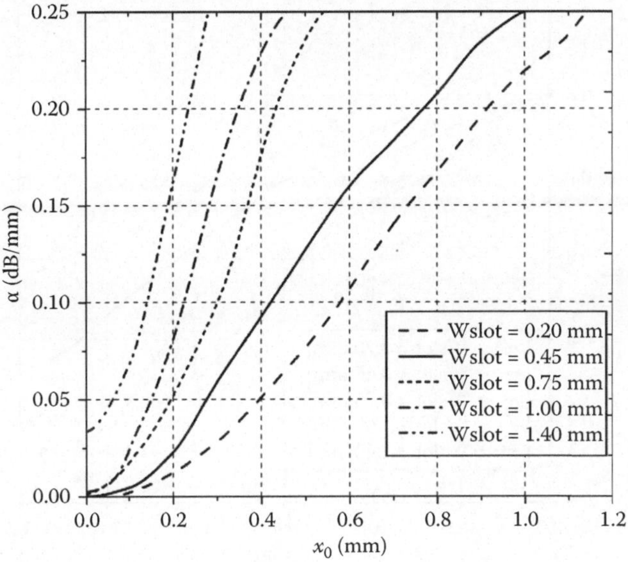

FIGURE 4.8 Relationship between α and x_0 for a long slot etched on the SIW.

a narrow slot is preferable to design a leaky-wave antenna with a desired radiation pattern.

According to Equation (4.3), accurate β/k_0 can be obtained from simulated radiation patterns as illustrated in Figure 4.9. The absolute variation of β/k_0 is 0.35 when x_0 increases from 0 mm to 1.2 mm. The average value of β/k_0 with different leakages corresponding to those used x_0 can be employed to estimate the beam direction. Here, the beam direction at 35 GHz is predicted to be 41°. In this study, $w_{slot} = 1.4$ mm is the optimum value. It provides a maximum $\alpha = 0.3$ dB/mm only with a variation $\Delta\beta/k_0 = 0.06$. It means that the normalized phase constant keeps almost unchanged along the length of the guiding structure. In this case, all sections of the antenna nearly radiate at the same pointing angle. The beam direction can be accurately designed.

Based on the overall consideration, w_{slot} is finally chosen to be 0.45 mm.

3. Determine x_0 at different position l. In this step, first, α at different position l, $\alpha(l)$, is calculated on the basis of the collapsed aperture distribution $A(l)$. Second, based on Figure 4.8, the offset x_0 is obtained at different position l, i.e., $x_0(l)$.

The experimental array is set to meet the Taylor distribution with –35 dB SLL. The desired aperture distribution is shown in Figure 4.10. There is a mathematical relationship between $\alpha(l)$ and $A(l)$ as described in [79],

$$\alpha(l) = \frac{\frac{1}{2}|A(l)|^2}{\frac{1}{1-R}\int_0^L |A(\zeta)|^2\, d\zeta - \int_0^l |A(\zeta)|^2\, d\zeta} \tag{4.5}$$

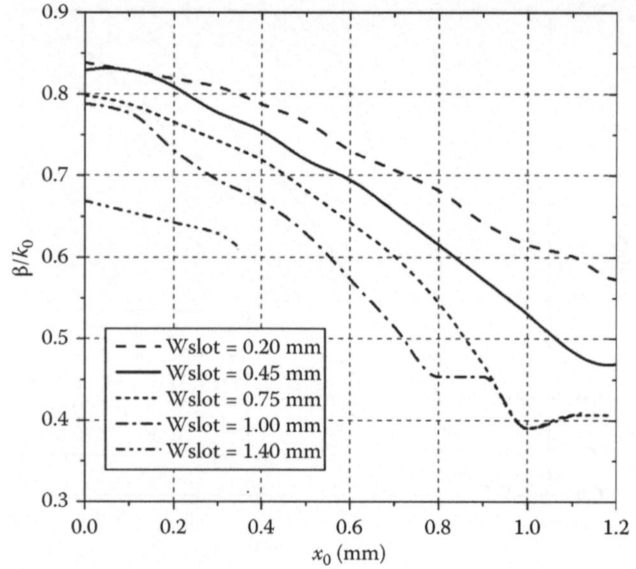

FIGURE 4.9 Relationship between β/k_0 and x_0 for a long slot etched on the SIW.

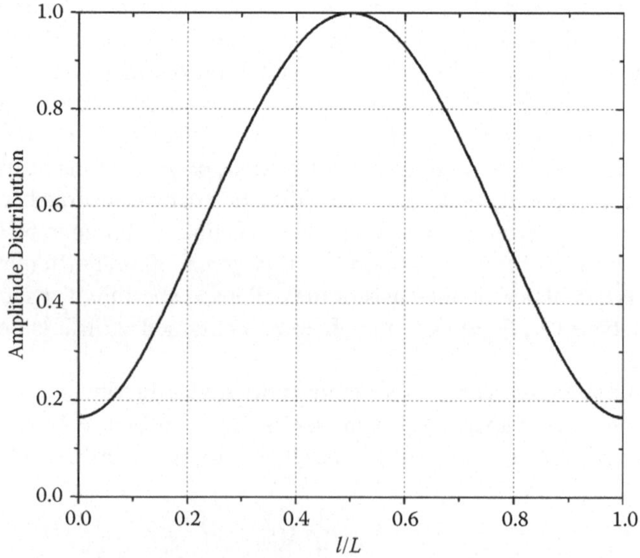

FIGURE 4.10 Aperture distribution of the Taylor distribution with –35 dB SLL.

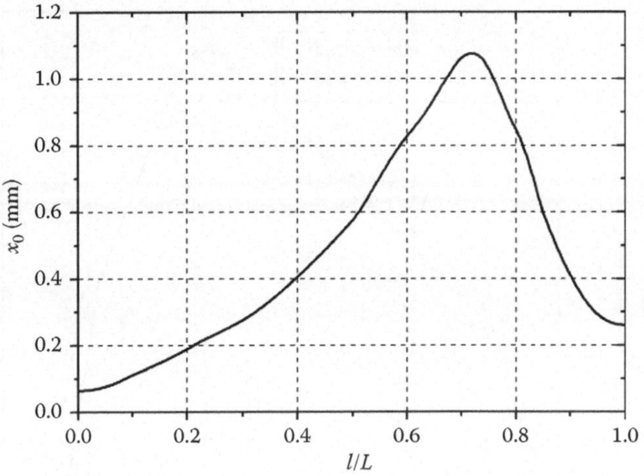

FIGURE 4.11 Design curve of the long slot leaky-wave antenna with –35 dB SLL.

The leaky-wave antenna is terminated with a load. It is assumed that a fraction, R, of the incident power is absorbed by the load, i.e., $R = P_{load}/P_{in}$. As a compromise between moderate peak value for $\alpha(l)$ and antenna efficiency, 5% of the incident power, i.e., $R = 0.05$, is set to be taken by the terminating load. For the same reason, the slot length, L, is set to be 115 mm. Now, $\alpha(l)$ can be calculated by Equation (4.5). Then, $x_0(l)$ is plotted to determine the shape of the meandering slot or the meandering sidewall as shown in Figure 4.11.

4. Modeling, simulation, and optimization. Up to the present, long slot SIW leaky-wave antennas can be modeled and optimized. It is very difficult to accurately model the meandering slot or the meandering wall based on the continuous curve as shown in Figure 4.11. To approximate the continuous curve, several straight lines are connected one-by-one. The approximation is acceptable when the number of segments is enough. In this design, there are seven discrete segments to approximate the meandering slot or the meandering lateral wall. The final dimensions of the meandering slot or the meandering sidewall are illustrated in Figure 4.12. It should be pointed out that some endpoints of line segments cannot be demonstrated in this figure. The straight SIW meandering long slot leaky-wave antenna and the meandering SIW long straight slot leaky-wave antenna were fabricated on single-layer Rogers 5880 substrates through a standard PCB process, respectively. Both of them are fed by the standard WR-28 waveguide with the configuration described in Section 2.4.3. Radiation from feeding discontinuity can be avoided in this topology to guarantee the performance of the fabricated antenna. These leaky-wave antennas have good matching over a wide band, even including the transition. Measured reflections of different two antennas are almost below –10 dB from 33 to 37 GHz. These leaky-wave antennas terminated with 50 Ω loads were characterized in an anechoic

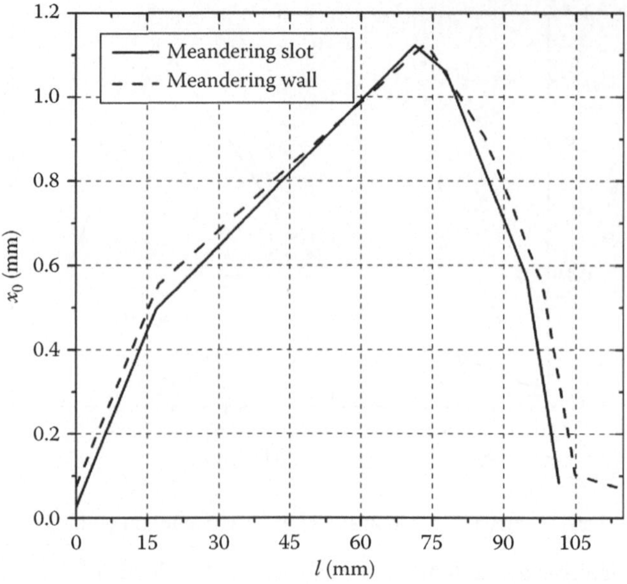

FIGURE 4.12 Dimensions of two types of leaky-wave antennas.

chamber. Figures 4.13 and 4.14, respectively, show the measured H-plane radiation patterns of different fabricated leaky-wave antennas at 35 GHz. Their beam directions are –38.1° and –37.9°, gains are 12.2 and 12.7 dBi, SLLs are –27.7 and –29.3 dB, cross-polarization levels are –27.8 and –38.9 dB, 3 dB beamwidths are 14.2° and 13.9°, and radiation efficiencies are 66.8 and 66.1% for type 1 and type 2, respectively. It is demonstrated that the meandering wall leaky-wave antenna has a much better cross-polarization characteristic. Both of the antennas were tested at a series of equidistant frequency points with spacing of 0.5 GHz from 33 to 37 GHz. As shown in Figure 4.15, their gains are in the range of 10.5–12.8 dBi and 11.3–13.5 dBi, respectively. As shown in Figure 4.16, their SLLs are in the range of –21.6 to –27.7 dB and –21.8 to –29.3 dB, respectively. As shown in Figure 4.17, the straight SIW meandering long slot leaky-wave antenna can cover an angular region of (–55.3°, –21.4°), and the meandering SIW long straight slot leaky-wave antenna can cover an angular region of (–55.7°, –20.4°) with 3 dB beamwidth. With the increase of the operating frequency, the beam is steered toward the propagation direction of the leaky wave.

4.2.1.2 Tolerance Analysis

As usual, substrate and fabrication tolerances could influence the SIW dimensions, the slot offset, and the slot length. Thus, electrical performances of the fabricated long slot leaky-wave antenna may be different from the designed one. A tolerance analysis of a leaky-wave antenna becomes necessary, especially in the low SLL antenna design.

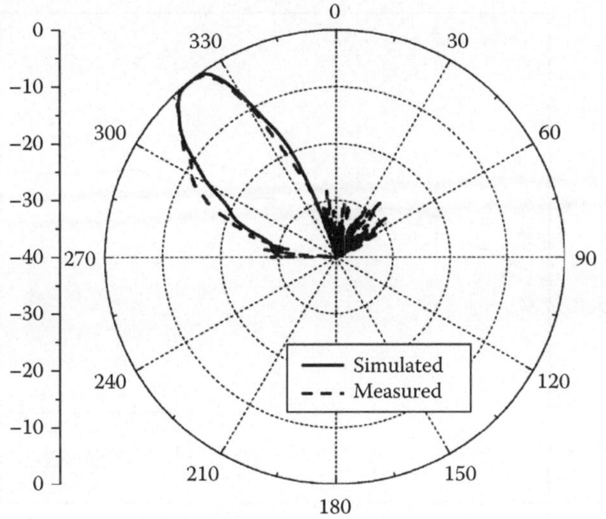

FIGURE 4.13 35 GHz H-plane pattern of the straight SIW meandering long slot leaky-wave antenna.

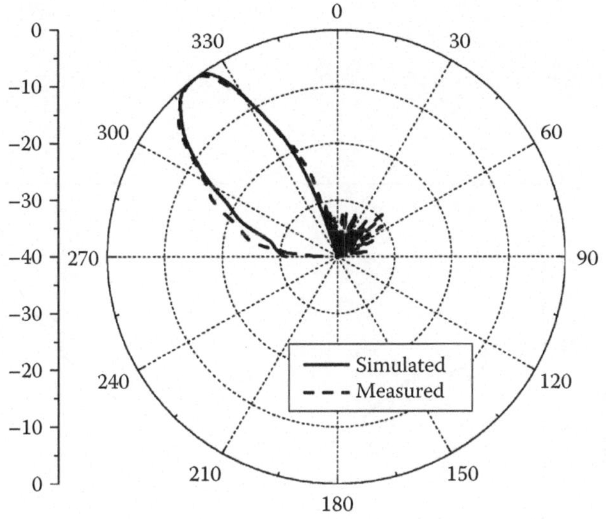

FIGURE 4.14 35 GHz H-plane pattern of the meandering SIW long straight slot leaky-wave antenna.

First, the fabrication tolerance is assumed to be ±20 μm with uniform distribution. Figure 4.18 gives the range of variation for 30 simulated radiation patterns for the meandering SIW long straight slot leaky-wave antenna. There is no visible difference in the aspects of beam direction, gain, and 3 dB beamwidth except SLL. However, the change in SLL is small. Thirty SLLs are all below –29.25 dB. The ideal SLL

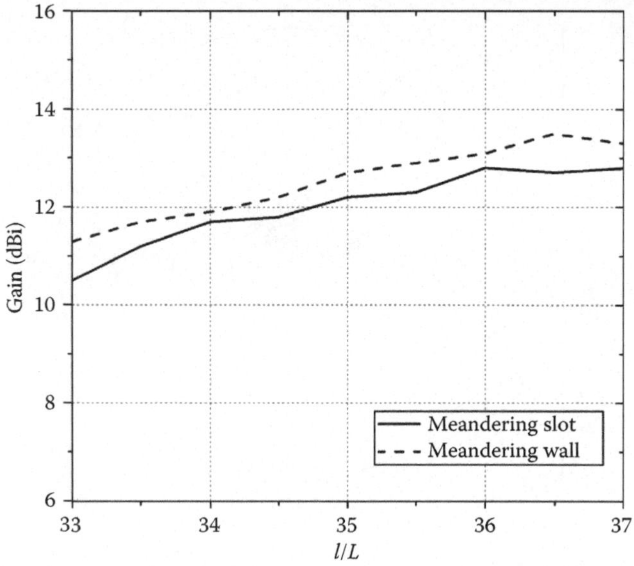

FIGURE 4.15 Measured gains of two types of long slot leaky-wave antennas at different frequencies.

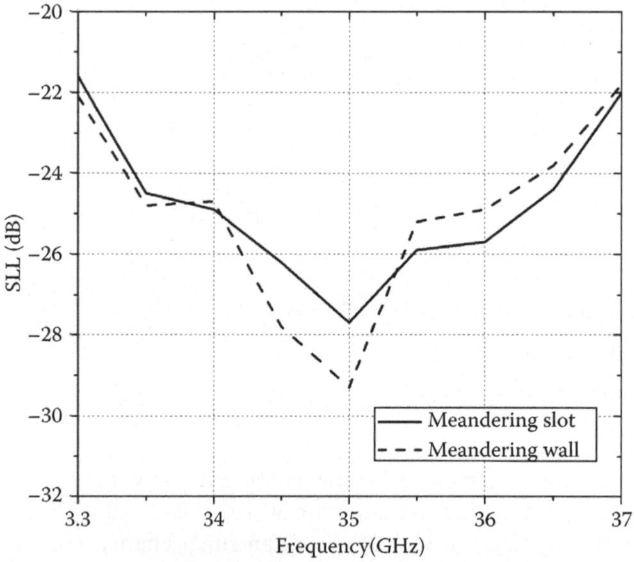

FIGURE 4.16 Measured SLLs of two types of long slot leaky-wave antennas at different frequencies.

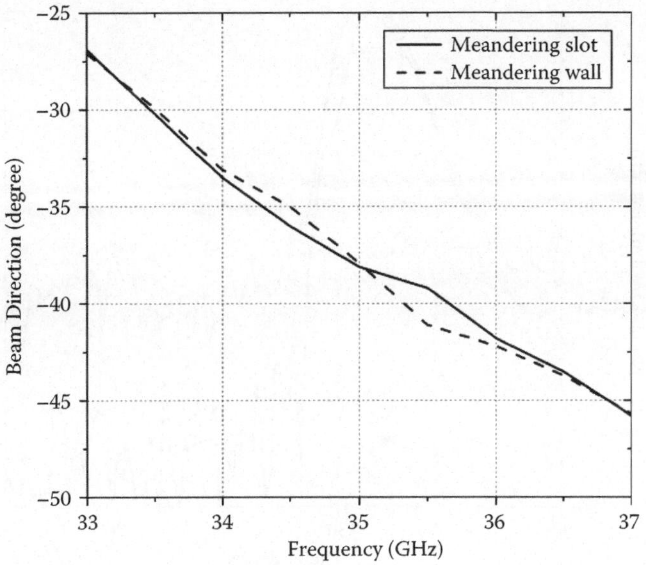

FIGURE 4.17 Measured beam directions of two types of long slot leaky-wave antennas at different frequencies.

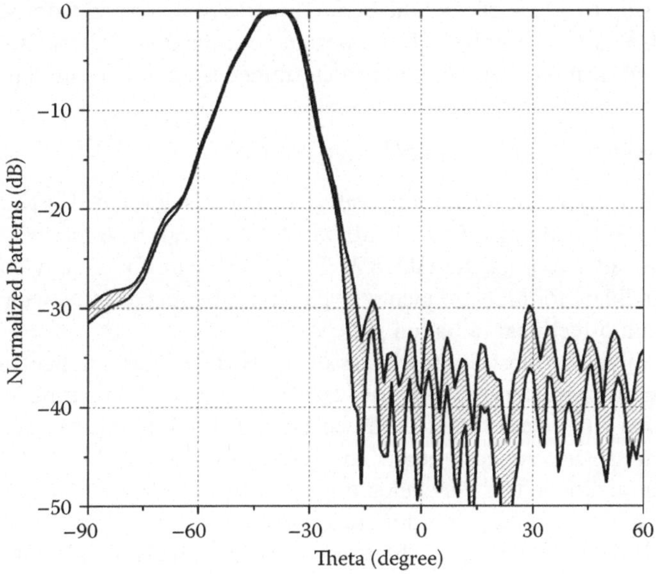

FIGURE 4.18 Upper–lower limits of the radiation pattern of the meandering SIW long straight slot leaky-wave antenna with a fabrication tolerance of ±20 μm at 35 GHz.

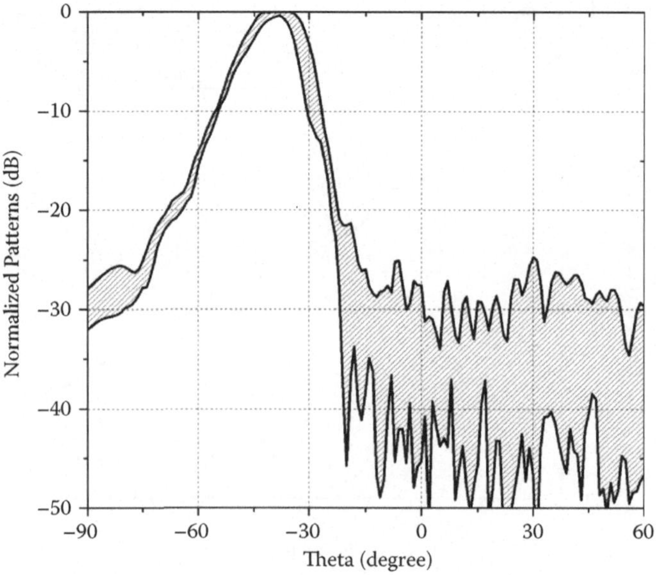

FIGURE 4.19 Upper–lower limits of the radiation pattern of the meandering SIW long straight slot leaky-wave antenna with a fabrication tolerance of ±100 μm at 35 GHz.

without tolerance is –30.9 dB. It is demonstrated that SLL is deteriorated just a little with a fabrication tolerance of ±20 μm. Then, the fabrication tolerance is assumed to be ±100 μm and a similar analysis is carried out. As shown in Figure 4.19, the worst SLL is up to –22.9 dB and the beam directions are varied from –35° to –42°. If so, it is difficult to comply with the design specifications. Through this study, a permissible fabrication tolerance can be determined to achieve desired performance.

4.2.2 HMSIW Uniform Leaky-Wave Antenna

In [140], a very simple uniform leaky-wave antenna was developed at Ka-band based on the HWSIW technology. It is found that the wave can be effectively leaked out from the open edge of a special HMISW. The HMSIW uniform leaky-wave antenna is a good candidate for high-frequency applications by means of its wide bandwidth and quasi-omnidirectional radiation pattern.

Through analysis, substrate thickness of the HMSIW has a noticeable influence on the radiation characteristic of the leaky-wave antenna. Another beam other than the main beam will emerge when a thin substrate is used. In this case, peak gains of two beams are so close to each other. They become imbalanced with the increase of the substrate thickness. It is concluded that the substrate thickness should be chosen carefully to achieve a quasi-omnidirectional pattern.

Besides, leakage capability of the leaky-wave antenna is mainly associated with the width of the HMSIW. A wider HMSIW can bind the electromagnetic and cannot radiate the energy out. The width of the HMSIW also determines the operating frequency range for good matching. There exists an optional region for the HMSIW width. Return loss is deteriorated below this range, while the efficiency is too low to serve as a leaky-wave antenna above this range.

It is obvious that a longer leaky-wave section leads to higher efficiency and narrower beamwidth of the HMSIW leaky-wave antenna. Thus, a trade-off should be made between acceptable antenna size and desirable specification.

4.3 SIC PERIODIC LEAKY-WAVE ANTENNA

In the periodic type leaky-wave antenna, there exists periodic modulation along the guided-wave structure. This periodicity is the major reason for wave leakage. To control the SLL, the periodic modulation has the small taper of the periodic properties.

4.3.1 SIW Periodic Leaky-Wave Antenna

SIW is comprised of periodically arranged metalized vias. Generally speaking, it has very low radiation loss. However, leakage loss of SIW structures can be increased with the distance between metalized vias. As shown in Figure 4.20, a periodic SIW with a large distance between metalized vias supports the propagation of leaky-wave modes. Thus, this structure can be employed in the design of a leaky-wave antenna as described in [141]. The leaky-mode phase and leakage rate can be simultaneously controlled by modifying the SIW width and the distance between vias.

Precise modeling of the SIW with different distances between metalized vias is essential in the design stage. A transverse equivalent network was developed in [142], and a corresponding parametric study was performed. Leaky-mode dispersion curves were obtained as a function of the main geometrical dimensions of the antenna. Additionally, the finite-difference frequency domain (FDFD) method can be employed to analyze the periodic SIW. Based on the leaky-mode analysis method, a periodic leaky-wave antenna at 28–34 GHz was designed and fabricated in [143].

This type of leaky-wave antenna can be synthesized with modulated geometry. In this case, it has the capability to flexibly tailor the radiation pattern as demonstrated

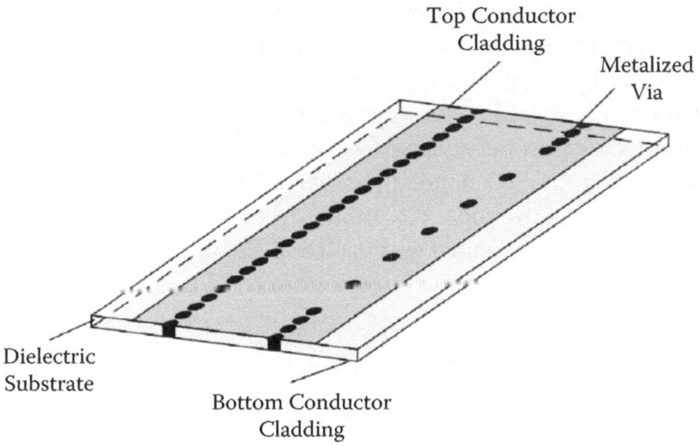

FIGURE 4.20 Configurations of the periodic SIW leaky-wave antenna.

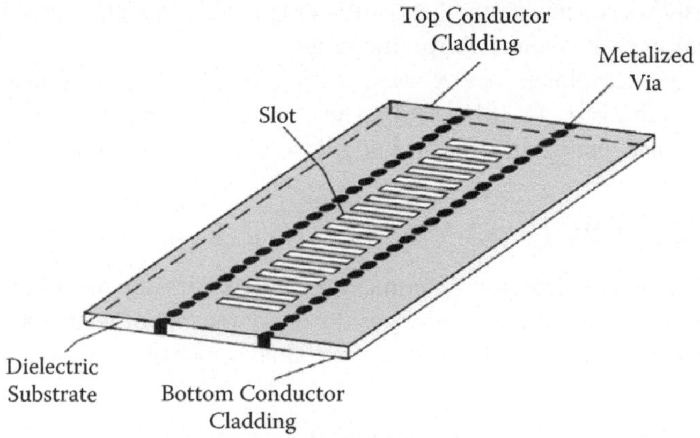

FIGURE 4.21 Configurations of the closely spaced slot SIW leaky-wave antenna.

in [144]. The synthesis technique is inspired in holographic concepts, which are related to the existence of modulated leaky waves. The requested modulation of the SIW width and distance between metalized vias can be obtained through the systematic design algorithm to synthesize the desired radiation pattern. Several design examples around 15 GHz have been reported and experimentally validated.

As another periodic leaky-wave antenna, an SIW leaky-wave antenna with closely spaced slots was proposed in [145]. As shown in Figure 4.21, a periodic set of transverse slots are etched on the top conductor cladding of the SIW. The current flow on the top wall of the SIW is interrupted to produce the leakage. There are three modes propagating in this structure, including a leaky mode, a waveguide mode, and a surface-wave mode. Their wavenumbers can be calculated theoretically and simulated numerically. After that, a uniform closely spaced slot SIW leaky-wave antenna is designed with a wide impedance bandwidth and a narrow beam that scans with frequency from near broadside to forward endfire. Based on this, a leaky-wave antenna has a butterfly-like configuration consisting of eight wings, as proposed in [146]. Better gain and SLL for lower-elevation angles are obtained. In the original design, these angles are scanned with poor radiation performances.

A double periodic composite right-left-handed (DP-CRLH) SIW proposed in [147] can also be employed in leaky-wave antenna design. In addition to the composite right-left-handed property, leaky-wave radiation is also observed at the low-frequency band below the left-handed passband. The radiation characteristics were tested to demonstrate the frequency scanning from backward to forward directions in the left-handed and right-handed regions from 12.6 to 17.8 GHz. Besides, it has the forward radiation in a new leaky-wave region from 5.2 to 5.8 GHz.

4.3.2 SIIG Periodic Leaky-Wave Antenna

4.3.2.1 SIIG Longitudinal Metal-Strip Leaky-Wave Antenna

A linear leaky-wave antenna fed by an SIIG at 94 GHz was presented in [148]. It can be regarded as the use of discrete leakage waves along the open traveling-wave path

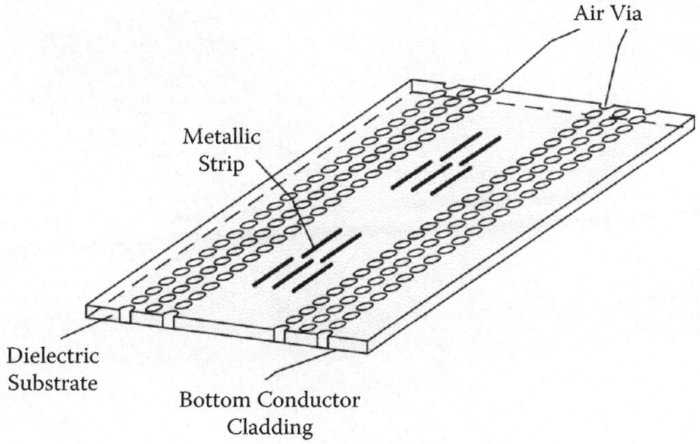

FIGURE 4.22 Configurations of the SIIG longitudinal metal-strip leaky-wave antenna.

of the SIIG structure. Five longitudinal metallic strips were grouped as one radiation element, which was fabricated on top of the SIIG as shown in Figure 4.22.

This scheme provides good linear polarization purity. The leaky-wave antenna usually suffers from a stopband near the broadside direction because of the distributed Bragg reflection. However, in this design, the beam of the leaky-wave antenna can be scanned with operating frequency through the broadside direction. The return loss can be effectively improved through an optimized reflection compensation technique. As an example, a 2-cm-long W-band SIIG longitudinal metal-strip leaky-wave antenna produces a fan beam with 11 dBi gain.

4.3.2.2 SIIG Transverse Metal-Strip Leaky-Wave Antenna

The low-temperature cofired ceramic (LTCC)-based SIIG can be employed in the design of a V-band periodic leaky-wave antenna, which has a periodic array of metal strips on the top surface of the SIIG, as shown in Figure 4.23 [149]. A CPW–SIIG transition as described in Section 2.4.4 is employed here to excite this antenna, offering a good solution to the integration onto chips. A V-band short linear leaky-wave antenna with six metallic strips was designed and fabricated as an example.

Here, the used material is Ferro A6-M with a relative permittivity of 5.9 and a loss tangent of 0.002. The dielectric strip has seven stacked ceramic tapes, while the isolating film has one ceramic tape. Each tape has a height of 0.1 mm. The thickness, h, is 0.7 mm. The SIIG strip width, w, is 1.8 mm. The CPW feedline is a 50 Ω grounded type with a substrate thickness of 0.1 mm.

The design process for this type of SIIG leaky-wave antenna is introduced as follows:

1. Radiation efficiency versus dimension. The single radiating element is modeled with different parameters and full-wave simulated to find the relationship between the dimension of the radiating element and the radiation efficiency. Conductor material is set to be the perfect electric wall, while dielectric material is considered lossless.

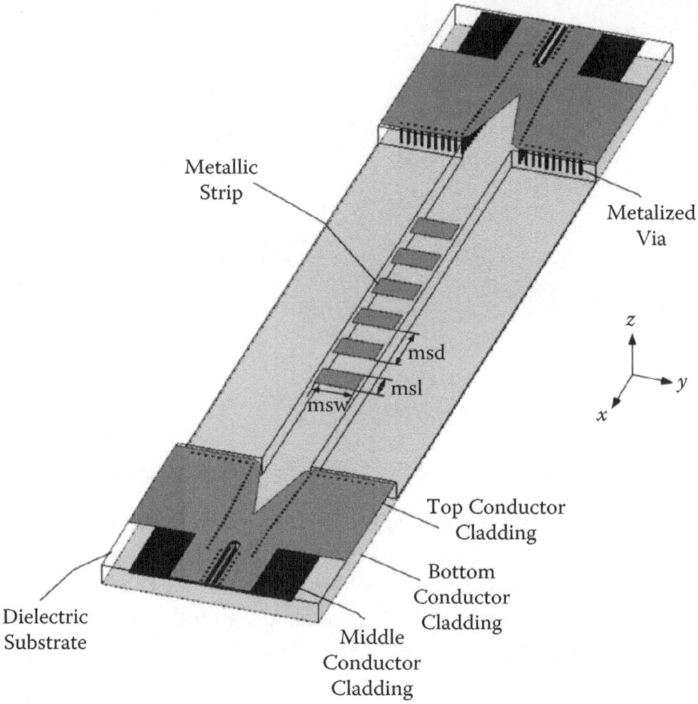

FIGURE 4.23 Configuration of the V-band six-element SIIG transverse metal-strip leaky-wave linear antenna.

Figure 4.24 shows the relationship between the radiation efficiency, the length of the metal strip, *msl*, and the SIIG thickness, *h*, at 60 GHz. The width of the metal strip, *msw*, keeps unchanged. As shown in this figure, the radiating efficiency decreases with the SIIG thickness. The electromagnetic wave is bounded much more in a thicker SIIG than in a thinner SIIG. Thus, a thick SIIG is not a good choice in leaky-wave applications. Besides, radiating capacity for a certain thickness is also weakened with the increase of the operating frequency. That stems from the fact that the same physical substrate thickness behaves electrically thicker at the higher-frequency band. A larger *msl* leads to higher radiating efficiency. That means the radiating efficiency can be controlled by *msl*.

In this design, the terminating load is designed to take in 10% of the incident power at 60 GHz, i.e., $R = 0.1$. The SIIG has a thickness of 0.7 mm. In this case, the required radiation efficiency for each element, *re*, can be estimated roughly by

$$re = 1 - \sqrt[n]{R} \tag{4.6}$$

In Equation (4.6), *n* is the number of the used metallic strips. Here, it is six. *msl* read from Figure 4.24 can be considered an initial value. In the next

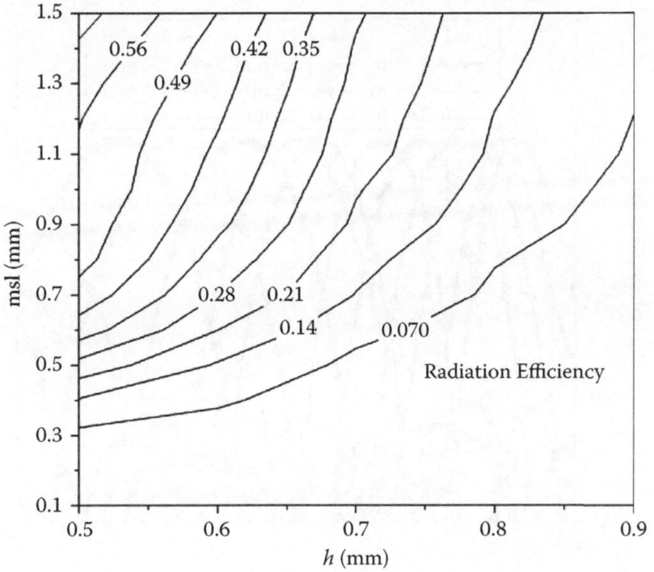

FIGURE 4.24 Radiating efficiency versus the dimension of the radiating element with $w = 1.8$ mm and $msw = 1.5$ mm at 60 GHz.

design step for a linear antenna, *msl* is optimized and changed to 1 mm to achieve the goal of $R = 0.1$.

2. Beam pointing versus patch distance and frequency. As shown in Figure 4.25, the beam of a six-element SIIG leaky-wave antenna operating at 60 GHz can be scanned from the backward quadrant into the forward quadrant. Beam direction, θ, of a periodic leaky-wave antenna can be controlled by a different period, *msd*. Their relationship can be described as follows [120]:

$$\theta = \arcsin\left(\frac{\beta}{k_0} - \frac{2\pi}{k_0 \times msd}\right) \tag{4.7}$$

By use of a full-wave simulator, β/k_0 is calculated to be 1.71 at 60 GHz for an SIIG with $w = 1.8$ mm and $h = 0.7$ mm. In our design, the required beam direction is 30° at 60 GHz; thus, the initial *msd* can be estimated at 2.25 mm. Through the optimization, *msd* is adjusted to 2.085 mm.

On the other hand, the leaky wave antenna can also be scanned with the operating frequency.

The final dimensions of the V-band SIIG periodic leaky-wave antenna are $w = 1.8$ mm, $h = 0.7$ mm, $msl = 1$ mm, $msd = 2.085$ mm, and $msw =$ 1.5 mm. The E-plane (*xoz*-plane) patterns at different frequencies are shown in Figure 4.26. The cross-polarization of such an antenna is lower than –35 dB at the beam direction of each operating frequency. Next, we pay

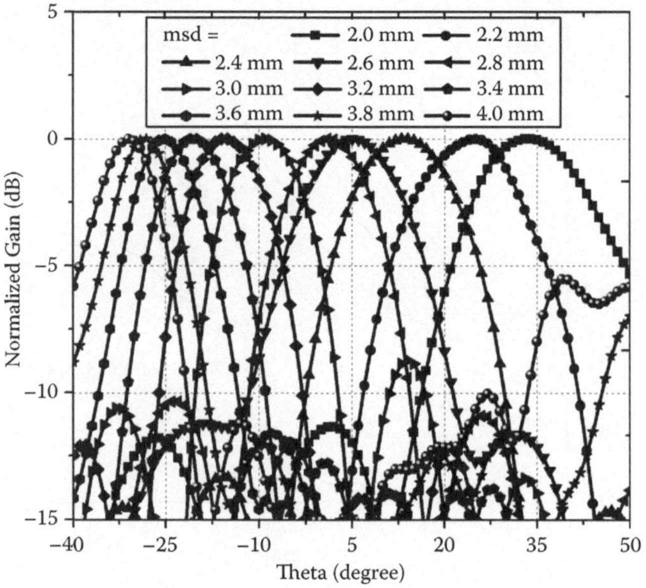

FIGURE 4.25 Simulated beam directions of the SIIG transverse metal-strip leaky-wave antenna with different periods at 60 GHz with w = 1.8 mm, h = 0.7 mm, msl = 1 mm, and msw = 1.5 mm.

attention to the SLL control of the SIIG periodic leaky-wave antenna. Because the SLL is usually high for a typical leaky-wave antenna, the practice is then to vary the periodic property slowly along the length while maintaining β unchanged, so as to control the aperture distribution to optimize the SLL performance.

In this design, the gratings can be tapered to control SLL, so that the width of the metallic strip should be very small at the beginning and increased with a tapered distribution. Two 19-element structures are used as examples. For the first uniform type, the periodic modulation itself is uniform along the length. The second one has the taper of the periodic properties along the length to control SLL. As shown in Figure 4.27, the nonuniform type is able to improve the SLL more than 9.7 dB. Certainly, this type of leaky-wave antenna has some difficulty in more accurate leakage control, and thus a further lower SLL is difficult to achieve.

The SIIG is suitable to act as antenna feeders on the strength of its extremely low loss. However, this semiopen structure is generally subject to leakage if there are discontinuities, bends, and corners. Thus, the SIIG divider is not a good choice in array applications. By use of the SIW–SIIG transition, the SIW can be employed in the design of a large-scale feeding divider to solve the array difficulty for SIIG linear antennas.

As an example, a grating of metallic strips with different widths and lengths was placed on the top surface of the SIIG to design a W-band antenna. A 19-strip linear antenna prototype was designed as described in

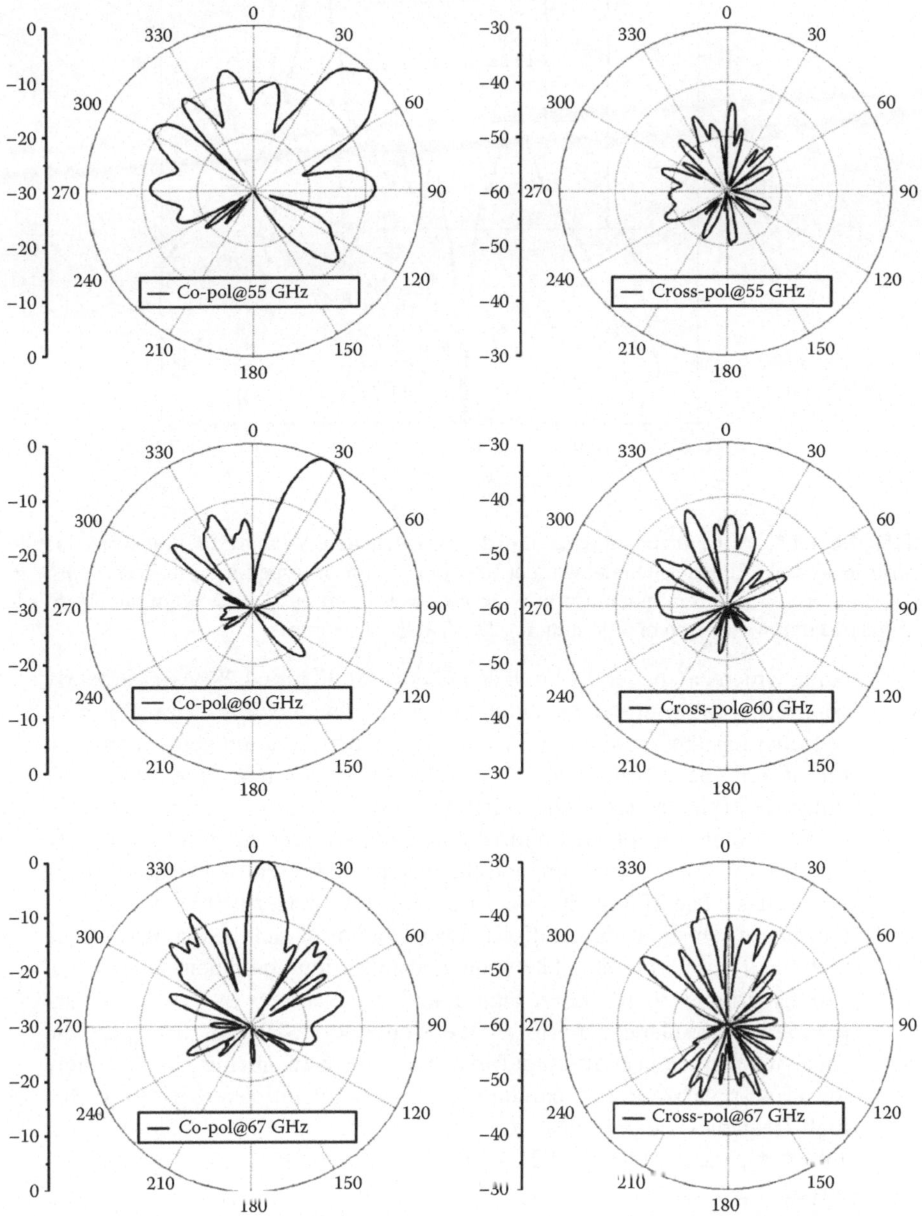

FIGURE 4.26 Simulated E-plane patterns of the V-band SIIG transverse metal-strip leaky-wave antenna at different frequencies.

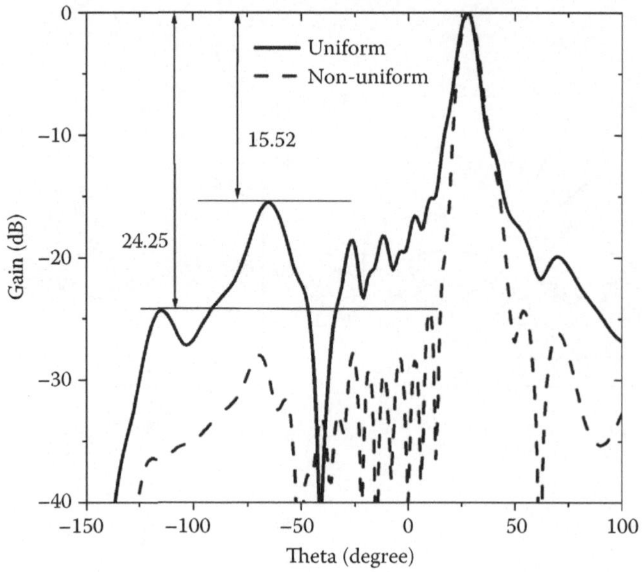

FIGURE 4.27 Normalized patterns of different 19-element V-band SIIG transverse metal-strip leaky-wave linear antennas without transition (uniform type: *msl* = 0.78 mm, *msw* = 1.5 mm, *msd* = 2.15 mm; nonuniform type: *msw* = 1.5 mm, *msd* = 2.15 mm, no. 1–16 *msl* 0.24–1.14 mm with a step of 0.06 mm, no. 17–19 *msl* = 1.08 mm).

[149], which can be fed by an SIW, as shown in Figure 4.28. After properly controlling the width and position of the first few metallic strips, this type of feeder has the possibility of achieving a better radiation performance. In this design, the metallic strip starts within the V-slot in order to weaken the influence of the feeder on the radiation pattern.

The dimensions of the optimized antenna are presented in Figure 4.29. The SIIG width is 1.155 mm, and the distance between two adjacent strips is 1.35 mm. The SIW width is 1.2 mm, the SIW height is 0.5 mm, the distance between adjacent metalized vias is 0.25 mm, and the diameter of the metalized via is 0.1 mm. The simulated reflection coefficient is less than –10 dB from 80 to 107 GHz. Figure 4.30 demonstrates the co- and cross-polarization E-plane patterns of the single-port SIW-fed SIIG periodic leaky-wave antenna at different frequencies. Such an antenna has extremely low cross-polarization. It maintains high radiation efficiency within such a high-frequency band. Its beam directions are 43°, 40°, 35°, 31°, 25°, 22°, 18°, 13°, 9°, 6°, and 2° at 88, 90, 92, 94, 96, 98, 100, 102, 104, 106, and 108 GHz, respectively.

Finally, eight of the proposed W-band linear elements, with the element distance of 2.5 mm, were grouped to form an array fed by an eight-way SIW parallel divider. Designed E-plane patterns at different frequencies are shown in Figure 4.31. The cross-polarization of the fabricated antenna array is lower than –35 dB at the beam direction. The beam direction of the

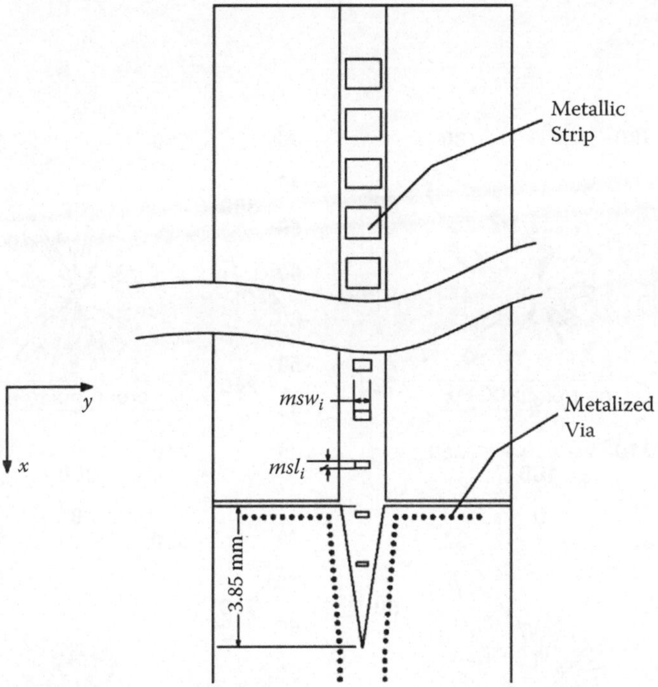

FIGURE 4.28 Configuration of the W-band SIW-fed SIIG transverse metal-strip leaky-wave linear antenna.

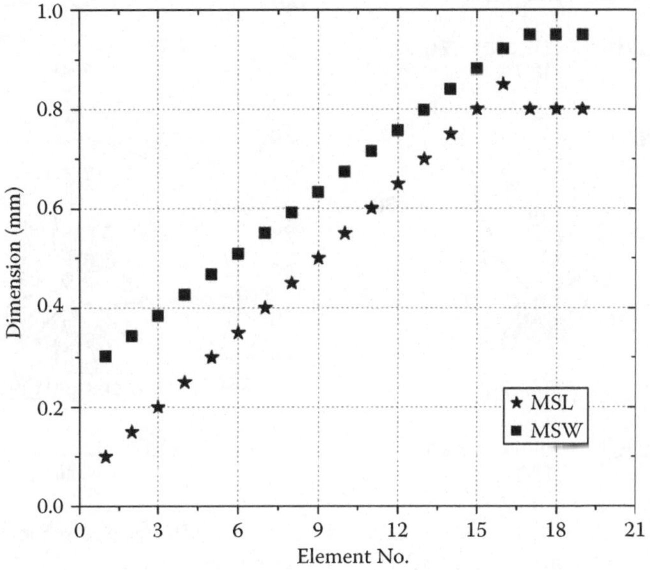

FIGURE 4.29 Dimensions of the W-band SIIG transverse metal-strip leaky-wave linear antenna.

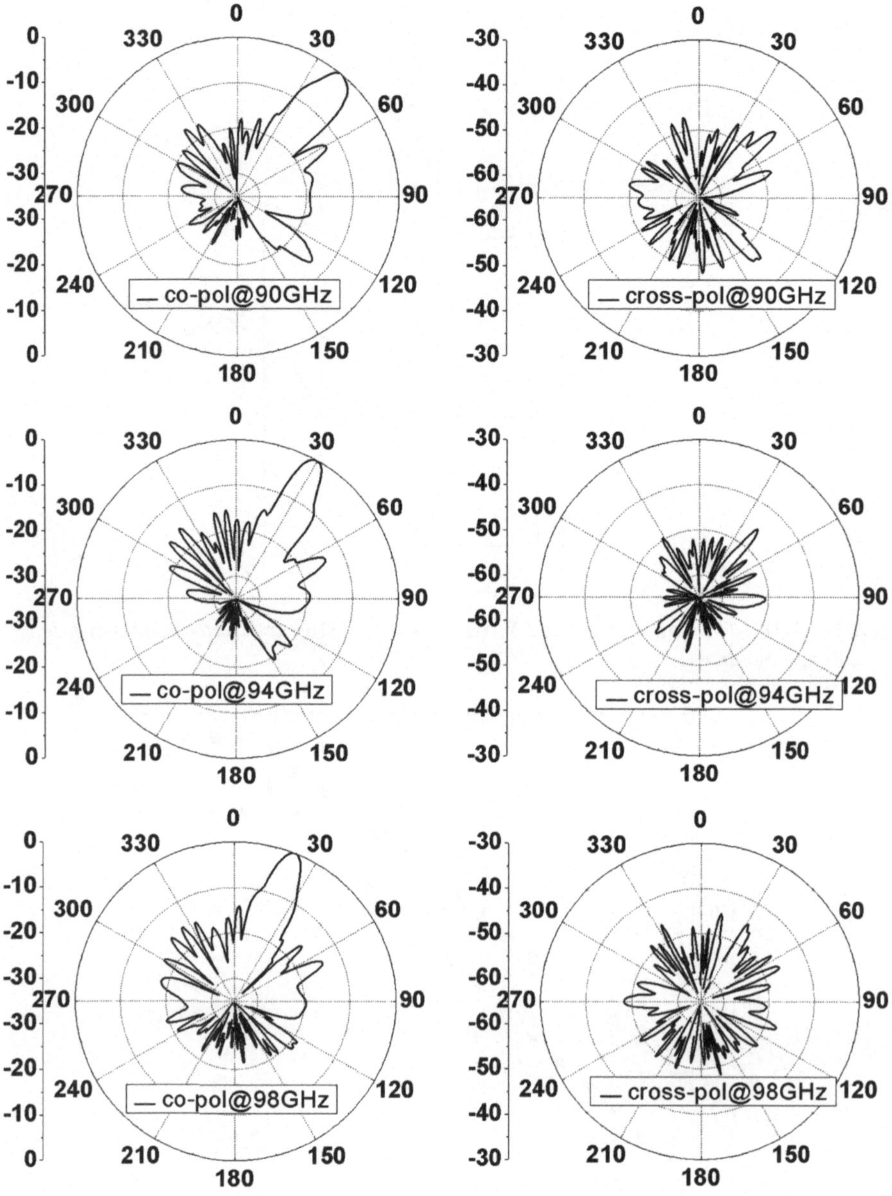

FIGURE 4.30 Simulated E-plane patterns of the W-band SIIG transverse metal-strip leaky-wave linear antenna, including the SIW feeder at different frequencies.

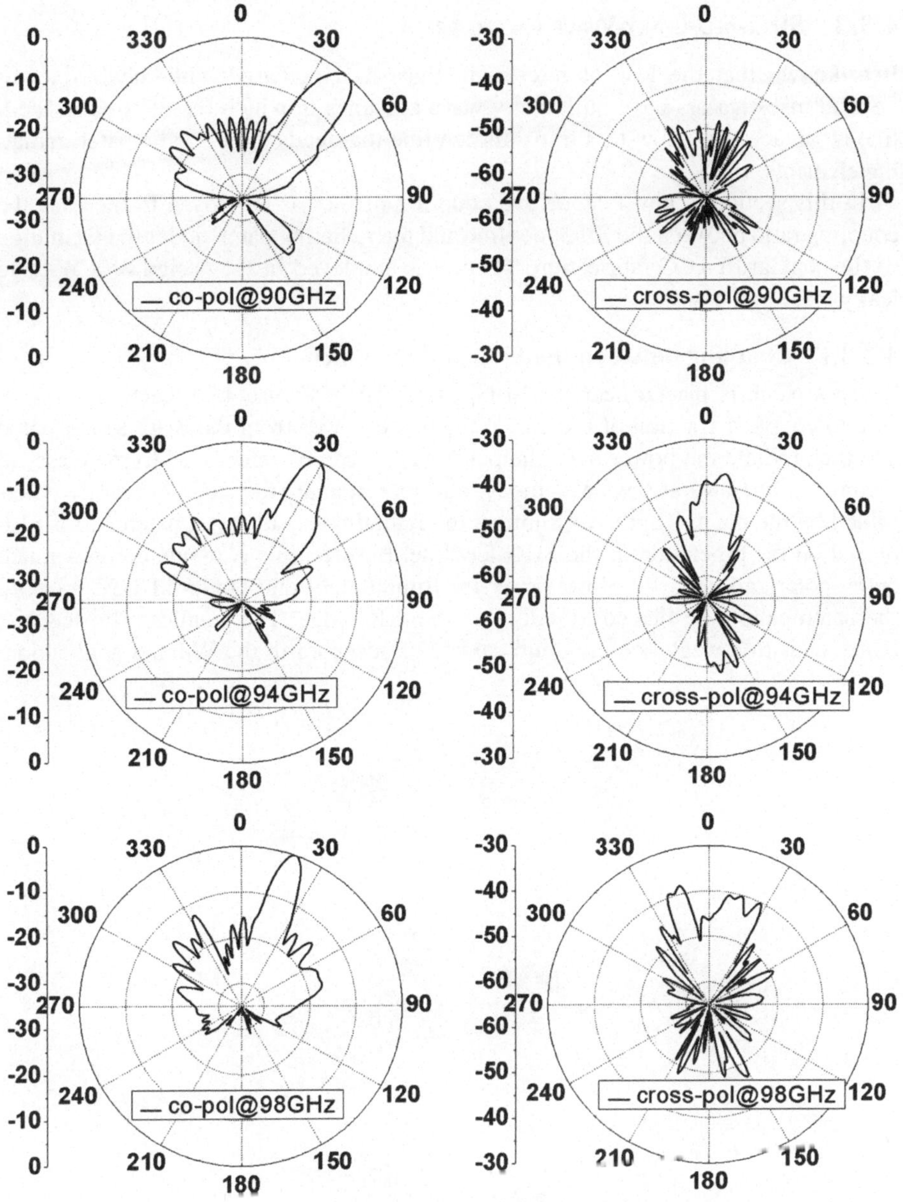

FIGURE 4.31 Simulated E-plane patterns of the W-band eight-element SIIG transverse metal-strip leaky-wave antenna array at different frequencies.

antenna scans from 23° to 45° when its operating frequency varies from 98 GHz to 90 GHz. Measured gain remains in the range of 20.5–22.5 dBi.

4.3.3 SIIG-Fed Leaky-Wave Patch Array

It is known that the loss of microstrip lines has increased with the frequency. Several investigators have built leaky-wave antennas in which the microstrip feedline is replaced by a low-loss transmission line that feeds the periodic patch radiating elements.

In this section, a novel capacitive coupling method is employed to increase the coupling ratio between the SIIG feedline and the radiating patch under the limitation of the standard LTCC fabrication. Then, it is employed in the design of a W-band leaky-wave antenna.

4.3.3.1 Coupling between the SIIG and the Patch

When a patch is placed near the SIIG, as shown in Figure 4.32, each unit cell is able to couple a fraction of the electromagnetic power from the SIIG to the patch and then radiate the power out. The patch design can be referenced to the classical method. It operates at the TM_{10} mode, and its resonant frequency is mainly determined by the patch length. To improve the radiating capability, a patch cell is also placed on the other side of the SIIG feedline. Besides, this two-side cell has much better polarization purity. However, as the limitation of the standard LTCC process, the patch cannot be placed too close to the SIIG edge. The smallest allowed distance leads to a weak coupling and cannot be accepted in the W-band application.

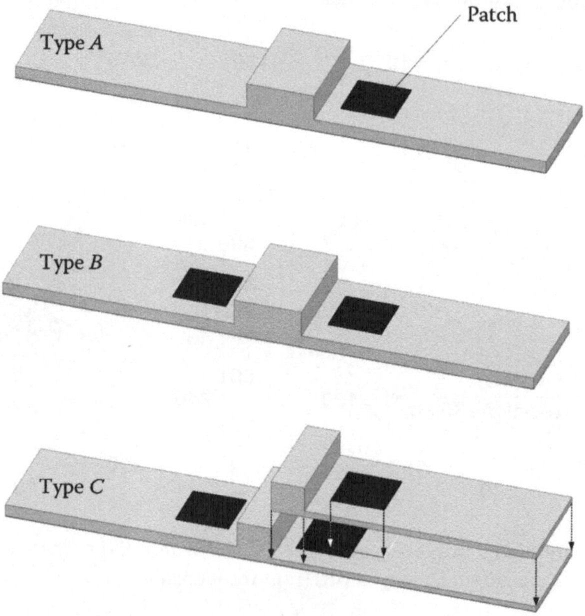

FIGURE 4.32 Different types of SIIG-fed patches.

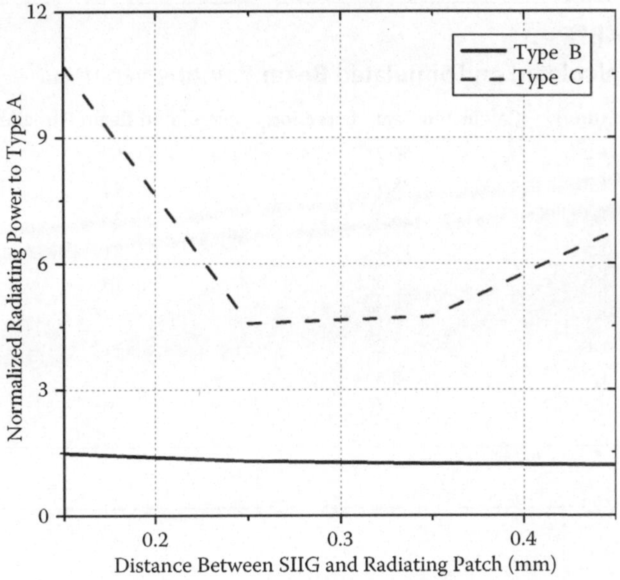

FIGURE 4.33 Normalized radiating power of different SIIG-fed patches.

Therefore, a feeding patch is added behind the SIIG and the radiating patch. It can be considered a capacitive coupling and increase the coupling significantly. Certainly, a wider coupling tuning range can be achieved, which is helpful to synthesize better radiation performance.

The used material is Ferro A6-M with a relative permittivity of 5.9 and a loss tangent of 0.02. For the SIIG feedline, the dielectric strip consists of four stacked ceramic tapes. Each tape has a height of 0.1 mm. Two additional tapes are employed to be the isolating film. The strip width is 1 mm. The used patch has a length of 0.8 mm. Simulated radiating power of types B and C with the same dimensions is normalized to that of type A at 94 GHz, as shown in Figure 4.33. There exists only one parameter for types A and B, i.e., the distance between the SIIG and the radiating patch. But there are several other parameters to optimize the performance of type C. The design freedom is increased as well.

4.3.3.2 Linear Array Design

1. Beam pointing versus patch distance. The investigated model is an SIIG-fed leaky-wave patch antenna with 19 two-side elements. Equation (4.7) can be used to calculate the beam pointing with the patch distance, *pd*, just by replacing *msl* with *pd*. Similarly, β/k_0 of the SIIG feedline can be obtained by simulation.

   Comparisons between the calculated results by Equation (4.7) and the simulated results are listed in Table 4.1. Meanwhile, simulated patterns with different *pd* are depicted in Figure 4.34. As shown, β will become a little smaller when leakage occurs. Equation (4.7) can certainly be used to predict the beam pointing versus frequency and the beam coverage, as presented in Table 4.2.

TABLE 4.1
Calculated and Simulated Beam Pointing versus *pd*

pd (mm)	Calculated Beam Direction	Simulated Beam Direction
1.2	56.2°	57°
1.3	38.7°	42°
1.4	26.8°	30°
1.5	17.4°	21°
1.6	9.5°	13°
1.7	2.7°	4°
1.8	−3.2°	−1°
1.9	−8.6°	−4°
2	−13.5°	−9°

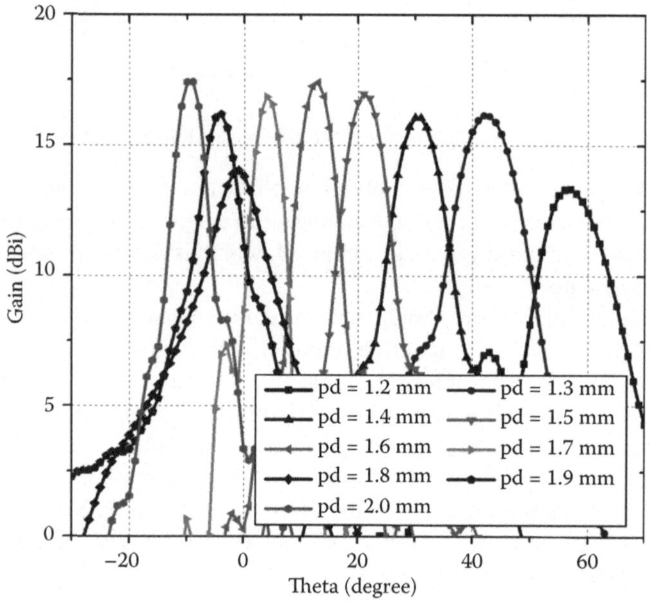

FIGURE 4.34 Simulated patterns of the 19-element SIIG-fed leaky-wave patch antenna with different periods.

2. Patch number versus residual power and radiating characteristic. For a conventional leaky-wave antenna, it is terminated with a load to absorb a fraction of the incident power. But in the high-frequency band, the matching load is expensive. In this design, the terminating load can be omitted when the leaky-wave antenna is long. The residual power is very small and the radiating influence of the backward wave can be ignored.

   First, the relationship between the element number of a two-port leaky-wave antenna and the residual power is investigated as shown in

TABLE 4.2
Calculated and Simulated Beam Pointing versus Frequency

Frequency (GHz)	Calculated Beam Direction	Simulated Beam Direction
91	40°	34.4°
92	37°	31.7°
93	33°	29.2°
94	30°	26.8°
95	28°	24.5°
96	25°	22.3°
97	22°	20.2°
98	20°	18.2°

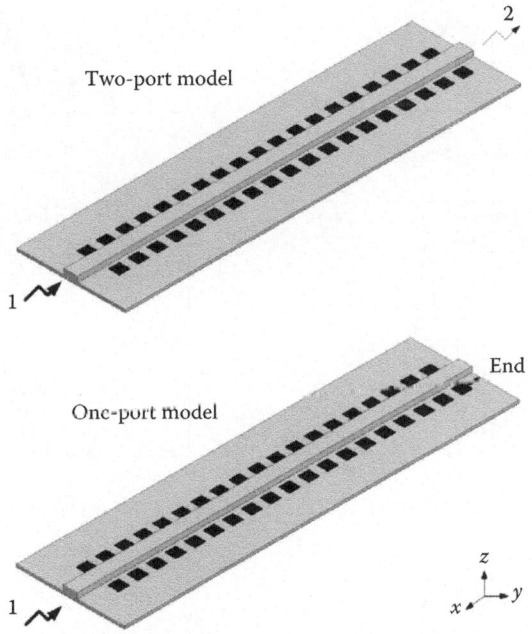

FIGURE 4.35 Models to investigate the relationship between patch number and residual power and radiating characteristic.

Figure 4.35. The simulated results are presented in Figure 4.36. Second, a one-port antenna with the open end is investigated, and its configuration is shown in Figure 4.35. When the residual power is lower than –19 dB, the SLLs of them are almost the same. Comparisons of radiating characteristics between two models are listed in Table 4.3. It is demonstrated that the beamwidth of the one-port model is narrow. Besides, their beam directions and gains are almost the same.

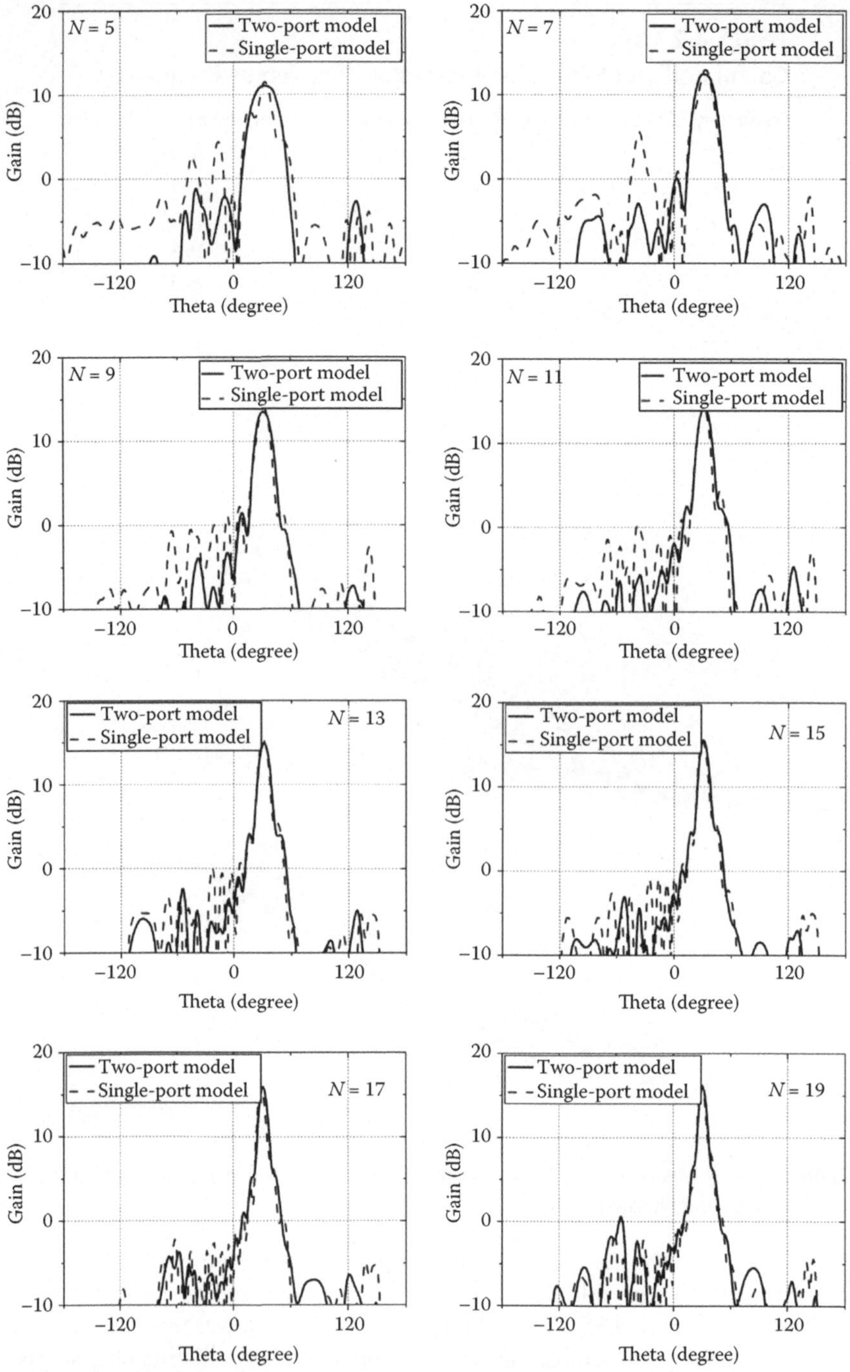

FIGURE 4.36 Radiation patterns of SIIG-fed patch arrays with different radiating element numbers, N, at 94 GHz.

TABLE 4.3
Comparisons of Radiating Characteristics between Two Models

Two-Port Model

N	Beam Direction	Gain (dBi)	Beamwidth	S21
5	32°	11.04	30.63°	–6.87
7	31°	12.49	22.13°	–9.13
9	31°	13.49	17.40°	–11.39
11	32°	14.30	14.01°	–13.80
13	31°	14.92	11.58°	–15.76
15	31°	15.59	9.89°	–17.78
17	31°	15.95	8.48°	–19.56
19	30°	16.06	8.22°	–21.46

One-Port Model

N	Beam Direction	Gain (dBi)	Beamwidth
5	33°	11.52	14.12°
7	33°	12.98	15.08°
9	32°	13.91	13.50°
11	31°	14.23	12.02°
13	31°	15.26	10.55°
15	30°	15.61	9.80°
17	29°	15.75	8.13°
19	30°	16.30	7.60°

3. End distance versus radiating characteristic. Next, the effect of the open end on radiation patterns is discussed. As shown in Figure 4.37, with 1–5 mm end length, the beam direction changes 2° and the gain changes 0.4 dBi. All of them have similar performances. After full-wave simulation and optimization, the final dimension of the proposed structure is shown in Figure 4.38. A transition between an SIIG and an SIW is also used here. For experimental purposes, an SIW interlayer transition is employed in front of the antenna input port. It can transfer the electromagnetic wave from the exciting standard WR-10 waveguide port to the SIW antenna port through five-layer SIWs, as shown in Figure 4.39. A coupled slot is etched on the middle conductor cladding between two SIWs. In practical applications, this transition can be omitted when the antenna is integrated with other printed circuits. Figure 4.40 presents radiated patterns at different frequencies.

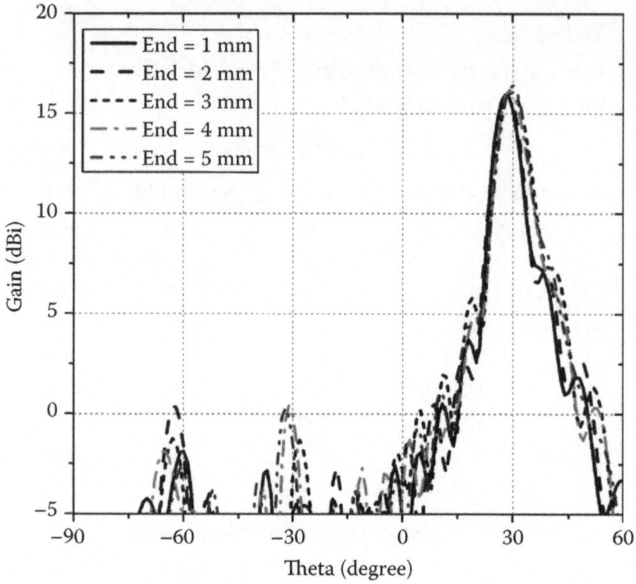

FIGURE 4.37 Relationship between the end distance and the radiating characteristic.

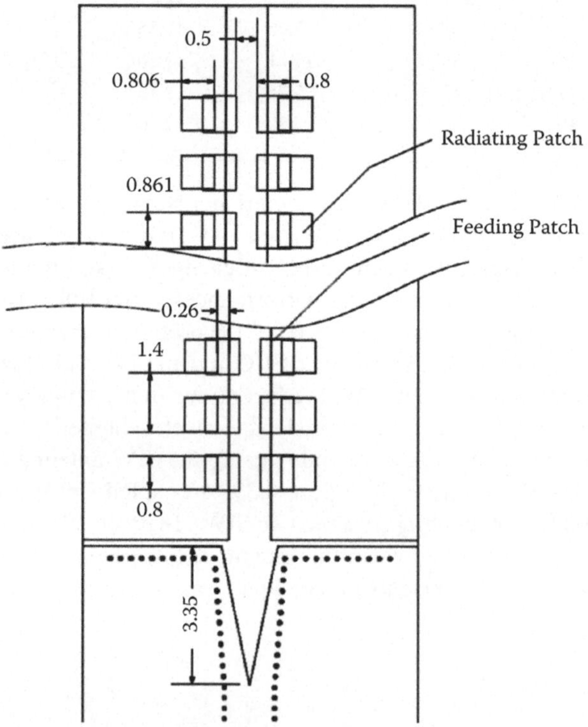

FIGURE 4.38 Configuration and dimension of the SIIG-fed leaky-wave patch antenna with an SIW feeder.

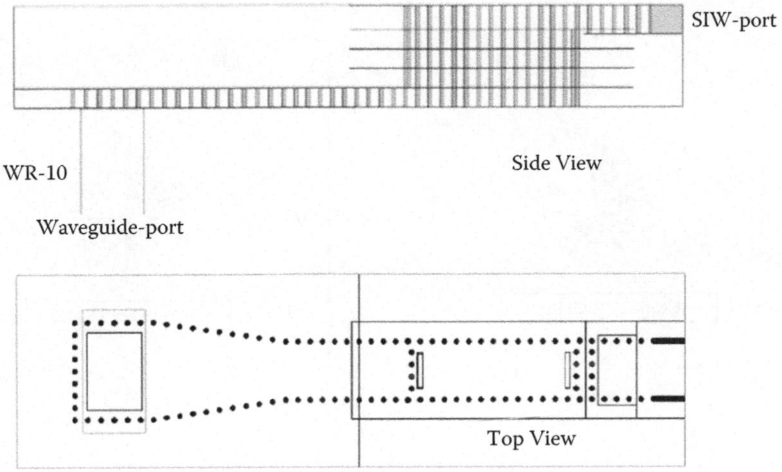

FIGURE 4.39 Configuration of the multilayer SIW–waveguide transition.

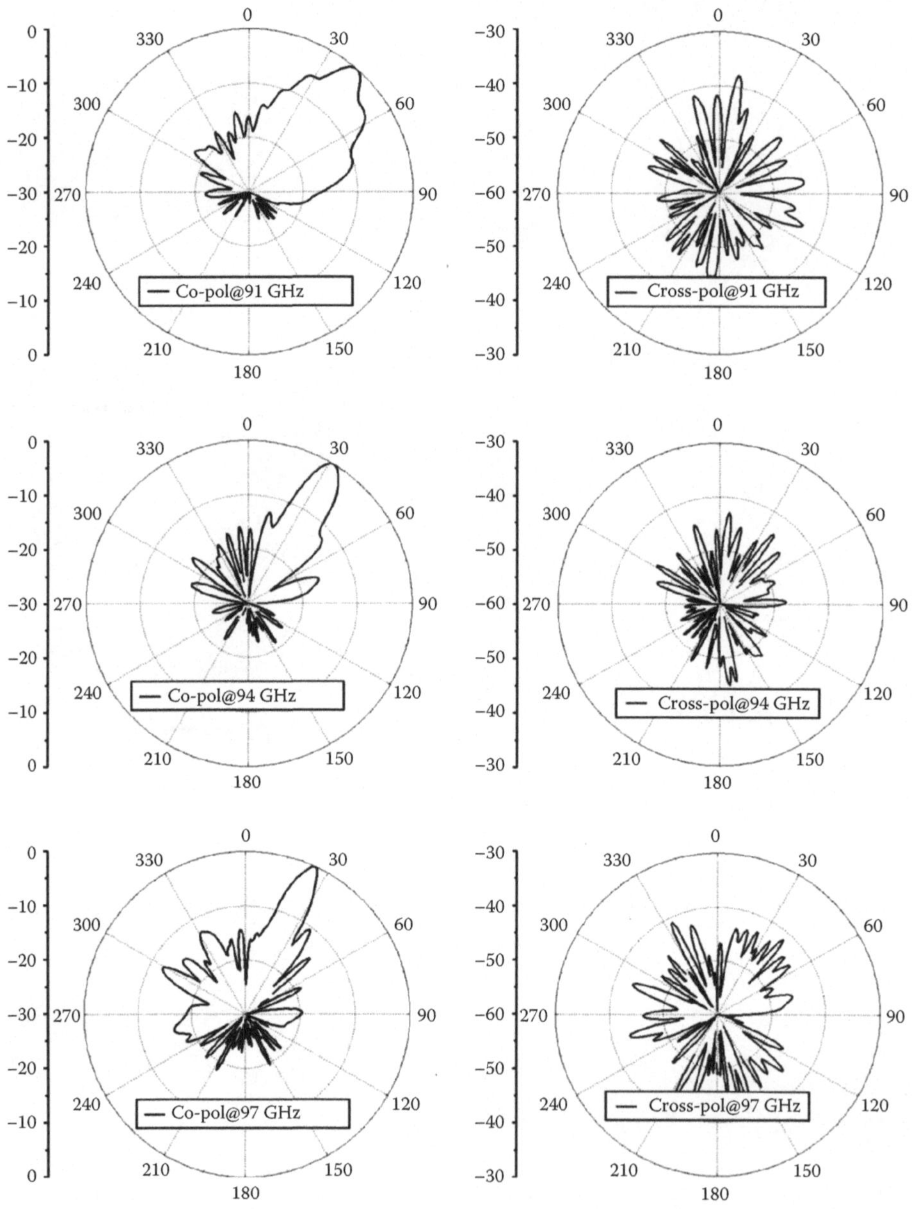

FIGURE 4.40 Radiation patterns of the SIIG-fed leaky-wave patch antenna at different frequencies.

5 Substrate Integrated Feeding Antennas

Antennas are one of the most important components responsible for the noise performance and power budget of systems. Bandwidth is an important specification for antennas, especially in frequency division duplexing (FDD) communication. Another critical specification for antennas is the realized gain. Conventional high-gain antennas most often used in the high-frequency band are parabolic reflector antennas and waveguide antennas, which need a precise and complicated machining process, resulting in a high production cost. On the other hand, these antennas usually have large size and big weight, leading to high shipping expenses and heavy manpower expenditure on equipment installation and adjustment.

These considerations impel us to develop a high-gain and wideband planar antenna as an alternative for these nonplanar antenna schemes. Microstrip-fed array antennas have been extensively used in the microwave frequency band. However, the feeding efficiency of microstrip lines drops significantly because of serious losses at bends and discontinuities when the operating frequency increases to the millimeter-wave frequency band. A low-loss, high-efficiency, planar feeding network is required.

The substrate integrated waveguide (SIW) technology combines the superior features of both the planar transmission line and nonplanar waveguide, and thus deserves more expectations in terms of millimeter-wave circuits at low-cost fabrication. There are little radiation leakage and coupling, even at SIW bends and discontinuities. Therefore, it is a good solution for millimeter-wave feeding network applications.

In this chapter, an overview of popular and representative substrate integrated feeding array antennas is introduced, including SIW-fed patch antenna array, SIW-fed cavity-backed patch antenna array, SIW-fed dielectric resonator antenna, SIW-fed open-ended cavity antenna array, and SIW-fed printed open-ended waveguide antenna array.

5.1 SIW-FED PATCH ANTENNA ARRAY

Until now, various types of patch antennas have been fed by the SIW slot series feeding topology.

Stacked multilayered patch designs were proposed and demonstrated in [150]. First, a single antenna element was designed at the V-band. It consists of four parasitic elements, which are driven by a circular patch. By properly selecting the element diameter, the distance between elements, the substrate thickness, and the dielectric permittivity, the center frequency, the gain, the beamwidth, and the front-to-back ratio of the antenna can be modified and optimized. After that, the array makes use of a four-way SIW parallel feeding divider in order to feed a 4 × 4 antenna array. The radiating elements are coupled through rectangular slots alternately spaced $\lambda_g/2$

apart etched on the top surface of a short-circuited feeding SIW branch. Each branch has four elements. Another E-band 4 × 4 circular patch array was demonstrated in [151] using the SIW technology for the design of its feeding network. Longitudinal slots etched on the SIW top metallic surface are also used to excite the radiating elements. To increase the gain of the circular patch elements, dielectric cubes with low permittivity were placed on top of each 1 × 4 linear antenna as described in [152]. Furthermore, two rectangular microstrip patch antennas were, respectively, fed by a transverse slot and a longitudinal slot etched on the conductor cladding of the feeding SIW at the V-band [153].

These resonant SIW feeding techniques usually lead to narrow bandwidth with the increase in the array size. Besides, the feeding network based on the tree-like parallel divider becomes very large. Therefore, it is difficult to fully meet the requirements of high gain, wide bandwidth, and miniaturization at the same time.

5.1.1 Microstrip-SIW-Fed Patch Antenna Array

In [154], compact substrate integrated parallel feeding technology was employed to feed a 32 × 32 patch array antenna for W-band applications. The contradiction between gain and bandwidth of conventional SIW array antennas can be solved. This antenna array consists of a stack of two dielectric substrates. The large-scale feeding network is completely hidden underneath the radiating aperture. In this case, the total size of the integrated array antenna is equal to the size of the radiating aperture. Thus, this configuration benefits greatly in miniaturization.

These 32 × 32 patches are divided into 16 × 16 subarrays, and each subarray is a 2 × 2 unit. Each patch fed by an SIW is impossible because the SIW width is larger than the spacing between neighboring patches. To solve this problem, the 2 × 2 subarray is fed by a four-way back-to-back microstrip divider first. Then, each 2 × 2 unit is fed through a longitudinal off-center slot etched on the top surface of an SIW branch. This method is more appropriate for W-band applications than using a metalized via, which is described in [155] due to its large electrical size.

The schematic of a 4 × 4 small integrated array antenna is shown in Figure 5.1. The bottom layer is used to place the H-type SIW feeding network, which is made of a 0.508 mm Rogers 5880LZ substrate. Its relative permittivity is 1.96. The thickness of each copper layer is 16 μm. The diameter of the metalized via is 0.3 mm. The top layer consists of 4 × 4 rectangular patches with an element spacing of $0.63\lambda_0$ on a 0.254 mm Rogers 5880LZ substrate. Each patch has a length of 1.35 mm and a width of 0.92 mm. Two such substrates are bonded by Rogers 3001 bonding film. The blind vias, which are laid only in the bottom dielectric layer, connect the middle and bottom conductor layers and form SIW branches.

To locate the slot at peak of the standing wave, the SIW branch is terminated by a short circuit that is a quarter of the guided wavelength away from the center of the last slot in order to align the slots at the peaks of the standing wave in the SIW. Three T-type dividers are placed back-to-back as an H-type divider, not only to connect the subarrays and excite the array antennas in phase with the broadband characteristic, but also to keep a compact configuration. The relatively small size leads to the nonuniform SIW width in such an H-type feeding network.

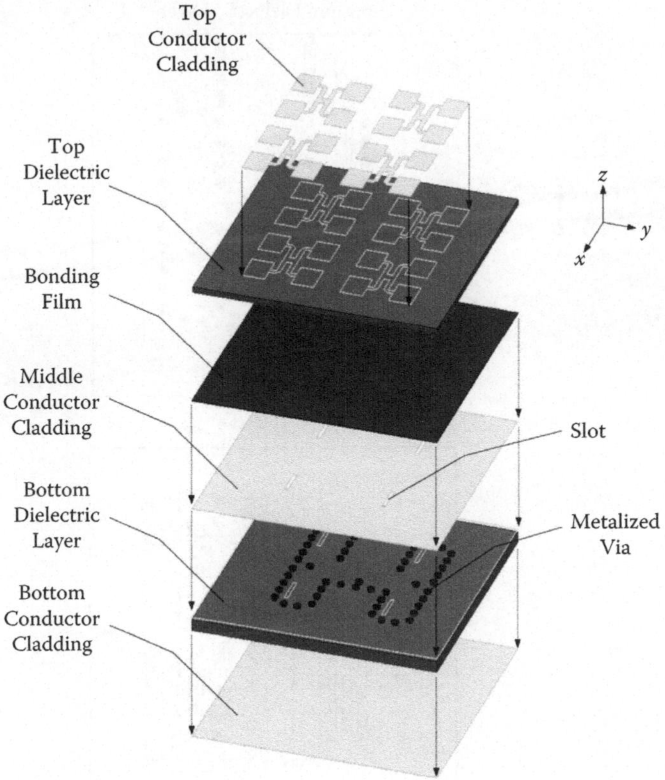

FIGURE 5.1 Configuration of the 4 × 4 microstrip-SIW-fed patch antenna array.

Depending on the topology, this two-layer antenna array prototype can be fabricated through the low-cost standard multilayer printed circuit board (PCB) process. First, the SIW feeding network is fabricated in the bottom substrate using the normal PCB process. After that, another dielectric substrate is attached by using the bonding film to combine the two substrates together as a new single substrate. Then, the patches on the top surface of the new substrate can be fabricated through the normal PCB process again. By virtue of its dual-layer configuration, the radiating and feeding layers can have different thicknesses and permittivity values to increase the design flexibility.

To explain the operating mechanism, a 4 × 4 small array is used by demonstrating the E-field associated with the topology, as shown in Figure 5.2. The E-field within the SIW near the input port is supposed in the z-direction, which produces a slot wave with the E-field in the y-direction. In this case, two microstrip paths on each side of the slot have two out-of-phase waves. This phase difference is independent of the operating frequency and can thus be used in wide-frequency band. Referring to the E-field in two-side microstrip lines lying in the opposite direction, the patches are placed back-to-back in the y-direction to get excited with the same phase.

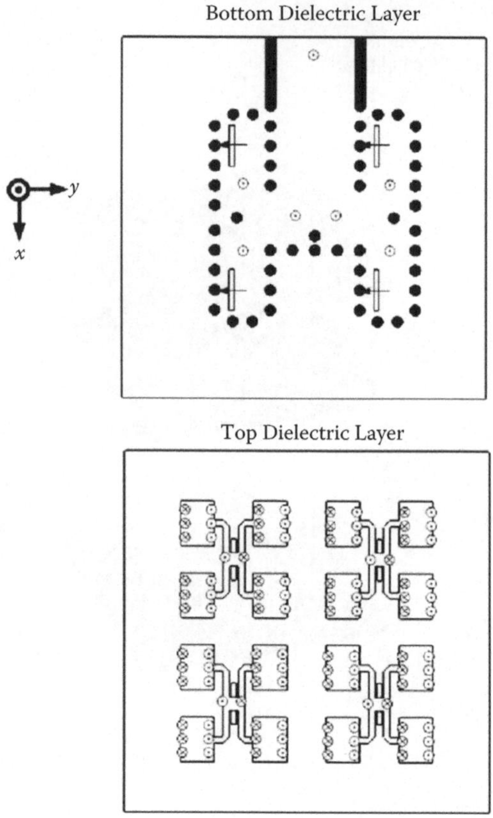

FIGURE 5.2 Feeding mechanism of the 4 × 4 microstrip-SIW-fed patch array antenna.

Now, a large-scale, high-gain planar integrated array antenna with 32 × 32 elements can be designed by connecting sixty-four 4 × 4 arrays. It is fed by the standard WR-10 metallic waveguide through a coupled aperture etched on the bottom conductor cladding. In practical applications, the feeder can be put at the center of the array. In this case, the size of the radiation aperture is only 63 × 63 mm^2.

A reflection coefficient below –10 dB is achieved within the frequency band from 89.5 to 98 GHz. The E-plane and H-plane radiation patterns at different frequencies are shown in Figures 5.3 and 5.4, respectively. The measured gains are in the range of 28.81–29.97 dBi from 91 to 97 GHz. Here, the insertion loss of the WR10 waveguide to SIW transition is calibrated. The gain of the antenna performs quite stably within the entire working bandwidth, which is important in wideband communications or the imaging systems. In the beam direction, the E-plane and H-plane cross-polarization levels are both less than –40 dB. Total radiation efficiencies that are calculated based on the measured gain and the simulated directivity of the antenna are in the range of 18.66–23.88%. The measured total radiation efficiency is 23% at 94 GHz.

The measured value is 29.9 dBi and the simulated gain of 94 GHz can be up to 32.3 dBi. To identify the reason, two SIWs with different lengths were fabricated and

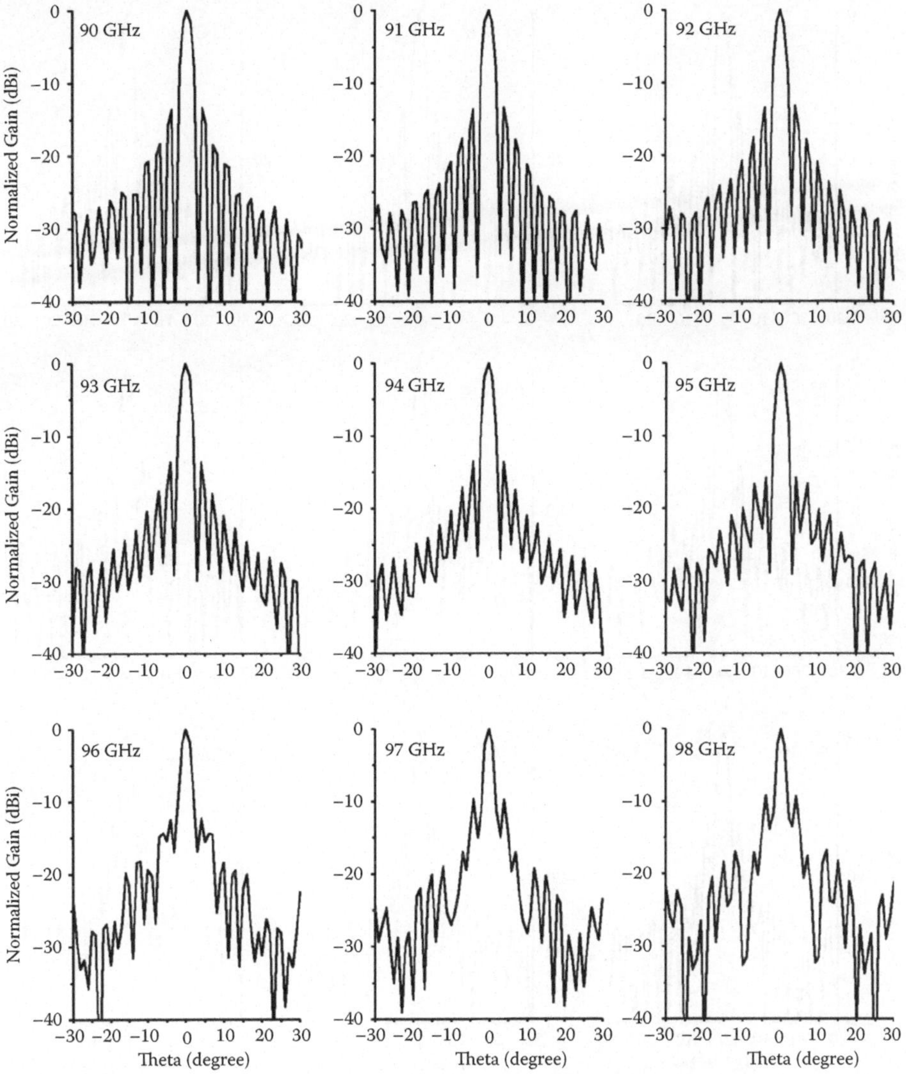

FIGURE 5.3 E-plane radiation patterns of the 32 × 32 microstrip-SIW-fed patch array antenna at different frequencies.

tested to estimate the actual dielectric loss tangent of the substrate Rogers 5880LZ. After experiment, the equivalent dielectric loss tangent of Rogers 5880LZ at 94 GHz can be calculated to be 0.0035. In this case, the realized gain decreases by 2.2 dBi in simulation and approaches the measured one. In conclusion, the inaccurate tan δ of the substrate and the bonding film at W-band leads to the discrepancy between the simulated and measured gains.

Moreover, this design is based on a dielectric-filled structure. Its radiation efficiency is decreased with the antenna size because of the larger loss from the longer feedline.

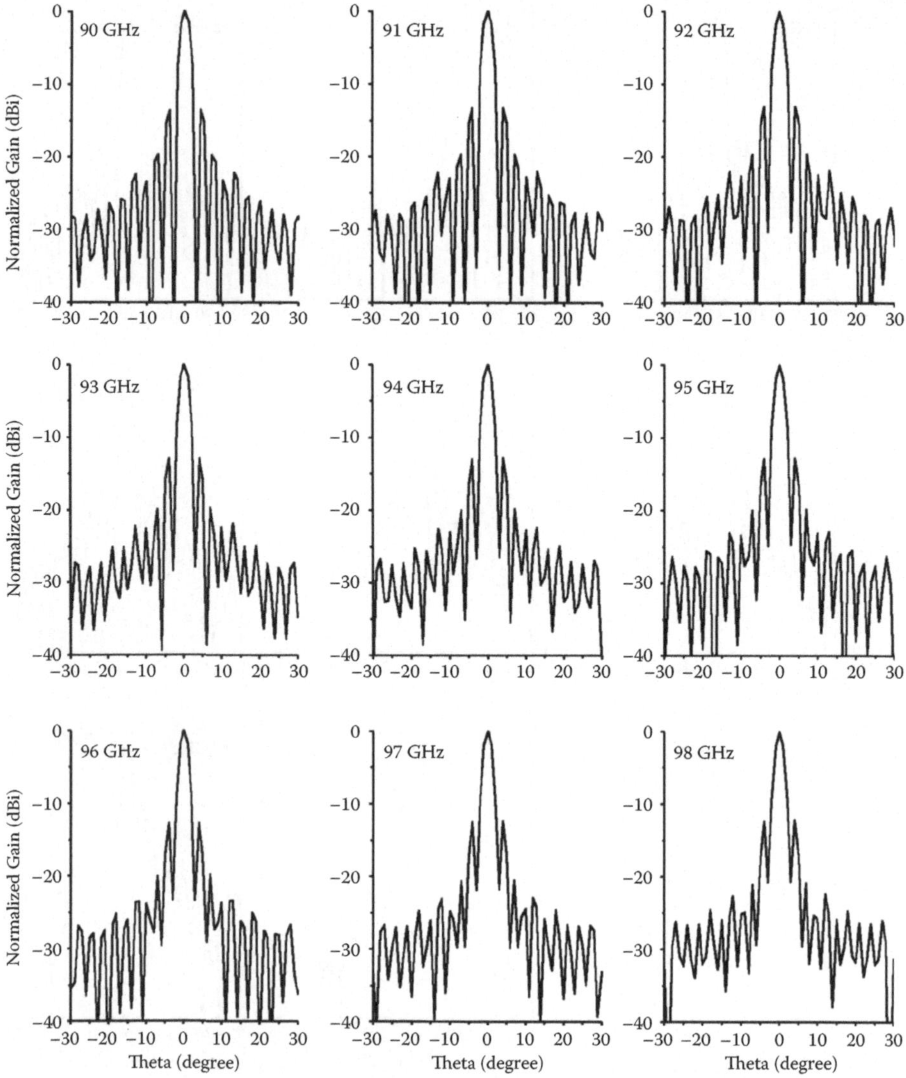

FIGURE 5.4 H-plane radiation patterns of the 32 × 32 microstrip-SIW-fed patch array antenna at different frequencies.

The radiation efficiency of this array is acceptable after taking account of the wide-bandwidth property. If this design is not a 32 × 32 case but a 4 × 4 case, the total length of the SIW feedline will be reduced by more than 84 mm. Besides, the loss from six 1-to-2 dividers can be excluded. The expected efficiency will be increased up to more than 90% at 94 GHz. Such a degree of efficiency is on the same level as indicated in [150–152]. A complete analysis of the loss in the feeding network with different scales at 94 GHz is provided in Figure 5.5.

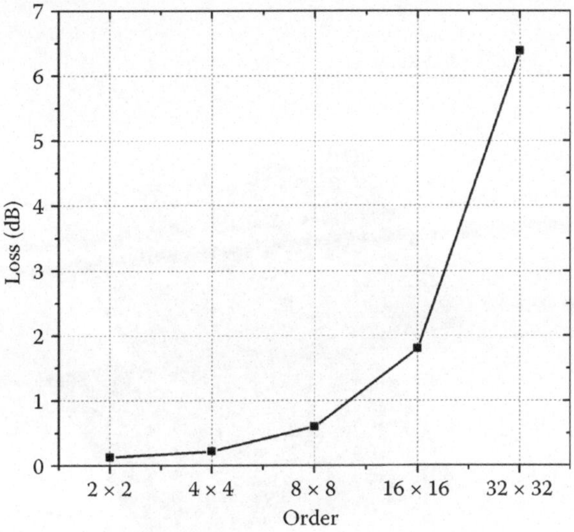

FIGURE 5.5 Loss of SIW feeding networks with different scales.

5.1.2 Microstrip-SIW-Waveguide-Fed Patch Antenna Array

The SIW feeding network is a good choice to increase the total efficiency of an antenna array to some extent. The sad news is the dielectric loss cannot be neglected when the array dimension becomes larger. It is found that the efficiency of the antenna decreases a lot when the array scale is increased to 32 × 32, as shown in Figure 5.5. Different sizes of SIW subarrays have been fabricated to investigate the relationship between the array scale and the efficiency. The optimal hybrid feeding topology can be recognized with the help of such an investigation. It is also found that the efficiency keeps a high level when the size of the SIW antenna is no larger than 16 × 16.

In this section, a V-band high-gain plate antenna is proposed for trying to replace the bulky parabolic antenna. Thus, both broadband and high efficiency are needed. In order to meet the requirement of high gain, high efficiency, compact configuration, and light weight, three kinds of transmission lines are used: microstrip, SIW, and waveguide. The pure SIW divider is replaced with a hybrid SIW-waveguide parallel divider in the application of a larger array in order to reduce the feedline loss. A better result may be obtained to carry out the replacement at a smaller array scale, such as 8 × 8. However, it is not realistic because of the limited space for the waveguide structure. There should be a trade-off between good performance and compact configuration. A special feeding structure from the SIW feeding slot to the microstrip subarray is proposed to expand the operating bandwidth. A broadband transition from the SIW power divider to the waveguide power divider is designed as well. Besides, the parallel divider is also used to keep the broadband characteristic.

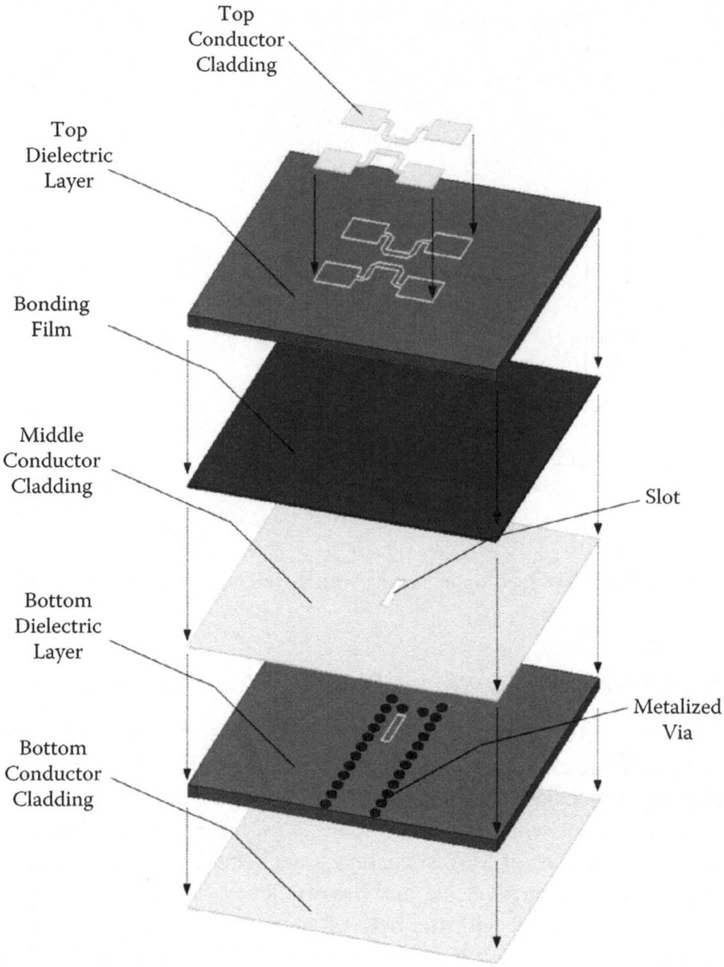

FIGURE 5.6 Configuration of the modified 2 × 2 microstrip-SIW-fed patch array antenna.

The basic subarray is 2 × 2 rectangular patches. Two patches are connected by a meandering microstrip line. Then, two parallel microstrip lines are fed by a narrow slot etched on the conductor cladding of the feeding SIW. Figure 5.6 shows the configuration of the 2 × 2 subarray. It is made of two dielectric substrates connected by a bonding film, which can be fabricated through a standard multilayer PCB process. Radiating patches and microstrip lines are printed on the top copper cladding. The thickness of the cladding is 35 μm. The used substrate is Taconic TLY-5 ($\varepsilon_r = 2.2$, tan $\delta = 0.0009$) with a thickness of 0.508 mm. Metalized vias are fabricated in the bottom dielectric layer connecting the middle copper cladding and the bottom copper cladding together to form the feeding SIW. The used bonding film is FR-28-0040-50 ($\varepsilon_r = 2.81$) with a thickness of 0.1 mm. The SIW width is fixed at 2.4 mm to make sure that only the fundamental mode propagates within the SIW.

TABLE 5.1
Measured Gain and Total Efficiency of the Fabricated 64 × 64 Microstrip-SIW-Waveguide-Fed Patch Array Antenna

Frequency (GHz)	Gain (dBi)	Efficiency
57	36.3	23%
58	37.2	30%
59	39.2	41%
60	38.8	37%
61	38.3	32%
62	38.7	33%
63	38.8	33%
64	39.0	34%
65	36.7	19%
66	37.5	22%

Simulated results of the above array show about 24.8% of the reflection coefficient bandwidth ($|S_{11}| < -10$ dB) in the frequency band of 50.8–65.2 GHz. The gain fluctuates less than 1 dB from 55 to 66 GHz. There is no distortion of the radiation pattern during the whole band.

The SIW H-type divider is used as the feeding network to realize a 16 × 16 microstrip-SIW-fed patch array antenna. Then, the compact waveguide power divider is processed on a brass block to assemble sixteen 16 × 16 arrays together. The total size of the proposed antenna is 240 × 240 × 13 mm. The radiating aperture size is 204.8 mm^2. Measured results show about 14.6% of the reflection coefficient bandwidth ($|S_{11}| < -10$ dB) in the frequency band of 57–66 GHz. Table 5.1 shows the measured gain and the total radiation efficiency of the fabricated 64 × 64 array. The 1 dB gain bandwidth is 8.1% within the frequency band of 59–64 GHz. The maximum efficiency is 41% at 59 GHz, and the maximum realized gain is more than 39 dBi.

5.2 SIW-FED CAVITY-BACKED PATCH ANTENNA ARRAY

5.2.1 SIW-Fed Cavity-Backed Linear Polarization Patch Antenna

As is well known, the surface wave is a major factor of deterioration for patch antenna systems. A cavity associated with a patch antenna can prevent the surface wave. The SIW technology helps in reducing the cost of the cavity-backed antennas.

In [156], a Ku-band cavity-backed microstrip patch array antenna was implemented using the SIW technology. As shown in Figure 5.7, each cavity is emulated using a circular array of metalized vias, and each patch is fed through a microstrip line. This antenna consists of two stacked substrates. The top dielectric substrate is used for fabricating radiating patches and microstrip feedlines. In this case, it should be thin enough to suppress the influence of the surface wave. The bottom dielectric layer is used for fabricating SIW cavities. Thus, the bottom substrate should be thick

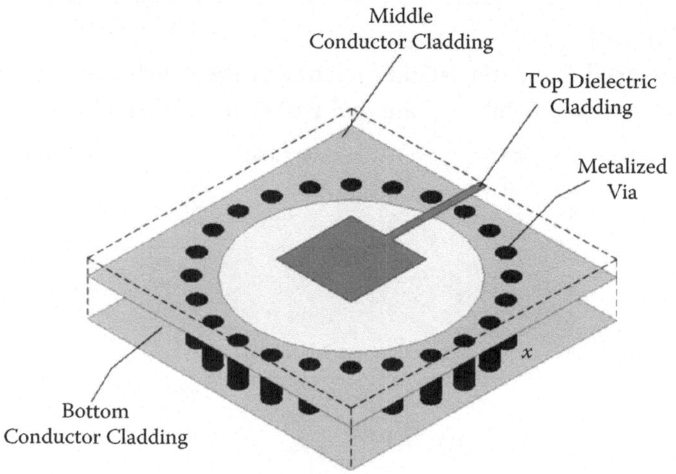

FIGURE 5.7 Configuration of a Ku-band two-layer cavity-backed patch antenna.

enough to increase the bandwidth. The proposed antenna arrays inherit the attractive features of the conventional metalized cavity-backed patch arrays. However, the microstrip feeding network limits its applications in the high-frequency band.

A single-layered SIW feeding network with T-junctions and cavity-backed patch antennas were employed to achieve high gain and wideband performance in the V-band [157]. As shown in Figure 5.8, the patch antenna consists of three dielectric substrates and an aluminum plate. An offset longitudinal slot etched on the third conductor cladding, which is the top surface of the feeding SIW, is used to couple the wave. Another coupling slot is on the second conductor cladding, which is the top surface of an SIW resonator. By adding the slot, one more resonance can be introduced to widen the bandwidth. Several circular holes are drilled in an aluminum plate to form cavities below patches. This multilayered antenna array can be fabricated by conventional single-layered PCB technology, and then all of the single layers are stacked and fixed together.

Another type of wideband SIW cavity-backed patch antenna and array for millimeter-wave applications were investigated in [158]. This Q-band antenna consists of a rectangular patch with a backed SIW cavity. In order to increase the bandwidth and radiation efficiency, the cavity resonates at its TE_{210} mode.

5.2.2 SIW-Fed Cavity-Backed Circular Polarization Patch Antenna

Some works try to realize the SIW-fed cavity-backed circular polarization (CP) patch antenna and array.

In [159], a 4 × 4 two-layer antenna array was designed and fabricated using the low-temperature cofired ceramic (LTCC) technology. As shown in Figure 5.9, a metallic topped via fence, as a cavity, is around the 45° rotated patch element. The mutual coupling between elements can be suppressed in this topology. A transverse

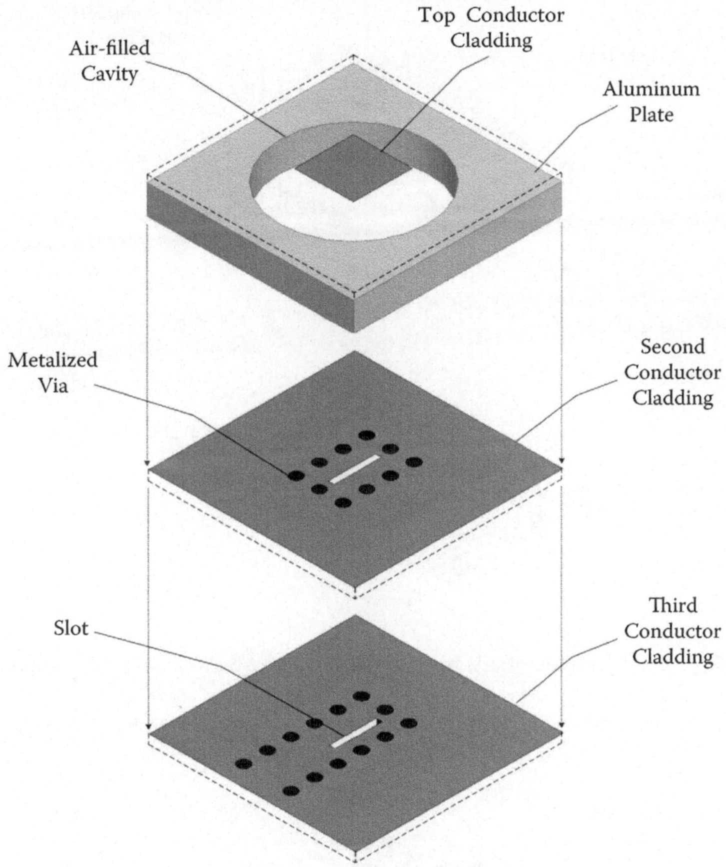

FIGURE 5.8 Configuration of a V-band four-layer cavity-backed patch antenna.

feeding slot, which is cut into the broad wall of the SIW, couples the energy into the cavity and then feeds the patch. The measured results show that a stable gain is greater than 12.5 dBi from 57 to 64 GHz. Besides, the axial ratio bandwidth is more than 7 GHz.

Another V-band CP design example was demonstrated in [160]. The proposed antenna is stacked in three dielectric layers as shown in Figure 5.10. A CP patch is selected as the radiating element, which is fabricated on the top conductor cladding. As the feedline, SIW is drilled on the bottom dielectric layer. A longitudinal slot with proper width and length is etched on the top surface of SIW, i.e., the third conductor cladding. The slot is offset from the edge wall and three-quarter wavelengths from the end wall. The slot position is optimized to match the input impedance of the antenna. Electric fields inside the slot are coupled to the patch antenna fabricated on the top layer. A square cavity constructed with metalized vias is placed between the top dielectric layer and the bottom dielectric layer. The dielectric material inside the cavity area is removed to leave air space between the slot and the patch.

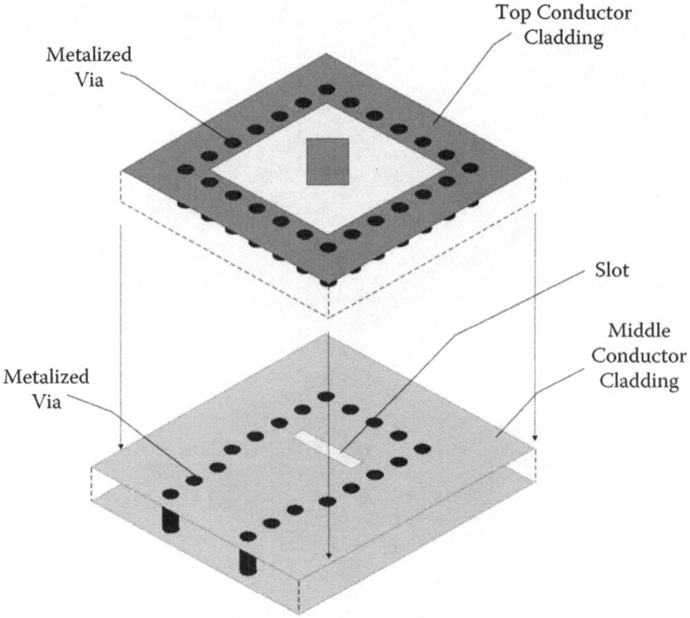

FIGURE 5.9 Configuration of a V-band two-layer cavity-backed CP patch antenna.

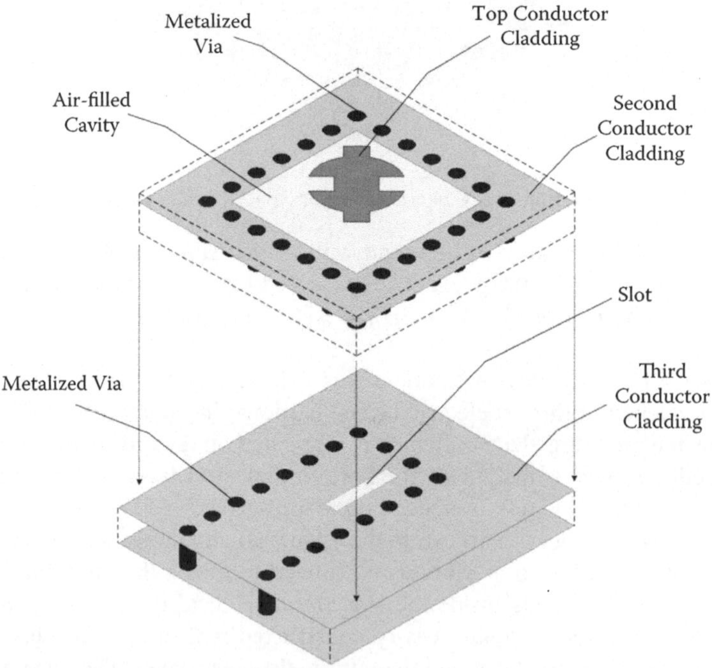

FIGURE 5.10 Configuration of a V-band three-layer air-filled cavity-backed CP patch antenna.

5.3 SIW-FED DIELECTRIC RESONATOR ANTENNA

Dielectric resonator antennas have been demonstrated to radiate efficiently at high frequency. They exhibit less conductor loss, high radiation efficiency, and larger bandwidth than microstrip patch antennas [161]. The first application of a cylindrical dielectric resonator as an antenna was described in [162]. Since then, there have been a number of excitation schemes for the dielectric resonator antennas.

Since the SIW is a high-quality structure and the dielectric resonator antenna is a low-loss radiator, the SIW-fed dielectric resonator antenna has high radiation efficiency [163]. As shown in Figures 5.11 and 5.12, a narrow slot is cut into the top conductor cladding of the feeding SIW with different orientations. It is used to excite the dielectric resonator antenna in its fundamental mode TE^{x}_{111} for radiating two different polarizations: the horizontal polarization and the vertical polarization. The measured gain for the SIW-fed single dielectric resonator antenna shows a gain of 5.51 dB, −19 dB maximum cross-polarization level, and calculated radiation efficiency of greater than 95%. Furthermore, an SIW series-fed dielectric resonator antenna array, formed by two different slot polarizations, was proposed in [164].

A dielectric resonator antenna fed by a half-mode substrate integrated waveguide (HMSIW) was studied in [165]. As shown in Figure 5.13, this investigated antenna consists of a dielectric resonator, which is surface mounted on an HMSIW. Energy is coupled into the dielectric cylinder through a transverse rectangular slot etched on the top surface of the HMSIW. Using this excitation scheme, an HMSIW-fed linear polarization (LP) dielectric resonator antenna was designed at 60 GHz. It exhibits a bandwidth of 24.2% for $|S_{11}| < -10$ dB in measurement. Measured gain is higher than 5.5 dBi over the whole operating frequency band. Measured radiation efficiency is between 80 and 92%. Besides, a CP dielectric resonator antenna fed by an HMSIW was designed, presenting a measured 3 dB axial ratio bandwidth of 4%. Through a

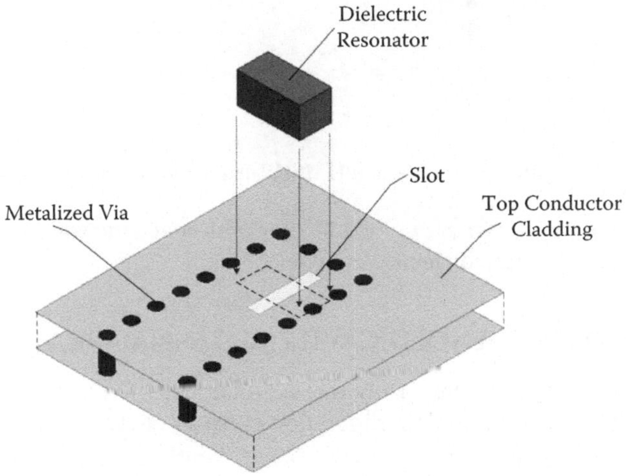

FIGURE 5.11 Configuration of a Ka-band SIW-fed dielectric resonator antenna using a longitudinal slot.

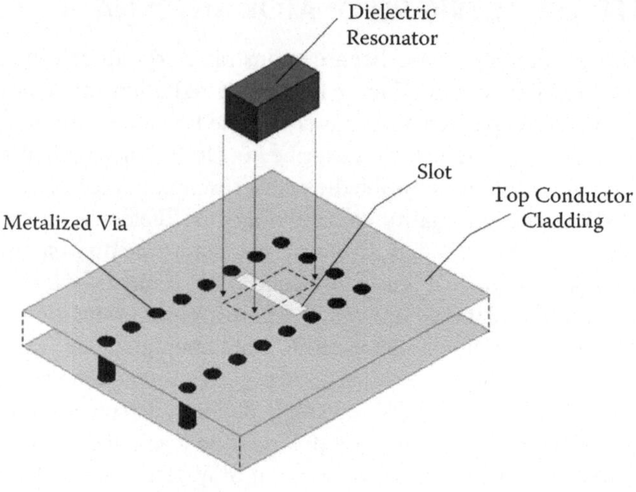

FIGURE 5.12 Configuration of a Ka-band SIW-fed dielectric resonator antenna using a transverse slot.

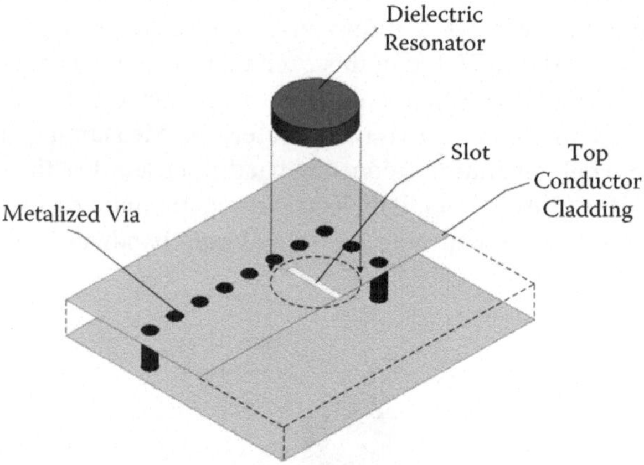

FIGURE 5.13 Configuration of a V-band HMSIW-fed LP dielectric resonator antenna.

pair of cross-slots, as shown in Figure 5.14, an electromagnetic wave is coupled from the interior HMSIW to the dielectric cylinder.

5.4 SIW-FED OPEN-ENDED CAVITY ANTENNA ARRAY

In [166], an 8 × 8 SIW-fed cavity array was symmetrically fed in parallel for 60 GHz applications. This multilayered configuration consists of 20 ceramic tapes, which can be fabricated by the LTCC process. A bandwidth of 17.1% and a gain up to 22.1 dBi can be achieved by properly implementing the radiating element, the feeding network, and the transition. As shown in Figure 5.15, the radiating element is a rectangular open-ended SIW cavity, which is fed by a feeding slot located on the

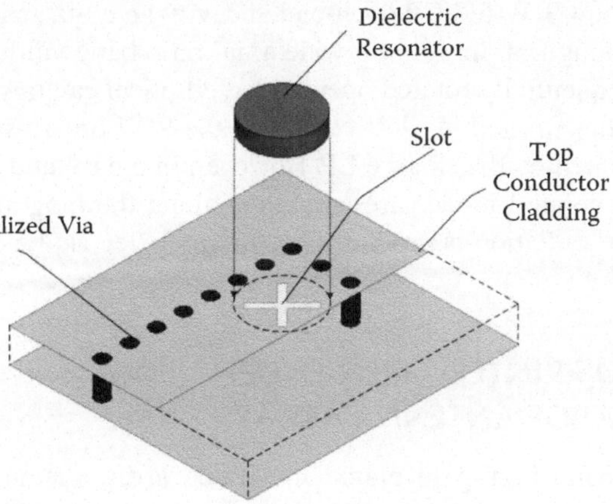

FIGURE 5.14 Configuration of a V-band HMSIW-fed CP dielectric resonator antenna.

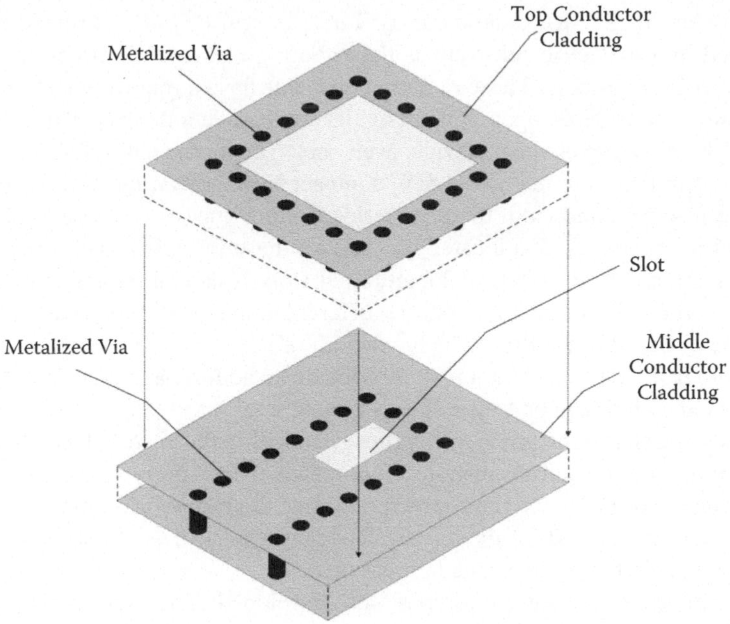

FIGURE 5.15 Configuration of a V-band SIW-fed rectangular open-ended cavity antenna.

bottom conductor layer of the cavity. The radiating aperture is realized on the top surface of the LTCC substrate through the screen-printing process. The sidewall of the SIW cavity is formed by metalized vias.

A millimeter-wave SIW cavity antenna fed by a slotted narrow wall was presented and investigated in [167]. The SIW feedline and the slotted narrow wall–fed high-permittivity dielectric loaded SIW cavity are integrated in a coplanar configuration. The prototypes operating at 35 and 60 GHz bands were fabricated to validate the concepts.

In [168], a 4 × 4 SIW-fed CP open-ended cavity antenna array was designed in a three-layered configuration. It has a wide axial ratio bandwidth. This antenna array consists of 16 sequentially rotated open-ended elliptical cavities acting as the radiating elements. They are fed by slots etched on the SIW broad wall. The electromagnetic waves radiated by the slot are LP. However, in the rotated elliptical cavity, two orthogonal degenerated modes are combined along the long and short axes of the cavity. Then, CP radiation is produced by the total electric fields in the aperture of the cavity.

5.5 SIW-FED PRINTED OPEN-ENDED WAVEGUIDE ANTENNA ARRAY

In [169], an E-band high-gain planar integrated array antenna made by low-cost PCB fabrication process was introduced with wide bandwidth or dual bandwidth and miniaturized size.

In this work, there are three methods to improve the bandwidth and gain performances of an SIW-fed antenna array. First, several cylindrical metalized holes are drilled in a dielectric substrate as the radiating elements through the standard single-layer PCB process. These radiating elements have a relatively large diameter, which can be considered open-ended circular waveguide antennas. Thus, the bandwidth of the radiating element is wide, even able to cover the entire E-band. Second, these wideband vias are fed by the SIW compact parallel feeding technology, which has a much wider bandwidth than the series feeding one, as described before. The used feeding divider is also located underneath these radiating vias, which aids in miniaturization. Third, a pair of longitudinal slots is etched on the top conductor cladding of the SIW. These two slots are different in length and offset away from the centerline of the SIW feedline. The bandwidth of the feeding structure is expanded much more from the consideration of introducing an additional resonance. High gain and wide bandwidth can be achieved at the same time.

Besides, the electromagnetic wave propagates within the air-filled metalized holes but not in the top dielectric substrate. That means the property of the top substrate material does not influence the characteristics of the radiating element. A cheaper substrate can be selected for the top dielectric layer regardless of the operating frequency, which further reduces the fabrication cost.

As examples, two array prototypes were developed. The first one is a 16 × 16 wideband array antenna, and the second one is an 8 × 8 dual-band array antenna. The proposed integrated array antenna consists of two dielectric layers and three conductor layers. As shown in Figure 5.16, the bottom dielectric layer is used to fabricate the SIW H-type divider, and the top dielectric layer is used to fabricate the radiating holes. The H-type SIW-based parallel feeding divider is employed to excite the radiating elements with constant amplitude and constant phase through a pair of longitudinal slots etched on the top conductor surface of the SIW feedline, i.e., the middle conductor cladding. To locate the slots at the peak of the standing wave, the SIW branch is terminated in a short circuit that is a quarter of the guided wavelength away from the center of the slot.

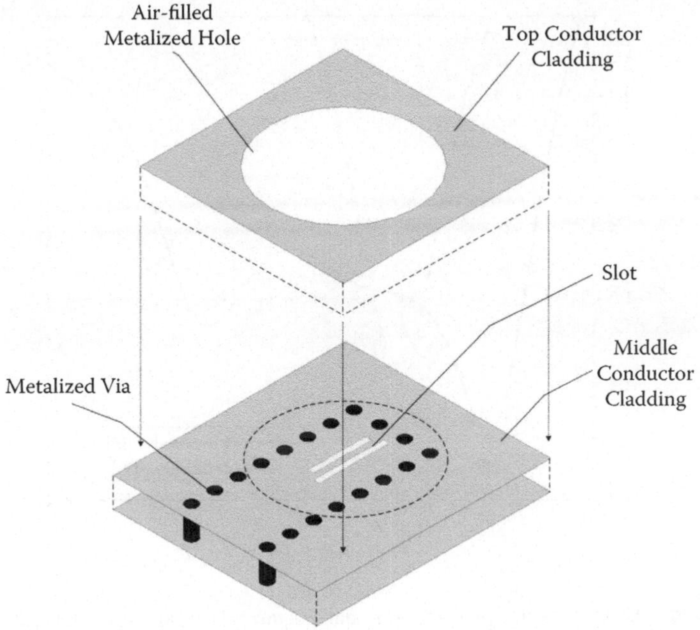

FIGURE 5.16 Configuration of an E-band SIW-fed printed open-ended waveguide antenna.

5.5.1 Parametric Study

First, a single radiating element shown in Figure 5.16 is investigated in detail. The thicknesses of the top and bottom dielectric substrates are h_1 and h_2, respectively. The width of the feeding SIW is a, the lengths of slots are sl_1 (the longer one) and sl_2 (the shorter one), the offsets away from the centerline of the SIW are so_1 and so_2, and the diameter of the radiating hole is D. The distance between two adjacent vias is 0.55 mm, and the diameter of the metalized vias in the bottom layer is 0.3 mm.

The double resonance can be generated by the dual-slot feeding structure, and the two resonance frequencies f_1 and f_2 determine the antenna bandwidth. To explain this, a printed open-ended waveguide antenna element fed by the dual-slot feeding structure, as shown in Figure 5.16, and another similar configuration fed by a single slot are modeled and simulated. The dimensions of the dual-slot-fed antenna unit are $a = 1.9$ mm, $h_1 = 1.52$ mm, $h_2 = 0.508$ mm, $sl_1 = 1.88$ mm, $sl_2 = 1.68$ mm, $so_1 = 0.1$ mm, $so_2 = 0.26$ mm, and $D = 3$ mm. The single-slot-fed antenna unit (without sl_2 and so_2) has the same geometrical dimensions as the first one except for $sl_1 = 1.7$ mm and $so_1 = 0.16$ mm. As shown in Figure 5.17, there exist two apparent resonance frequencies, $f_1 = 73$ GHz and $f_2 = 80.6$ GHz for the dual-slot case, while only one resonance frequency at 75.3 GHz for the single-slot case. The bandwidth of VSWR < 2 is expanded by 4.2 times. By properly controlling the resonance frequencies f_1 and f_2, the wideband or dual-band operating mode can be obtained.

The length of two slots, sl_1 and sl_2, mainly determines the bandwidth of the antenna. For a set of fixed sl_1 and sl_2, the slot offset so_1 and so_2 can be optimized

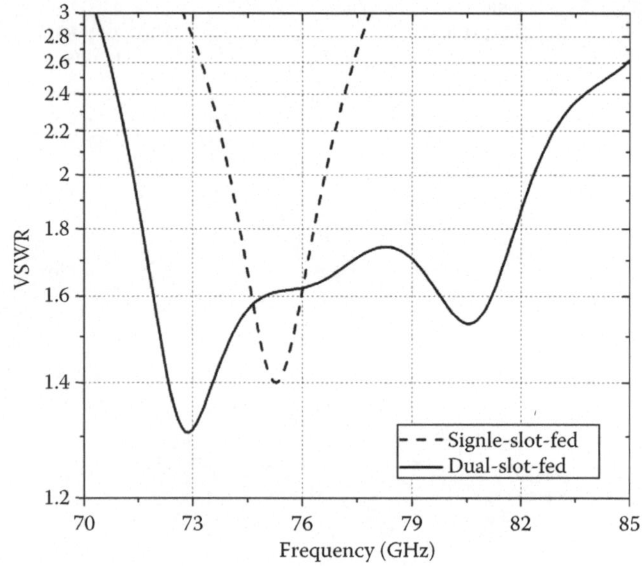

FIGURE 5.17 VSWR of the printed open-ended waveguide antenna element fed by the dual-slot and single-slot structures.

to achieve a better matching. The relationship between the bandwidth and the slot length sl_1, sl_2 is presented in Figure 5.18. For a fixed sl_1, a smaller sl_2 leads to a higher upper resonance frequency f_2 and then a larger bandwidth. That means the interval between two resonance frequencies is expanded. In addition, a larger sl_1 increases the bandwidth when sl_2 keeps unchanged.

The reflection coefficient and the gain of the open-ended waveguide antenna are influenced by the diameter of the radiating hole. The radiating hole is used not only for the energy radiation, but also for the mode conversion. The TE_{10} mode propagating within the SIW is converted to the TE_{11} mode propagating within the cylindrical via. Therefore, the equivalent circular waveguide should satisfy the single-mode condition, i.e., $0.587\lambda < D < 0.766\lambda$. Table 5.2 presents the relationship between the reflection coefficient, the gain, and the diameter of the hole at 73.5 GHz. When D increases from 2.4 mm to 3.2 mm, the antenna gain becomes larger. When D = 3.2 mm, the best reflection coefficient can be achieved. When D becomes larger, higher gain can be achieved. However, the reflection is deteriorated with an oversized radiating via. Besides, the processing limitations, such as the mechanical strength of the substrate, should be considered. Too large D is not permitted.

Table 5.3 presents the relationship between the reflection coefficient, the gain, and the thickness of the upper substrate h_1, i.e., the height of the radiating via, at 73.5 GHz. When $h_1 = 0$, the antenna becomes a slot antenna, and it has the lowest gain. When h_1 increases from 0 mm to 2 mm, the antenna gain becomes larger. However, it is not worthwhile to use an excessively thick substrate considering volume and cost.

The thickness of the bottom substrate affects the operating band of the antenna. With the decrease of h_2, the operating frequency band becomes higher. When

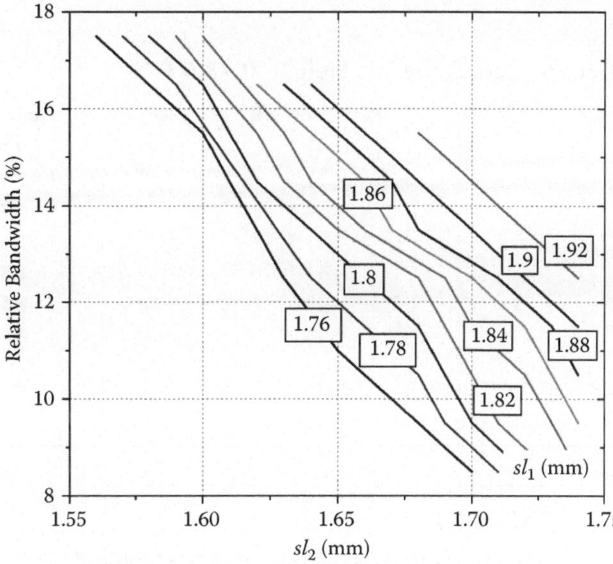

FIGURE 5.18 Relative bandwidth versus the slot length (a = 1.9 mm, D = 3 mm, h_1 = 1.52 mm, and h_2 = 0.508 mm).

TABLE 5.2
Gain and Reflection versus the Via Diameter at 73.5 GHz

D (mm)	Reflection (dB)	Gain (dBi)
2.4	–6.2	7.39
2.6	–8.6	7.82
2.8	–9.9	8.16
3	–11.5	8.66
3.2	–13.8	8.97
3.4	–12.6	8.18
3.6	–9.8	9.42

Note: a = 1.9 mm, sl_1 = 1.9 mm, sl_2 = 1.7 mm, so_1 = 0.1 mm, so_2 = 0.26 mm, h_1 = 1.52 mm, and h_2 = 0.508 mm.

h_2 = 0.381 and 0.508 mm, the double resonances can be obtained within the desired band. Thus, these two thicknesses are appropriate in E-band design.

Element spacing should be selected carefully to prevent the grating lobe formation. In this work, the selection of element spacing is also limited by the dimensions of the feeding divider and the radiating element. Table 5.4 presents the gain and SLL of a 16-element linear array with different element distances. The operating

TABLE 5.3
Gain and Reflection versus the Via Height at 73.5 GHz

h_1 (mm)	Reflection (dB)	Gain (dBi)
0	−16.6	7.28
0.5	−21.4	8.12
1	−14.8	8.5
1.5	−11.5	8.62
2	−11.7	8.72
2.5	−13.1	8.54
3	−16	8.68

Note: a = 1.9 mm, sl_1 = 1.9 mm, sl_2 = 1.7 mm, so_1 = 0.1 mm, so_2 = 0.26 mm, D = 3 mm, and h_2 = 0.508 mm.

TABLE 5.4
SLL and Gain versus the Element Spacing

Element Spacing (mm)	SLL (dB)	Gain (dBi)
3	−13.2	19.6
3.2	−13.6	19.9
3.4	−13.2	20.2
3.6	−13.3	20.4
3.8	−13.8	20.6
4	−13.4	20.7
4.2	−13	20.6
4.4	−10.6	20.5
4.6	−9	20.4
4.8	−7.7	20.4
5	−6.4	20.3

frequency is set to be 73.5 GHz and the element spacing is changed from 3 mm to 5 mm, i.e., from $0.735\lambda_0$ to $1.23\lambda_0$. The gain increases with the element spacing to a maximum value at 4 mm, and then decreases slightly. When the element spacing is less than 4.2 mm, the SLL stays at a normal level.

Furthermore, a linear array with an element spacing of 3.8 mm is employed to investigate the mutual coupling between radiating elements. The input port of the center radiating element is no. 1, and the ports of other radiating elements are no. 2 to no. 8 from proximal to distal. As shown in Table 5.5, the mutual coupling between different radiating elements is weak in our design.

5.5.2 Design Results

A low-cost multilayer PCB process was used to fabricate such wideband and dual-band dual-layer antenna prototypes. The SIW feeding network placed in the bottom

TABLE 5.5
Mutual Coupling between Different Radiating Elements at 73.5 GHz

Port	2	3	4	5	6	7	8
Coupling (dB)	−22.17	−28.40	−36.27	−49.02	−51.58	−50.25	−58.53

layer was fabricated in a 0.508 mm Rogers 5880 substrate. Its permittivity and loss tangent of the substrate are 2.2 and 0.0009 at 10 GHz, respectively. The metalized radiating holes placed in the top layer are fabricated in a 1.52 mm Taconic RF-35 substrate. Two such substrates are bonded together by use of the bonding film.

A standard metallic rectangular waveguide to SIW transition was designed for measurement. A reduced-height waveguide is used here to connect the metallic waveguide and the feeding SIW. This transition has the potential to cover the whole E-band, and thus can be used for both the wideband and dual-band cases.

The final dimensions of the 16 × 16 wideband SIW-fed printed open-ended waveguide antenna array are a = 1.9 mm, sl_1 = 1.9 mm, sl_2 = 1.7 mm, so_1 = 0.1 mm, so_2 = 0.26 mm, h_1 = 1.52 mm, h_2 = 0.508 mm, and D = 3 mm. The element spacing is 3.8 mm in E-plane and 4.4 mm in H-plane, respectively. A reflection coefficient below −10 dB is achieved within 70.5–78.3 GHz.

The radiation patterns of the fabricated 16 × 16 wideband antenna array were measured at different frequencies as shown in Figure 5.19. Figure 5.20 presents the simulated and the measured gain of the proposed array antenna. The loss of the waveguide–SIW transition was calibrated. The measured gain contains the reflection loss, the conductor loss, and the dielectric loss. So it is the realized gain. Measured maximum realized gain is 30.7 dBi at 76 GHz, while the minimum one is 29 dBi at 71 GHz. It should be noted that the gain of the proposed antenna is almost constant within the desired bandwidth. Figure 5.20 also shows the efficiency of the fabricated array antenna. This efficiency is defined as the ratio of the measured realized gain to the simulated directivity. So it is the total radiation efficiency that includes the mismatch factor. The measured maximum total radiation efficiency is 61.7% at 76 GHz.

The final dimensions of the 8 × 8 dual-band SIW-fed printed open-ended waveguide antenna array are a = 1.8 mm, sl_1 = 2.2 mm, sl_2 = 1.76 mm, so_1 = 0.12 mm, so_2 = 0.25 mm, h_1 = 1.52 mm, h_2 = 0.508 mm, and D = 2.6 mm. The element spacing is 3.6 mm in the E-plane and 4.6 mm in the H-plane, respectively. A reflection coefficient below −10 dB is achieved within the frequency bands of 70.6–72.4 GHz and 80.8–82.2 GHz. The corresponding bandwidths are 1.8 and 1.4 GHz for such two bands.

The radiation patterns of the fabricated 8 × 8 dual-band antenna were measured at different frequencies, as shown in Figures 5.21 and 5.22. Figure 5.23 presents the simulated and the measured gain of the antenna, as well as the total radiation efficiency of the antenna. Measured maximum realized gains are 25.2 dBi at 72 GHz and 25.8 dBi at 82 GHz, while the minimum ones are 24.6 dBi at 71 GHz and 24.7 dBi at 82.25 GHz, respectively. Measured maximum total radiation efficiency is 61.7 and 72.4% for dual bands, respectively.

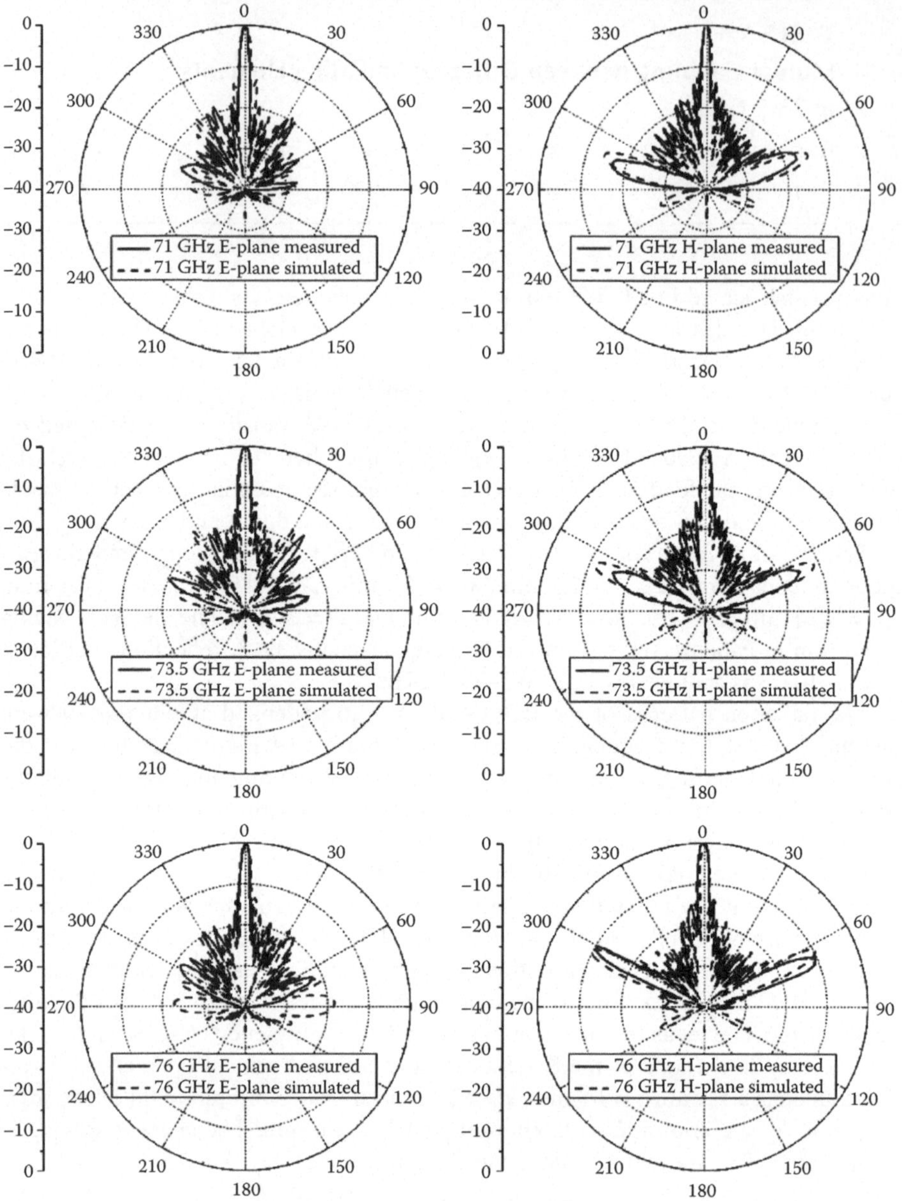

FIGURE 5.19 Radiation patterns of the 16 × 16 wideband SIW-fed printed open-ended waveguide antenna array at different frequencies.

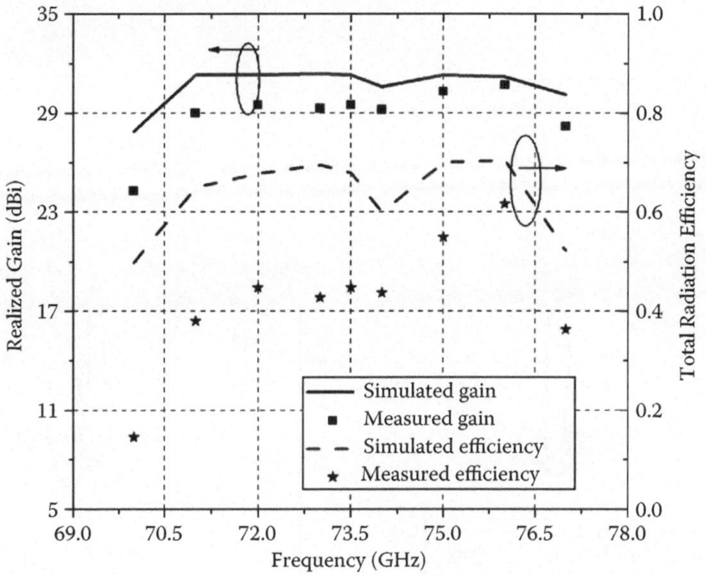

FIGURE 5.20 Simulated and measured gain and efficiency of the 16 × 16 wideband SIW-fed printed open-ended waveguide antenna array.

Three SIW resonant cavities operated at different frequencies, and two SIW transmission lines with different lengths were designed, fabricated, and measured to estimate the actual relative permittivity and the SIW loss within the E-band. According to the simulated and measured resonant frequencies, the relative permittivity of the used substrate is about 2.185. Moreover, the SIW loss per unit length was calculated to be 0.03 dB/mm. Compared with the results based on the substrate properties from the manufacturer's data sheet, there exists 2.2 dB additional loss for the 16 × 16 wideband array and 1.05 dB additional loss for the 8 × 8 dual-band array.

5.6 COMPARISON OF SUBSTRATE INTEGRATED FEEDING ANTENNAS

Table 5.6 lists comparative results among different substrate integrated feeding antennas, including frequency band (FB, GHz), number of substrate layer (NL), normalized thickness (NT, λ_0), number of element, realized gain within the frequency band (dBi), peak total radiation efficiency within the frequency band (TRE, %), operating polarization, cross-polarization level, used dielectric substrate, and fabrication process.

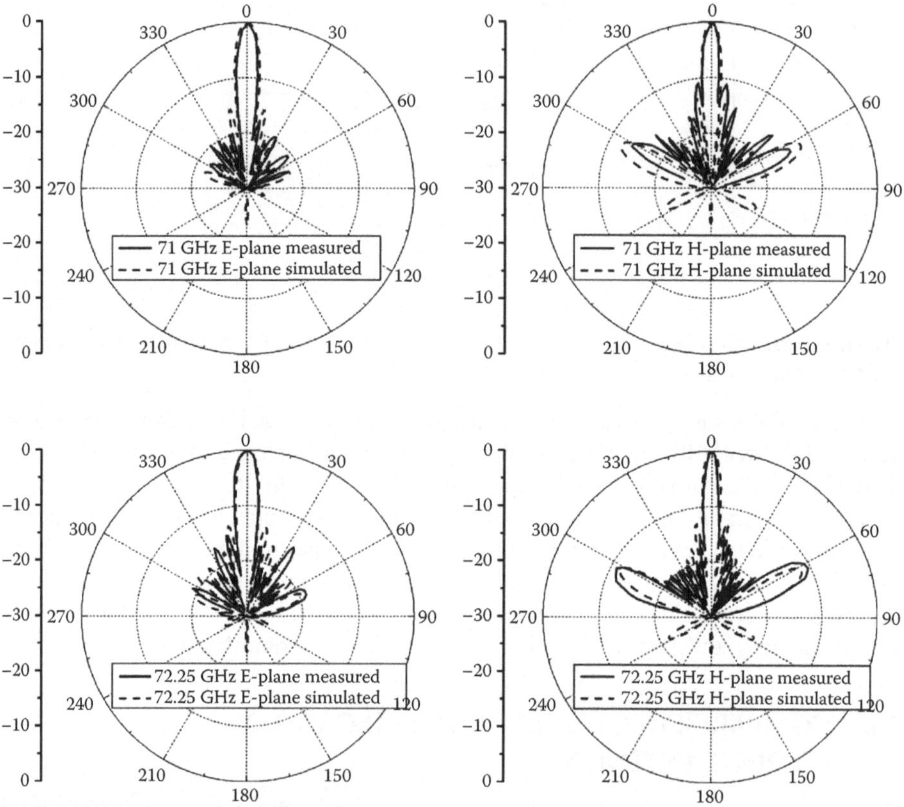

FIGURE 5.21 Radiation patterns of the 8 × 8 dual-band SIW-fed printed open-ended waveguide antenna array at the lower-frequency band.

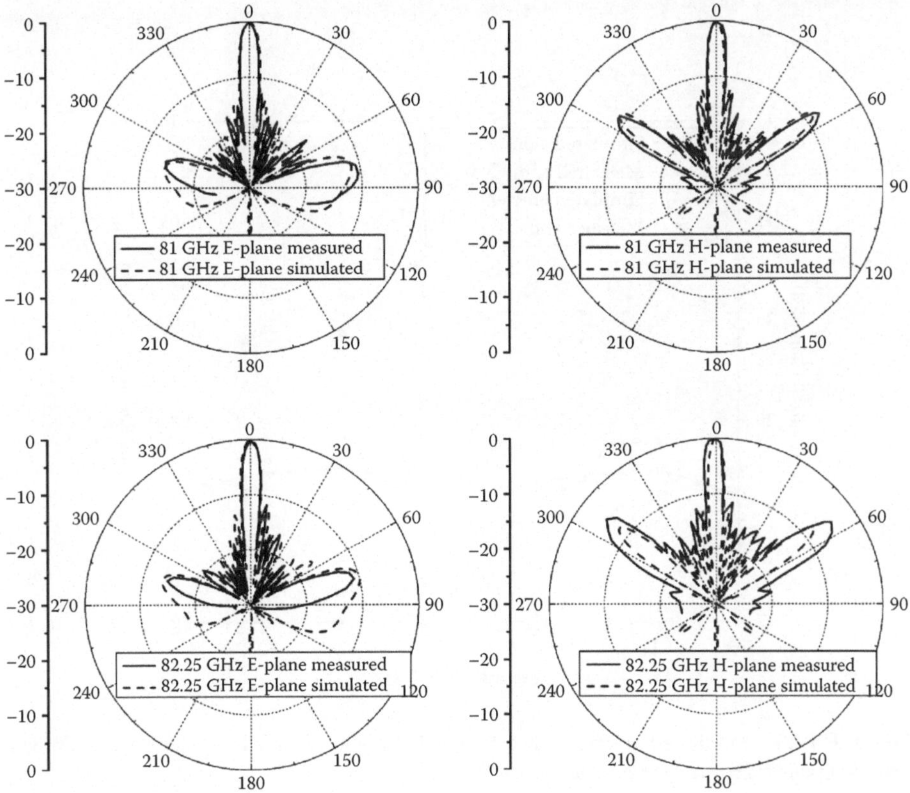

FIGURE 5.22 Radiation patterns of the 8 × 8 dual-band SIW-fed printed open-ended waveguide antenna array at the higher-frequency band.

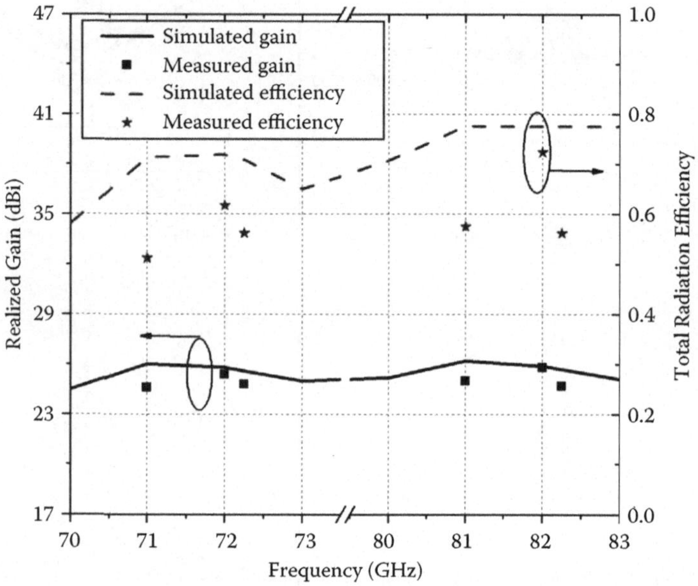

FIGURE 5.23 Simulated and measured gain and efficiency of the 8 × 8 dual-band SIW-fed printed open-ended waveguide antenna array.

TABLE 5.6
Compared Results between Different SIW-Fed Antennas

	FB (GHz)	NL	NT (λ_0)	Element No.	Gain (dBi)	TRE (%)	Polarization	Cross-polarization (dB)	Substrate	Fabrication
[150]	57.6–62.1	6	0.48	4 × 4	≤17.7	—	LP	>22	Rogers 6002, 5880	Nonstandard PCB process
[151]	81–86	2	0.32	4 × 4	17.5–18.5	90	LP	>20	Rogers 6002, 5880LZ	Nonstandard PCB process
[152]	94.2–101.8	3	0.9	4 × 4	17–19	81	LP	>23	Rogers 6002, 5880	Nonstandard PCB process
[154]	91–97	2	0.26	32 × 32	28.8–30	24	LP	>40	Rogers 5880LZ	Standard PCB process
Section 5.1.2	57–66	3	2.6	64 × 64	36.4–39.2	41	LP	>38	Taconic TLY-5, copper	Standard PCB process, machining operation
[156]	12.1–12.8	2	0.08	8 × 8	23–24.5	50	LP	>23	Rogers 5880	Standard PCB process
[157]	55.5–67	4	0.4	4 × 4	16.6–19.6	54.5	LP	>25	Rogers 5880, 6006, Aluminum	Nonstandard PCB process, machining operation
[158]	41.5–42.5	1	0.07	4 × 4	17.4–17.8	74.9	LP	>21	Rogers 5880	Standard PCB process
[159]	60–64	20	0.4	4 × 4	15–17.5	—	CP	<3.3[a]	Ferro A6-M	LTCC process
[160]	59–61	3	0.18	2 × 2	12–12.5	82	CP	<3.1[a]	Rogers Ultralam, 6002, 3006	Nonstandard PCB process
[164]	36.9–38.7	2	0.29	1 × 4	<10.6	93	LP	—	Rogers 5870, 6010	Nonstandard PCB process
[166]	54.9–65.1	20	0.4	8 × 8	19.6–22.1	44.4	LP	—	Ferro A6-M	LTCC process
[167]	54.5–60.3	1	0.13	2 × 2	9–11.8	—	LP	<17	Rogers 3006	Standard PCB process
[168]	22–24	3	0.44	4 × 4	15.3–17.9	—	CP	<3[a]	—	Nonstandard PCB process
[169]	71–77	2	0.5	16 × 16	27.5–30.7	61.7	LP	—	Rogers 5880, Taconic RF35	Nonstandard PCB process

[a] Axial ratio.

6 Substrate Integrated Monopulse Antennas

The monopulse antenna, which is a key technique for angle estimation [170], is widely used in rapid direction-finding systems, radar systems, and communication systems. It is usually developed from antenna arrays with even and odd excitations to generate a pair of sum and difference beams. These sum and difference radiation patterns can be obtained through a pair of interleaved antenna arrays with different beam squints [171]. The sum and difference patterns can be also generated by antenna arrays that operate in either of even and odd modes of excitation with separate hybrid ports for each mode [172]. The poor performance of the gain and the angular sensitivity are the main drawbacks of the first overlapped beam technique. That is because the excitations of the equivalent sum and difference antenna arrays are far away from the ideal independent amplitude tapers. With proper excitations, two closely spaced antenna arrays were interleaved to obtain the monopulse in dual plane [173]. Cassegrain parabolic antennas or lens antennas are also commonly applied in monopulse antennas, but they are usually heavy and complicated. Some printed arrays have been utilized to realize monopulse antennas described in [174, 175]. However, the microstrip feedline may cause a large radiation loss from feeding paths, especially in the millimeter-wave frequency band.

Consequently, the SIW technology is used for the design of monopulse antennas. Easy connection between printed radiating elements and an SIW comparator is secured. These features are important for the development of a monopulse system with both compact planar configuration and promising performance, especially in the high-frequency band. In this chapter, an overview of popular and representative substrate integrated monopulse array antennas is introduced, including one-dimensional monopulse antenna array, multimode monopulse antenna, and two-dimensional monopulse antenna array.

6.1 ONE-DIMENSIONAL SIW MONOPULSE ANTENNA

6.1.1 SIW-Fed TSA Monopulse Antenna Array

As shown in Figure 6.1, a compact planar E-plane monopulse antenna was proposed and built on the SIW topology at 36.5 GHz [176]. An array of four tapered slot antenna (TSA) radiators was designed and developed as a monopulse antenna with a simple and effective feeding mechanism. This configuration is able to generate the sum and difference beams along the substrate instead of in the vertical plane.

A 180° directional coupler, also used as a planar magic-T, was developed in the design of an integrated feeding network to provide the required excitations for the sum and difference channels. First, an H-plane SIW coupler was designed to

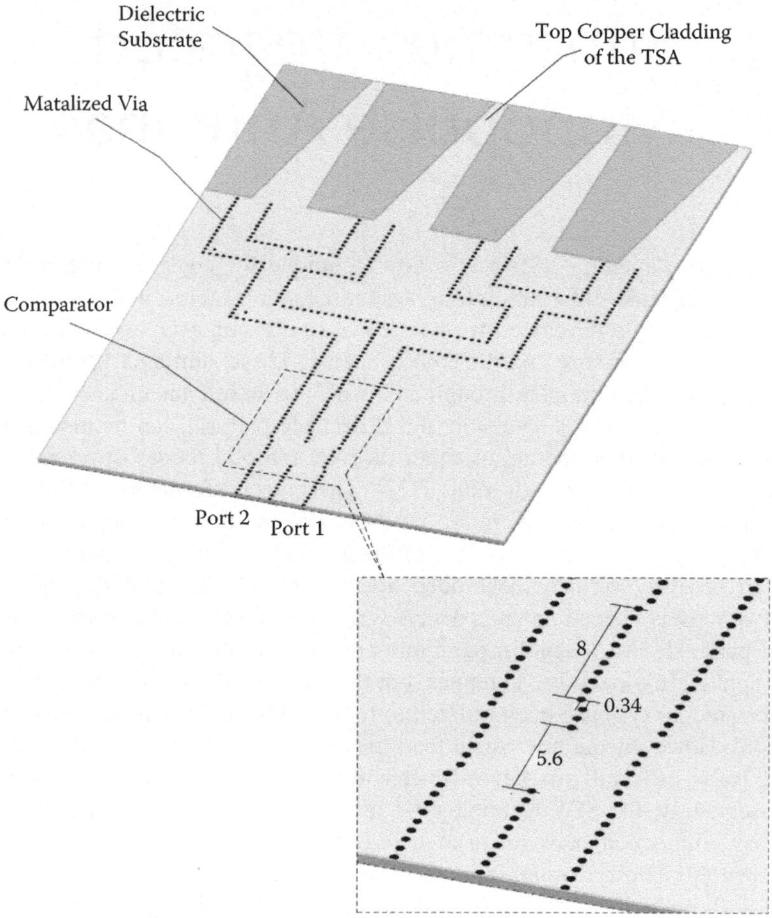

FIGURE 6.1 Configuration of the one-dimensional SIW-fed monopulse antenna with four integrated TSA elements. (Unit: mm).

operate at 36.5 GHz. There exists a 90° phase difference between output ports. Then, the 180° coupler was realized by combining the 90° coupler structure and another 90° fixed phase shifter, as shown in Figure 6.1. Such a phase shifter has equal length and unequal width for two SIW paths. Therefore, excited at different input ports, the equal amplitude and in-phase (or out-of-phase) excitations can be generated at output ports.

Rogers 5880 substrate is used in this design with a thickness of 0.508 mm and permittivity of 2.2. Because the 180° coupler is employed to feed the monopulse antenna, two input ports should be respectively investigated. Measured results are shown in Figures 6.2 and 6.3. When excited at port 1, ±0.25 dB power equality is achieved within 6.7% relative bandwidth. Measured return loss is better than 17.5 dB, and the isolation is better than 19.2 dB over the whole operating frequency range. When excited at port 2, ±0.25 dB power equality is achieved within 6.1% relative bandwidth. Measured return loss is better than 17.2 dB. Measured phase difference

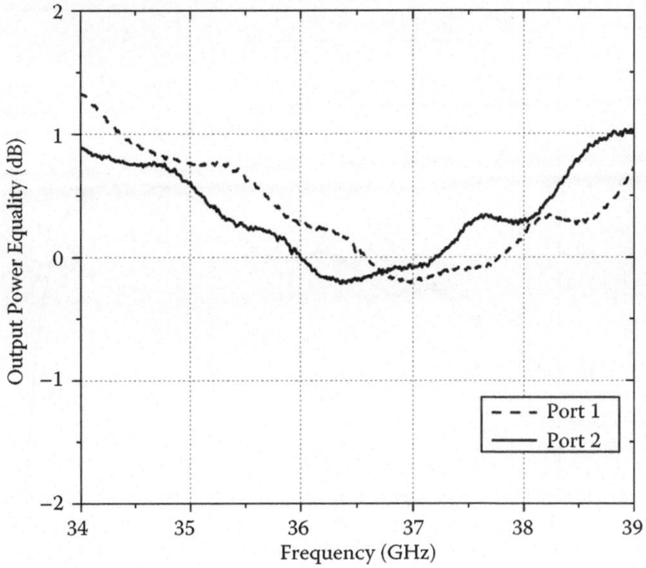

FIGURE 6.2 Measured output power equality of the 180° SIW coupler excited at two different input ports.

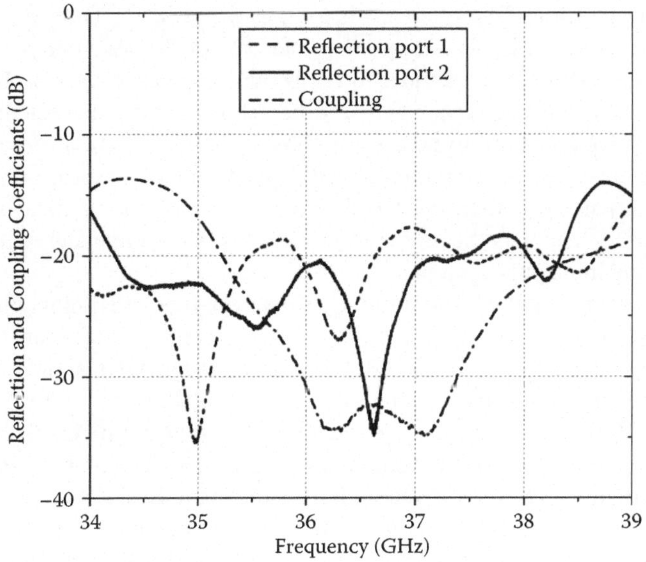

FIGURE 6.3 Measured reflection and coupling coefficients of the 180° SIW coupler.

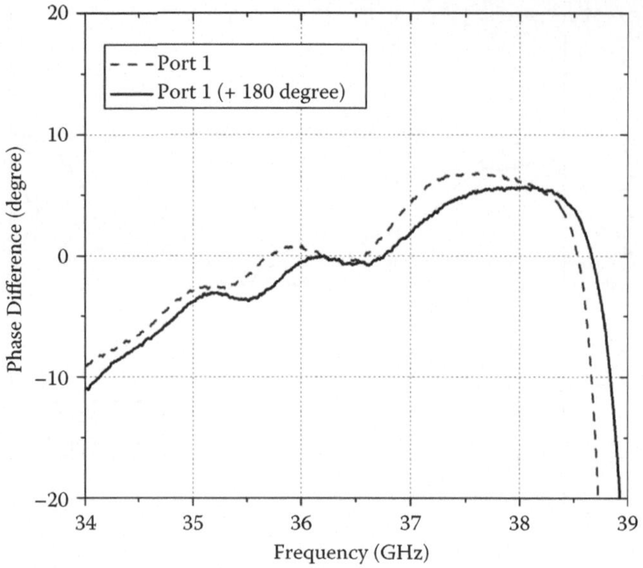

FIGURE 6.4 Measured phase differences between two output ports of the 180° SIW coupler excited at two different input ports.

between two output ports is shown in Figure 6.4. The phase difference is 0° ± 5° over 34.3–37.5 GHz for port 1 and 180° ± 5° over 34.6–37.3 GHz band for port 2.

As described in Section 4.1.1, the SIW-fed TSA has been proposed and experimentally demonstrated with excellent performances. This type of antenna would be a good partner to the monopulse antenna at the millimeter-wave frequency band. An array consisting of four TSA elements is used as the radiator. They are divided into two groups, each of which is fed by an output port of the 180° coupler connecting with a two-way SIW parallel divider to realize the in-phase or out-of-phase feeding mechanism. Absolutely, a higher gain can be achieved with more elements. When the incident wave comes from port 1, the SIW structure is excited in phase. Waves are guided right up to the radiating aperture with an even-term illumination. The sum radiation pattern is then produced. Similarly, when the incident wave comes from port 2, the difference beam is produced. This is in agreement with the design principle of the monopulse antenna.

Measured return losses for two input ports of the fabricated monopulse antenna are both better than 13 dB. The isolation between them is better than 17.5 dB over the frequency range of 33–40 GHz. Measured sum and difference radiation patterns are given in Figure 6.5. The E-plane sum radiation pattern has a 3 dB beamwidth of 7.2°. Table 6.1 presents the measured gain of the sum beam versus the operating frequency. The peak gain of the sum beam is 16.13 dBi. The gain of the difference beam is less than 2.95 dB compared with the sum value. Besides, the null depth is at −38.1 dB. Those results suggest that this monopulse antenna is able to operate over a relative wideband with this type of SIW comparator that has an influence-limited error in generating the required 0° or 180° phase difference.

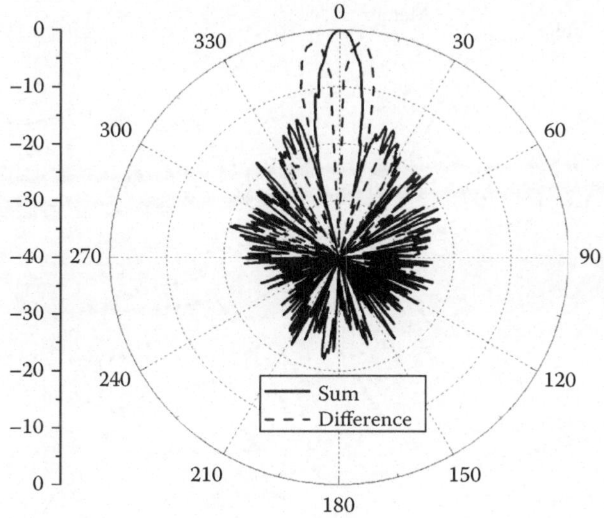

FIGURE 6.5 Measured radiation patterns of the fabricated SIW-fed TSA monopulse antenna array at 36.5 GHz.

TABLE 6.1
Measured Frequency-Dependent Gain of the Sum Pattern of the Fabricated SIW-Fed TSA Monopulse Antenna Array

Frequency (GHz)	34	34.5	35	35.5	36	36.5
Gain (dBi)	15.9	16	15.7	15.8	16.1	16.1
Frequency (GHz)	37	37.5	38	38.5	39	
Gain (dBi)	16.1	16.1	15.8	15.9	15.4	

6.1.2 SIW-Fed Horn Monopulse Antenna Array

The horn antenna is a very popular endfire antenna with wide bandwidth. An elliptical-shaped dielectric loaded SIW H-plane sectoral horn antenna was proposed in [177]. The horn and the loaded dielectric can be integrated in the same single-layer substrate resulting in easy fabrication and low cost.

As illustrated in Figure 6.6, such an antenna with loaded dielectric is used to form a one dimensional monopulse antenna array at the center frequency of 27 GHz. This monopulse antenna array is formed by two 1 × 4 subarrays. A 90° coupler and 90° phase delay line are also used to form a comparator to supply odd and even excitations. The structure of this comparator is similar to the above design, but they are different in their phase shifters. A substrate with a relative permittivity of 4.8 and a thickness of 2.5 mm is employed. Measured gain of the sum pattern is 15.65 dBi. Measured null depth of the difference pattern is –25 dB.

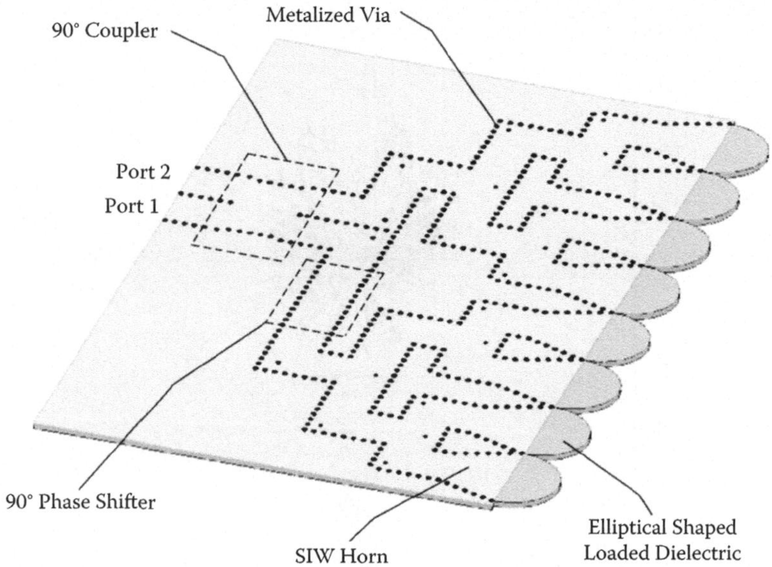

FIGURE 6.6 Configuration of the one-dimensional SIW-fed horn monopulse antenna array.

6.1.3 SIW-Fed Yagi Monopulse Antenna Array

Integrated with an SIW hybrid ring coupler, the developed X-band SIW-fed Yagi-Uda antenna described in [133] can also be employed in the design of a one-dimensional monopulse array. Measured gain of the sum pattern is larger than 13 dBi, whereas the maximum null depth of the difference pattern is lower than 20 dB in 9.4–10.05 GHz.

6.2 ONE-DIMENSIONAL SIW MULTIMODE MONOPULSE ANTENNA

6.2.1 Dual V-Type Linearly Tapered Slot Monopulse Antenna

In many applications of millimeter-wave integrated circuits, an antenna element would prefer to be integrated with the rest of the circuit on a single substrate. The uniplanar TSA has been widely implemented in the development of millimeter-wave integrated circuits and systems for its superior salient features. This type of antenna can be grouped as an array and fed by an SIW comparator to obtain the monopulse ability. However, a single TSA element cannot be employed as a monopulse antenna according to its structural characteristic. Therefore, a dual V-type linearly tapered slot antenna (DVLTSA), which is a modified type of TSA, was proposed in [180] to realize a miniaturized endfire monopulse antenna. Using the multimode feeding technology, one-dimensional monopulse capability can be realized in the E-plane only by a single element of this type of antenna.

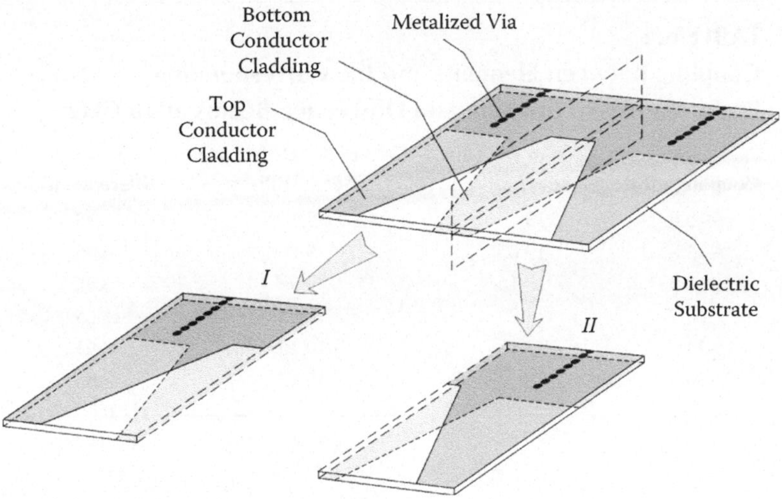

FIGURE 6.7 Configuration of the DVLTSA element.

6.2.1.1 DVLTSA Element

As shown in Figure 6.7, a DVLTSA element has some tapered slots etched in the opposite conductor claddings on a substrate. The appropriate analytical model is based on a TSA array with two antisymmetric elements. Each element is similar to a conventional TSA element. It should be noted that the DVLTSA just demands the symmetry of the whole configuration, not each TSA part.

The equivalent model, i.e., an ALTSA array consisting of two antisymmetric elements, can be used to interpret the monopulse principle. When the elements are fed out of phase, the DVLTSA produces the sum beam according to its antisymmetric configuration. When inputs of elements are in phase and amplitude, the difference beam can be generated.

Parameters of the DVLTSA, such as shape, total length, dielectric thickness, and permittivity, should be selected correctly. The mutual coupling between elements of a DVLTSA should be investigated to achieve a good performance. The odd and even mode analysis can be employed to recognize the influence of the coupling. As shown in Table 6.2, the gain of the sum beam almost stays the same value around 13 dBi. Only the gain of the difference beam is worsened by the coupling. Therefore, the gain of the difference beam can be considered the criterion to measure the coupling. For a monopulse antenna, the general gain difference between the sum and difference beams is known to be 3 dB. The coupling below –20 dB has no effect on the performances of a DVLTSA. Therefore, values of fw_1 and fw_2, as labeled in Figure 6.8, should not be selected too small to avoid the strong coupling.

From these considerations, a DVLTSA was designed at 36 GHz with parameters: $fw_1 = 1$ mm, $fw_2 = 7$ mm, $bw_1 = 15$ mm, and $bw_2 = 2$ mm. The length of the DVLTSA is 17.39 mm. Rogers 5880 substrate was used in this design with a thickness of

TABLE 6.2
Coupling between Elements and the Corresponding Performances of the Sum and Difference Beams at 36 GHz

Coupling (dB)	Gain of the Sum Beam (dBi)	Gain of the Difference Beam (dBi)	Difference (dB)
–10.6	13	7.15	5.85
–13.8	13.05	9.2	3.85
–18.5	13.15	9.9	3.25
–20	13.2	10.15	3.05
–24	13.1	10.15	2.95
–30	13	10.15	2.85

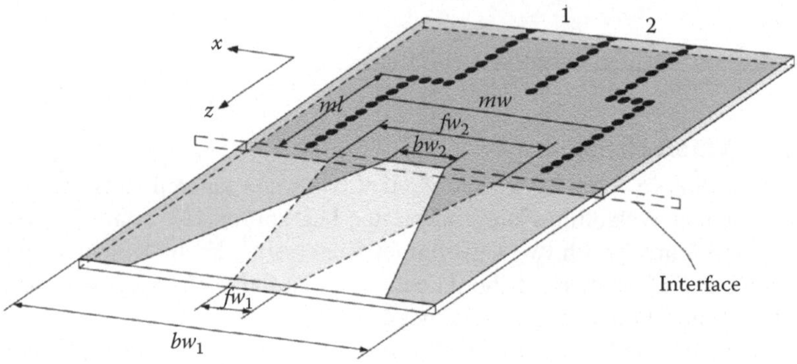

FIGURE 6.8 Configuration of the DVLTSA fed by a multimode SIW.

0.508 mm and a permittivity of 2.2. By use of the odd and even mode analysis, the coupling is below –20 dB over the whole frequency band from 34 to 39 GHz, which guarantees the good performances of the antenna.

6.2.1.2 Multimode SIW

Multimode technology is a very useful and effective way to feed a monopulse antenna [178, 179]. Thus, an SIW multimode feeding network is developed and employed in this design. The multimode zone is $ml \times mw$, as shown in Figure 6.8. It has two input ports. Here, ml and mw are supposed to be approximately equal to the corresponding equivalent length and width of the SIW multimode section.

To guarantee the propagation of the initial four modes in the multimode SIW, the width mw should be chosen by Equation (6.1):

$$\frac{4}{2}\lambda < mw < \frac{5}{2}\lambda \tag{6.1}$$

That means TE_{10}, TE_{20}, TE_{30}, and TE_{40} can be excited by the moder. In Equation (6.1), λ is the wavelength in the substrate. If just the first three modes are permitted to propagate in multimode SIW, lengths of fw_1 and fw_2 may be too small, which deteriorates the performances of the DVLTSA owing to the strong coupling.

When two input SIWs are excited out of phase, the modes excited by the discontinuity of the moder only have odd transverse components. The nonevanescent TE_{20} and TE_{40} modes propagate along the multimode SIW right up to the interface at their respective phase velocities and amplitudes. In this case, it is the odd illumination law. The equation of the said law is

$$S_{odd}(x) = \sin\left(\frac{2\pi x}{mw}\right)e^{-j\beta_{20}z} + c_1 \times \sin\left(\frac{4\pi x}{mw}\right)e^{-j\beta_{40}z} \tag{6.2}$$

When two input SIWs are excited in phase, only even-term modes are excited by the discontinuity of the moder. All of them evanesce except TE_{10} and TE_{30}. They propagate to the interface, resulting in an even illumination law. In this case, the equation of the said law is

$$S_{even}(x) = \cos\left(\frac{\pi x}{mw}\right)e^{-j\beta_{10}z} + c_2 \times \cos\left(\frac{3\pi x}{mw}\right)e^{-j\beta_{30}z} \tag{6.3}$$

In Equations (6.2) and (6.3), c_1 and c_2 denote the mode ratios. The coefficients β_{10}, β_{20}, β_{30}, and β_{40} are the phase constants of the TE_{10} to TE_{40} modes, respectively. They are expressed as follows:

$$\beta_{10} = \frac{2\pi}{\lambda}\sqrt{1-\left(\frac{\lambda}{2\times mw}\right)^2} \tag{6.4}$$

$$\beta_{20} = \frac{2\pi}{\lambda}\sqrt{1-\left(\frac{2\lambda}{2\times mw}\right)^2} \tag{6.5}$$

$$\beta_{30} = \frac{2\pi}{\lambda}\sqrt{1-\left(\frac{3\lambda}{2\times mw}\right)^2} \tag{6.6}$$

$$\beta_{40} = \frac{2\pi}{\lambda}\sqrt{1-\left(\frac{4\lambda}{2\times mw}\right)^?} \tag{6.7}$$

Neglecting the effects of the reflected and evanescent modes, the illumination law at the interface can be simplified by equating z and ml in Equations (6.2) and (6.3) as follows:

$$S_{odd}(x) = \sin\left(\frac{2\pi x}{mw}\right) + c_1 \times \sin\left(\frac{4\pi x}{mw}\right) e^{-j(\beta_{40}-\beta_{20})ml} \quad (6.8)$$

$$S_{even}(x) = \cos\left(\frac{\pi x}{mw}\right) + c_2 \times \cos\left(\frac{3\pi x}{mw}\right) e^{-j(\beta_{30}-\beta_{10})ml} \quad (6.9)$$

As shown in Figure 6.9, the even and odd transverse component of the E-field can be realized at the output of the multimode SIW section, respectively, in theory.

The illumination law at the center point of the interface equaling zero is another key factor in the design of the DVLTSA. At the center point of the interface, the odd illumination law always equals zero. Thus, the field of the even term at the interface has more influence on the reflection of the DVLTSA. If the respective phases of the TE_{10} and TE_{30} modes are adjusted to the same value at the interface, the approximate null-center-field distribution can be achieved for the even illumination law. By altering *ml*, one is able to control the phase of each mode at the interface.

First, adjust *ml* to make TE_{10} and TE_{30} in phase at the interface. That is,

$$ml = \frac{2\pi}{(\beta_{30} - \beta_{10})} \quad (6.10)$$

Then, the width *mw* satisfying Equation (6.1) is also regulated to get proper mode ratios c_1 and c_2, especially c_2, to satisfy the condition of null field at the center of the interface for the even illumination law.

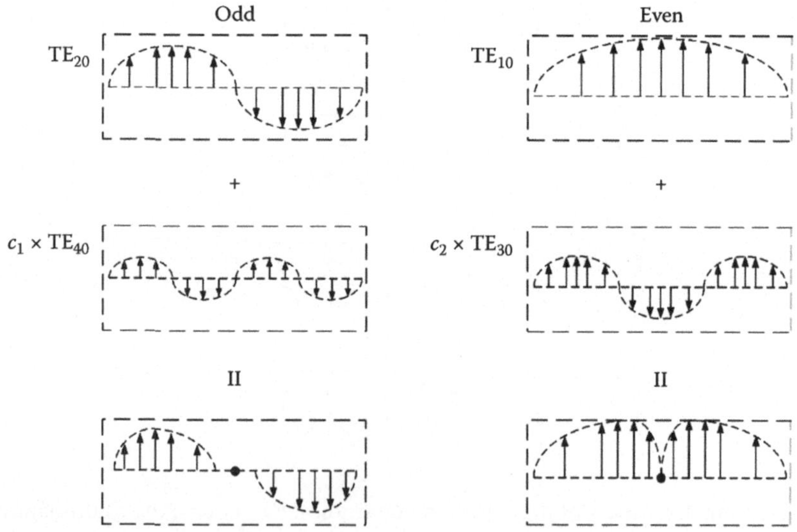

FIGURE 6.9 Even and odd illumination laws at the output of the multimode SIW section in theory.

At last, the even illumination equation can be rewritten as

$$S_{even}(x) = \cos\left(\frac{\pi x}{mw}\right) + c_2 \times \cos\left(\frac{3\pi x}{mw}\right) \tag{6.11}$$

In this design, the width of the input SIW is 4 mm, the distance between the adjacent metalized vias is 0.8 mm, and the diameter of the metalized vias is 0.4 mm. According to Equation (6.1), the range of choice for mw is from 10.5 to 13 mm. When $ml = 9.13$ mm, the case of $mw = 12$ mm satisfies the requirement of null center field best.

6.2.1.3 Design Result of the DVLTSA Monopulse Antenna

After balancing the efficiency of the sum and difference beams, as well as optimizing the performance, a monopulse antenna was designed based on a single DVLTSA at 36 GHz. As shown in Figure 6.10, it is fed by a multimode SIW cascading with a 180° directional coupler. The 180° directional coupler, worked as a comparator, is used to excite the multimode SIW in phase or out of phase. A 3 dB coupler offers the first 90° phase difference, and another 90° phase difference is provided by variable SIW widths.

When the incident wave comes from port 1, the multimode SIW is excited out of phase and provides an odd-term illumination to excite the DVLTSA. In this case, the sum beam is generated. Similarly, when the incident wave comes from port 2, the difference beam is produced.

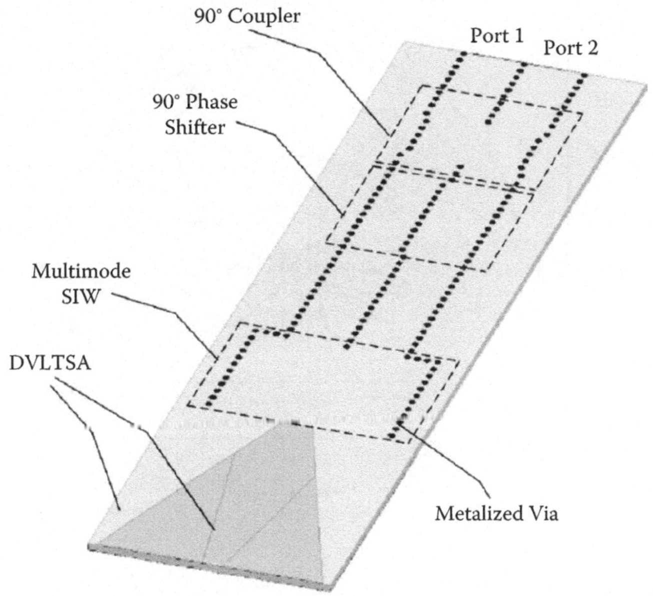

FIGURE 6.10 Configuration of the DVLTSA monopulse antenna.

Measured return losses of two input ports are both better than 12.5 dB, and the isolation between input ports is better than 17.5 dB from 33 to 40 GHz. The E-plane and H-plane radiation patterns measured at 36 GHz are given in Figures 6.11 and 6.12. The gain of the sum beam is 10.6 dBi. The corresponding radiation efficiency is 57.5%. The 3 dB beamwidth of the sum beam is 22.5° in the E-plane. The gain of the

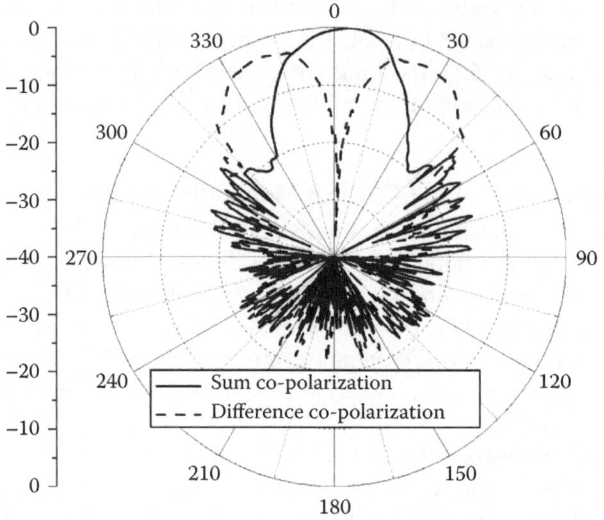

FIGURE 6.11 Measured E-plane radiation pattern of the DVLTSA monopulse antenna at 36 GHz.

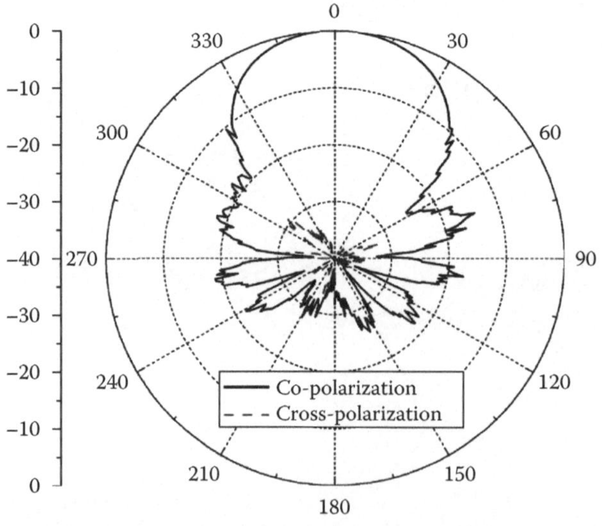

FIGURE 6.12 Measured H-plane radiation pattern of the DVLTSA monopulse antenna at 36 GHz.

TABLE 6.3
Measured Gains of the Sum Beam versus the Operating Frequency

Frequency (GHz)	34	35	36	37	38	39
Gain (dBi)	9.4	9.8	10.6	10.6	10.7	10.8

difference beam is less than 3 dB compared with the sum one. The null depth of the difference pattern is at –35 dB. Both E- and H- planes have good cross-polarization. Table 6.3 shows the measured gain of the sum beam versus the operating frequency.

6.2.2 Multimode SIW Monopulse Horn

For purposes of tracking and searching, monopulse capability is often included in communication systems and radars. Therefore, the design of a monopulse feeder for a reflector antenna or a lens is a significant work. In the case of a monopulse antenna, there are usually a sum mode and a difference mode, and the requirements for the sum and difference modes are not the same. It has to design a feeder for simultaneous optimum performance in both modes. There are several works focusing on this area. Among these designs, the feeding technology [179, 181, 182] has been considered to be a good compromise. A multimode horn and an appropriate feeding network were combined to balance the efficiency of the sum and difference beams. A rectangular waveguide is usually used to obtain the optimum feeder.

As shown in Figure 6.13, an optimum monopulse horn was presented at the Ku-band fed by a multimode SIW [183]. A 180° directional coupler was employed to realize the in-phase or out-of-phase two-way excitations. An H-plane moder excited by the 180° coupler was employed to provide even and odd illumination laws at the aperture. Only the first three modes are permitted to propagate in the multimode SIW. The H-plane SIW horn antenna cannot radiate energy effectively because of its very large width-to-height ratio. To address this problem, the dielectric substrate applied to fabricate the SIW and three pieces of metallic plates with required shapes were welded together as a horn antenna to radiate energy effectively. It is worthwhile to note that there is no conductor cladding on the top surface of the substrate. Here, good performance is realized and the complicated fabrication is avoided.

Following is a brief introduction of the operation principle. When two input SIWs of the moder are excited in phase and amplitude, modes within the multimode SIW have even transverse components. All these modes are evanescent except the TE_{10} and TE_{30} modes. An even illumination law can be realized at the radiating aperture, and optimized by properly controlling the respective amplitudes and phase constants of the TE_{10} and TE_{30} modes. Thus, the sum beam can be generated with optimal performance. When two input SIWs of the moder are excited out of phase, different odd-term modes are excited in the multimode SIW because of the discontinuity. The only nonevanescent mode is the TE_{20} mode. In this case, the difference beam is provided according to an odd illumination law at the radiating aperture.

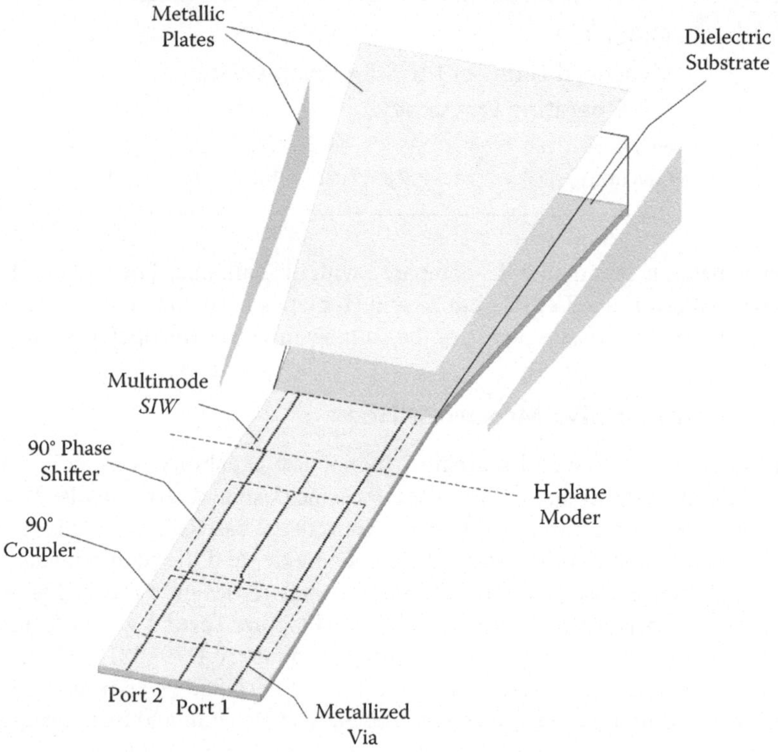

FIGURE 6.13 Configuration of the multimode SIW monopulse horn.

The whole circuit was fabricated through a printed circuit board (PCB) process. Rogers 5880 dielectric substrate was used in this design with a permittivity of 2.2 and a thickness of 1.575 mm. The isolation is better than 15 dB over 13–18 GHz. The measured return loss of port 1 is better than 12 dB over 14.2–18 GHz, and the return loss of port 2 is better than 11.5 dB over 13–18 GHz. Figure 6.14 presents the sum and difference radiation patterns in the H-plane measured at 15.5 GHz. The peak gain of the sum beam is measured to be 9.9 dBi. Compared with the sum beam, the gain of the difference beam is less than 2.15 dB. The measured null depth of the difference pattern is at –44 dB. The 3 dB beamwidth of the sum beam is 28°. Both H- and E-planes have good cross-polarization.

6.3 TWO-DIMENSIONAL SIW MONOPULSE ANTENNA ARRAY

The two-dimensional monopulse antenna, which has been developed in a variety of technologies, is much more attractive for high-resolution tracking applications. Narrow beamwidths for high angular accuracy are possible in the millimeter-wave band with a relatively small aperture size. Conventional waveguide, which provides coupling-free and low-loss features, may be the proper choice for the design of a

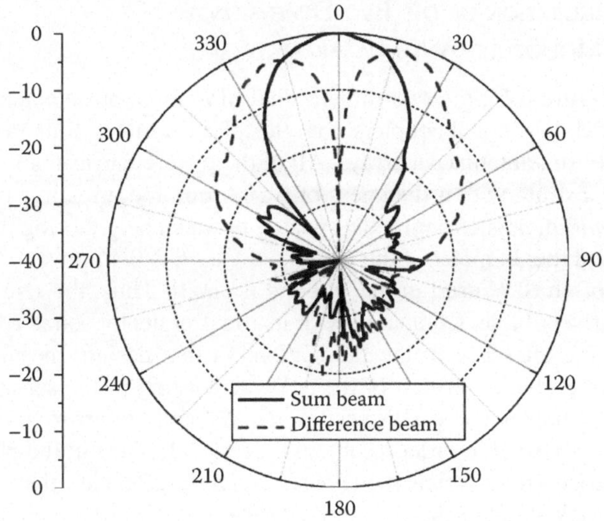

FIGURE 6.14 Measured H-plane radiation pattern of the multimode SIW monopulse horn at 15.5 GHz.

millimeter-wave monopulse antenna array. However, integration problems between those planar and nonplanar configurations have to be considered in the system design. To deal with those problems, the cost may be prohibitively high.

As an alternative, the SIW technology is very appropriate to construct the high-frequency planar monopulse antenna array. The two-dimensional comparator, divider, and radiator can all be integrated in a single-layer dielectric substrate through the low-cost PCB process. A Ka-band single-layer SIW monopulse slot array antenna was designed, fabricated, and experimented with in [184]. The total size of the antenna is 140 × 130 mm. The relative bandwidth ($|S_{11}| < -10$ dB) of the antenna is 7.39% from 30.80 to 33.14 GHz. The peak gain is 18.74 dBi at 31.5 GHz, and the maximum null depth is −46.3 dB.

Besides, an SIW comparator network was investigated and demonstrated in [185] for Ku-band tracking applications. The feeding network and the monopulse comparator implemented by the SIW technology are integrated into a substrate with a size of $8.2\lambda_0 \times 9\lambda_0$ at a center frequency of 17.525 GHz. Thirty-two output ports are welded perpendicularly for connecting active transmitting and receiving modules. The relative bandwidth ($|S_{11}| < -10$ dB) of the network is 6%, in which the phase imbalance of all ports is less than 5.5° and the power imbalance is about 0.5 dB. The isolations between the three channels are all better than 25 dB. The insertion losses are about 3–5 dB.

In [186], a planar W-band monopulse antenna array was designed based on the SIW technology. Compared with the Ku-band and Ka-band versions developed in [184, 185], there are several new challenges and design considerations in this higher-frequency and larger-scale design.

6.3.1 Configuration of the Two-Dimensional SIW Monopulse Antenna Array

As shown in Figure 6.15, the two-dimensional SIW monopulse antenna array consists of four 3 dB directional couplers, four 90° phase shifters, four 16-way dividers, and four 16 × 16 slot antenna subarrays. All of them were fabricated in a single-layer Rogers 5880 substrate with a thickness of 0.508 mm and a relative permittivity of 2.2. The SIW width, a, is 1.76 mm, the spacing between neighboring metalized vias, s, is 0.6 mm, and the diameter of the metalized via, d, is 0.4 mm.

The width of an SIW used in the W-band is small. Thus, the size of the metalized via is relatively large. Compared with its low-frequency counterpart, it is more susceptible to the tolerance in the fabrication. In this design, the most important factor may be the phase constant of the SIW. According to the sensitivity analysis as described in Chapter 1, the SIW used in the W-band prefers to have a larger a, a larger s, and a smaller d in order to obtain a better tolerance in the phase constant. The existing processing restricts the choices of d and s. Besides, the wave propagating along the SIW is more difficult to be restricted within two rows of metalized vias by the operating frequency. Thus, a more rigorous design rule is required for the SIW used in the W-band. The radiation loss of an SIW can be predicted when the SIW is constructed with the perfect conductor and the ideal dielectric material. In simulation, at 94 GHz, s should be less than 0.65 mm to keep affordable leakage, i.e., $\alpha_r/k < 10^{-4}$, as described in [15]. It is evident that the tolerance must be combined with the radiation leakage in the selection of s.

Four 3 dB couplers and four 90° fixed phase shifters are employed to construct the two-dimensional SIW comparator. The 3 dB coupler is the H-plane coupler, which has the continuous coupling aperture and matching steps. Two input and two output ports of the coupler are connected with 90° right-angled corners. A metalized via is

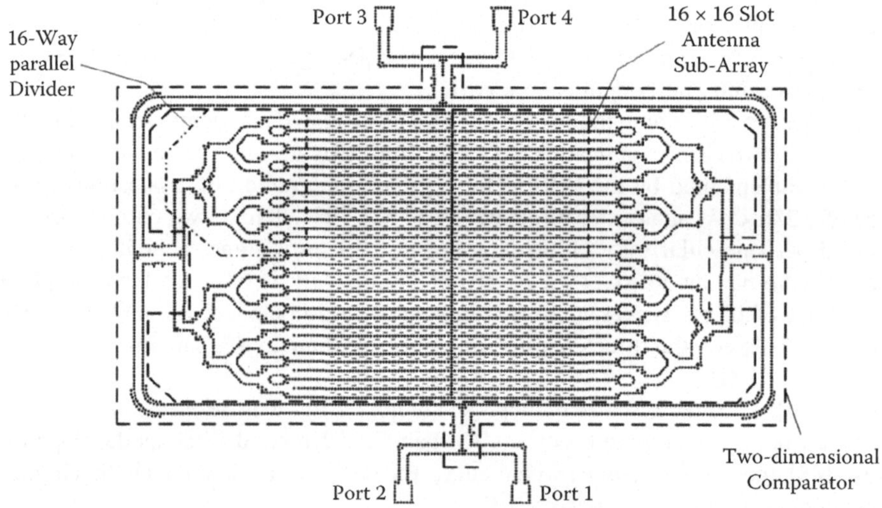

FIGURE 6.15 Configuration of the two-dimensional SIW monopulse antenna array.

placed at a proper position in the corner to achieve good matching. Ninety degree SIW fixed phase shifters are employed after 90° couplers to provide the in-phase and out-of-phase feeding mechanism. There are two different types of phase shifters in the comparator. One is the delay line, which is used to produce the 90° phase shift in H-plane. The other is the equal-length unequal-width phase shifter to realize the 90° phase shift in the E-plane. These couplers are connected by four 90° round bends and straight SIW feedlines. There are four rows of metalized vias to construct the round bend to avoid the leakage at such high frequency. Now, the SIW comparator has been constructed. It reserves enough area for dividers and slot arrays.

The input ports of the SIW comparator were fed by WR-10 waveguides to avoid the direct radiation from the feed discontinuity. The SIW–waveguide transition has a configuration similar to that shown in Figure 2.21. It is able to make the reflection coefficient less than –16 dB from 90 to 98 GHz.

Four-stage two-way dividers can be grouped together to realize the 16-way divider as presented in Figure 2.3. Such a large-scale divider is used to divide the power from each output port of the comparator to an antenna subarray consisting of 16 linear SIW slot arrays. Each linear SIW slot array includes 16 slots. Two linear slot arrays are placed back-to-back and designed together. If the slot array is set to satisfy a special aperture distribution, the number of slots will be 32, but not 16 in the design. The back-to-back array should be modeled and optimized together. The sum and difference beams of the designed back-to-back array are simulated at 94 GHz as shown in Figure 6.16. Because the back-to-back array itself owns a 180° phase difference between elements, the out-of-phase excitation generates a sum beam and the in-phase excitation generates a difference beam.

Figures 6.17 and 6.18 give simulated H-plane results of the back-to-back array at 94 GHz to investigate influences of amplitude and phase imbalances. The errors

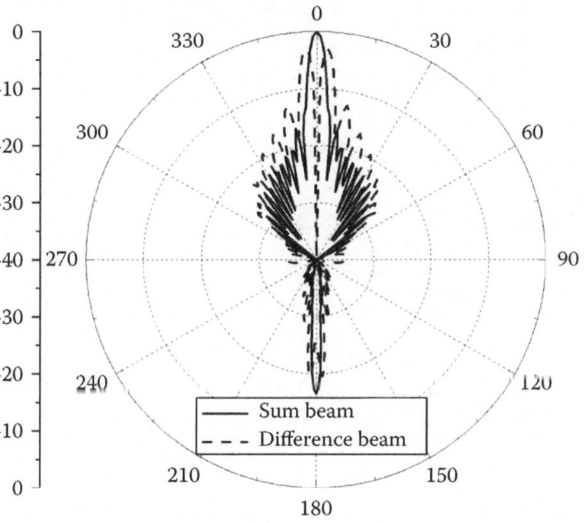

FIGURE 6.16 Simulation radiation patterns of the back-to-back SIW slot antenna array at 94 GHz.

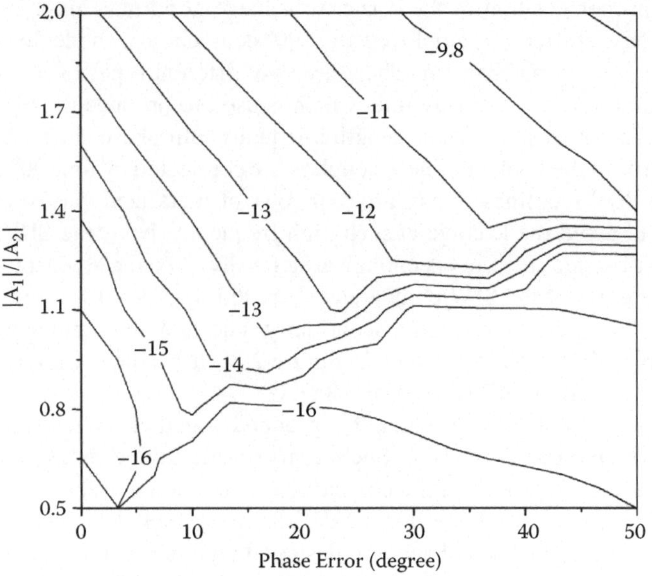

FIGURE 6.17 SLL of sum beams in the H-plane at 94 GHz versus amplitude and phase errors.

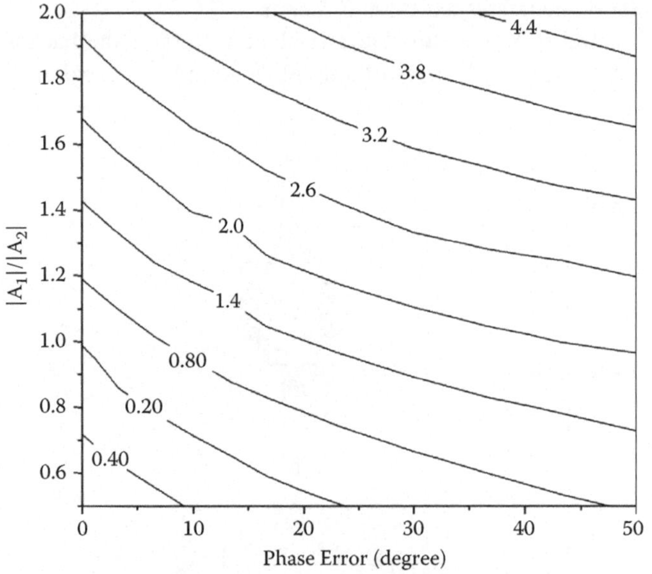

FIGURE 6.18 Amplitude imbalances of difference beams in the H-plane at 94 GHz versus amplitude and phase errors.

of two incident waves (A_1 and A_2) in amplitude and phase deteriorate sidelobe level (SLL) of the sum beam. The errors also lead to amplitude imbalances of the difference beam. For example, when $|A_1|/|A_2| = 2$, a phase error of 50° causes 8.3 dB deterioration of SLL for the sum beam and 4.7 dB amplitude imbalance for the difference beam.

Loss is also important for the W-band SIW monopulse antenna array. The used substrate has a tan δ of 0.0009 at 10 GHz according to the manufacturer's data sheet. However, the actual tan δ at 94 GHz should be known to evaluate the W-band circuit loss accurately. Through experiment, tan δ = 0.0012 can be set as the equivalent value. In this case, the efficiency of the SIW monopulse antenna array can be predicted when all simulations are carried out with tan δ = 0.0012. For example, excited at port 1, the average loss of the SIW comparator is 4.66 dB, the average loss of the 16-way divider is 1.32 dB, and the loss of the SIW–waveguide transition is 0.65 dB at 94 GHz. After considering that the radiation efficiency of the slot array is 75.14%, the total radiation efficiency is estimated to be 16.3% for the SIW monopulse array.

6.3.2 Operation Principle of the Two-Dimensional SIW Monopulse Antenna Array

The basic operation principle of the monopulse array excited at port 1 can be explained as shown in Figure 6.19.

When excited at port 1, the incident wave is divided into two equal portions (signal 1 and signal 2) with 90° phase difference by a 3 dB directional coupler at first.

On one side, signal 1 enters into coupler II, and then generates two equal outputs (signals 11 and 12) with 90° phase difference again. Signal 11 is input into a 16-way divider and directly fed subarray I. Signal 12, which is added with a 90° phase shift after passing a phase shifter, will be input into a 16-way divider and then fed subarray II. Now, the phase of signal 12 is the same as that of signal 11.

On the other side, signal 2 passes through a 90° phase shifter at first. Then coupler III divides it into two parts: signals 21 signal 22. They have equal amplitude but 90° phase difference. Subarray III is fed by signal 21 through a 16-way divider. Signal 22 is added with 90° phase shift through a phase shifter at first. Here, it has the same phase as signal 21. Then, it is input into a 16-way divider and fed subarray IV.

Finally, four ideal feeding signals prior to entering into dividers are decided. They are (–6 dB, 0°), (–6 dB, 0°), (–6 dB, 180°), and (–6 dB, 180°). Considering the 180° phase difference between subarrays I/II and III/IV, the sum beam will be generated in this case.

In a similar way, when excited at port 2, four ideal feeding signals prior to entering into dividers are (–6 dB, 0°), (–6 dB, 0°), (–6 dB, 0°), and (–6 dB, 0°). The difference beam in the *xoz*-plane will be generated. When excited at port 3, four ideal feeding signals prior to entering into dividers are (–6 dB, 0°), (–6 dB, 180°), (–6 dB, 180°), and (–6 dB, 0°). The difference beam in the *yoz*-plane will be generated.

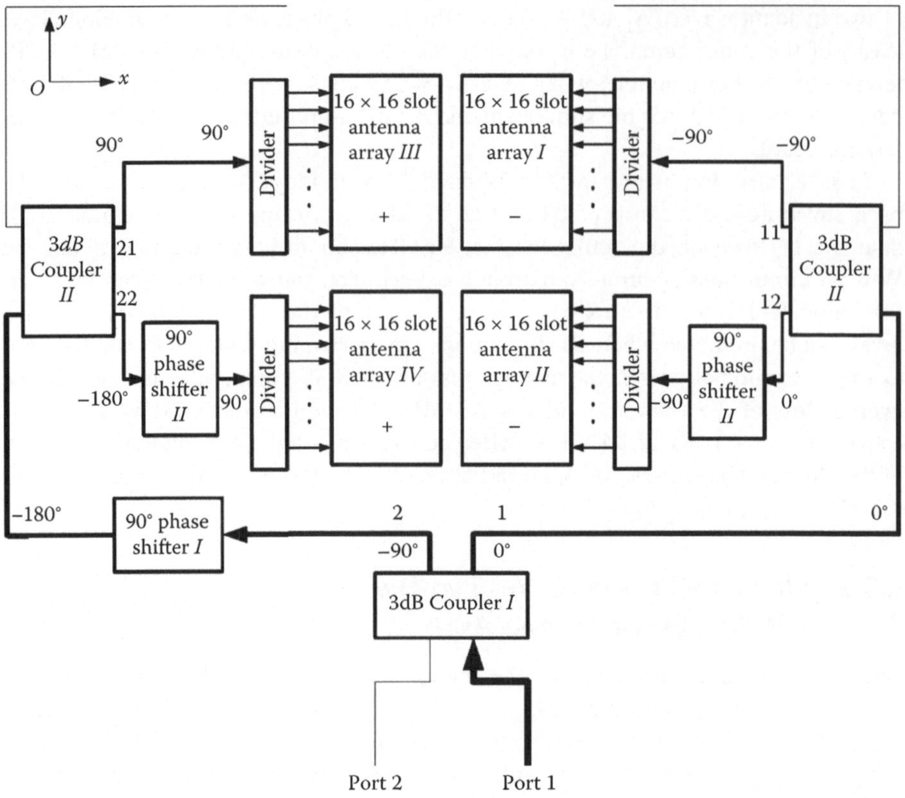

FIGURE 6.19 Basic principle of the two-dimensional SIW monopulse antenna array excited at port 1.

6.3.3 Design Result of the Two-Dimensional SIW Monopulse Antenna Array

The fabricated SIW monopulse antenna array has the size of 130 × 125 mm. When excited at port 1, the sum beam is generated by the monopulse antenna array in both the E-plane (the *yoz*-plane) and the H-plane (the *xoz*-plane). Measured radiation patterns at different frequencies are shown in Figures 6.20 to 6.21. The 3 dB beamwidths are in the range of 2°–3° in the E-plane and 3°–4° in the H-plane. The SLLs are in the range of 9.82–12.14 dB in the E-plane and 5.95–19.74 dB in the H-plane. As shown in Figure 6.22, the measured maximum gain is 25.75 dBi at 93.6 GHz, and the minimum one is 21.29 dBi at 96 GHz. The measured total radiation efficiency is 15.2% at 94 GHz, which agrees well with our expected value.

When excited at port 2, the sum beam in the E-plane and the difference beam in the H-plane can be generated by the monopulse array. Measured radiation patterns at different frequencies are shown in Figure 6.23. As shown in Figure 6.24, the maximum null depth is −45.81 dB at 93.8 GHz. The best amplitude balance of the difference beam is 2.12 dB at 95.2 GHz.

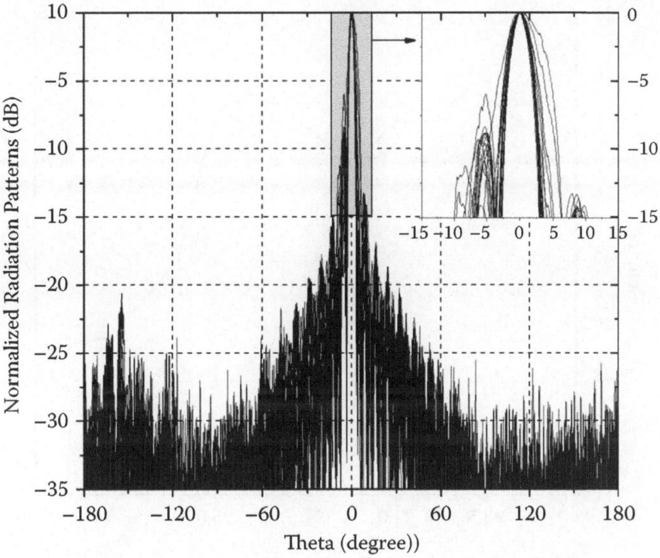

FIGURE 6.20 Measured H-plane radiation patterns of the two-dimensional SIW monopulse antenna array at different frequencies excited at port 1.

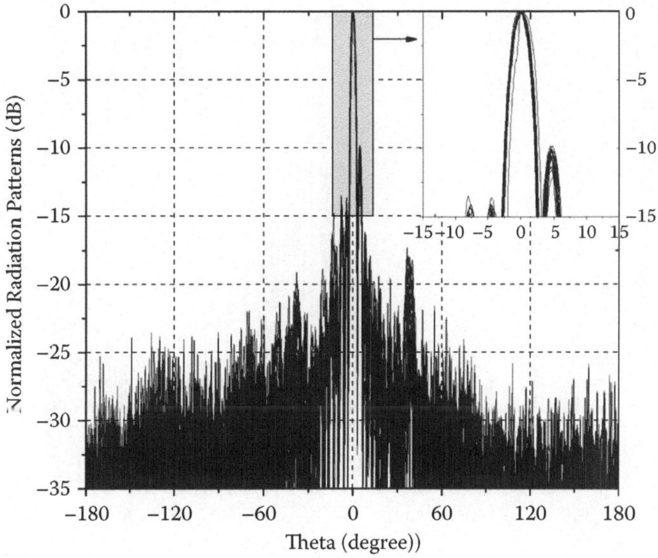

FIGURE 6.21 Measured E-plane radiation patterns of the two-dimensional SIW monopulse antenna array at different frequencies excited at port 1.

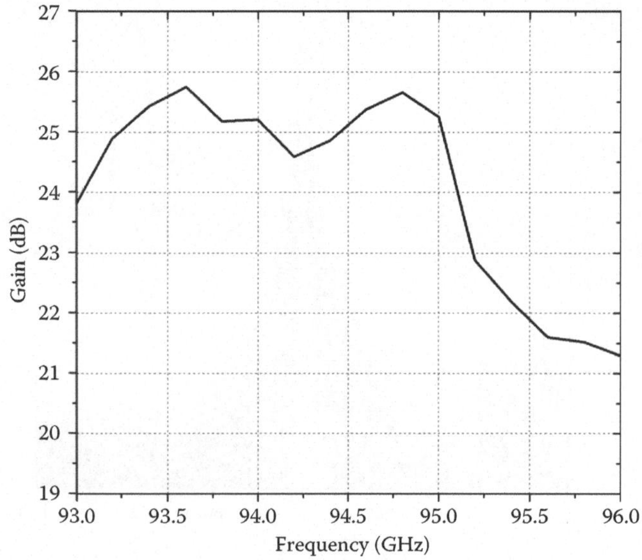

FIGURE 6.22 Measured gains of the two-dimensional SIW monopulse antenna array at different frequencies excited at port 1.

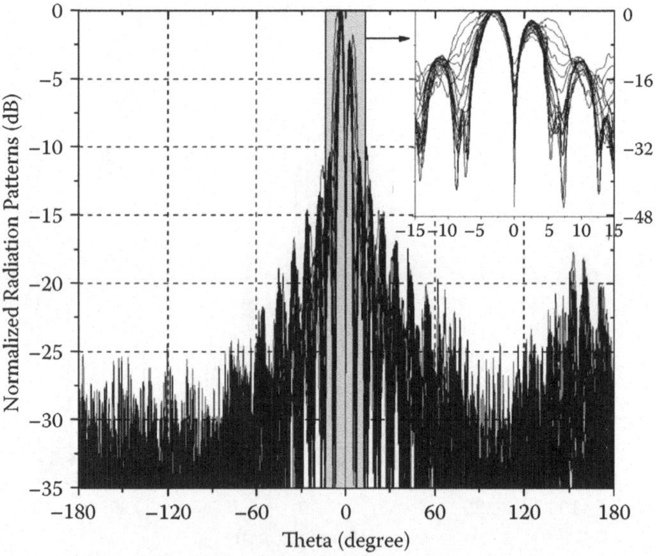

FIGURE 6.23 Measured H-plane radiation patterns of the two-dimensional SIW monopulse antenna array at different frequencies excited at port 2.

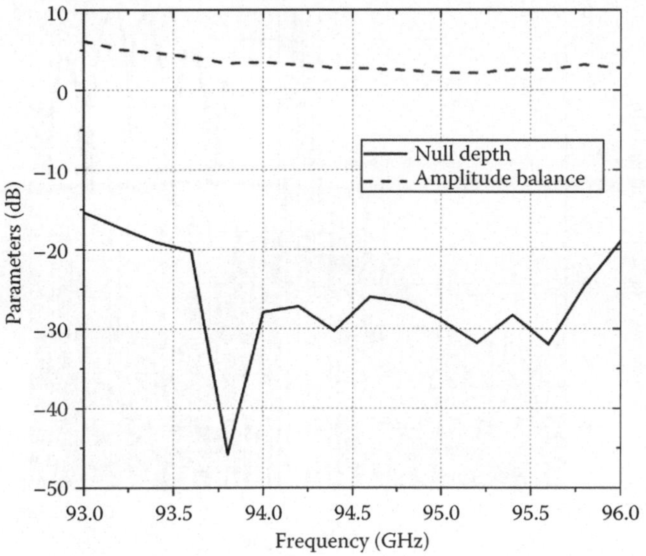

FIGURE 6.24 Measurements of the two-dimensional SIW monopulse antenna array at different frequencies excited at port 2.

When excited at port 3, the difference beam in the E-plane and the sum beam in the H-plane can be generated by the monopulse array. Measured radiation patterns at different frequencies are shown in Figure 6.25. As shown in Figure 6.26, the maximum null depth is –37.71 dB at 95.2 GHz. The best amplitude balance of the difference beam is 0.4 dB at 93.6 GHz.

The total loss of the fabricated two-dimensional SIW monopulse array in the W-band is a little high. The total radiation efficiency is relatively low, even though it is better than that of other planar counterparts. The multilayer configuration is effective in shortening the path length of the comparator, and thus able to improve the efficiency. To avoid the use of the large-scale parallel divider, each center of top-layer slot array antennas can be fed in turn by a crosswise SIW using series slots. The reliability of the multilayer fabrication process in the W-band is the most key factor to such a revised design.

Such an SIW monopulse array has better radiation characteristics in the E-plane than in the H-plane. Because of different types of phase shifters, the phase shifting in the H-plane changes much more versus the operating frequency than that in the E-plane. In the E-plane, the phase error is –3° at 93 GHz and 8.1° at 96 GHz. In the H-plane, the phase error is 9.5° at 93 GHz and –24.2° at 96 GHz, which lead to larger variation of radiation characteristics in the H-plane. Besides, as a resonant type antenna, the H-plane performance of the slot array antenna is more sensitive to the operating frequency.

Compared with the measured and designed results, the radiation characteristics of the antenna array are deteriorated, especially in aspects of SLL of the sum beam and null depth and amplitude imbalance of the difference beam. The properties of the feeding network deviate from the designed specifications because of unavoidable

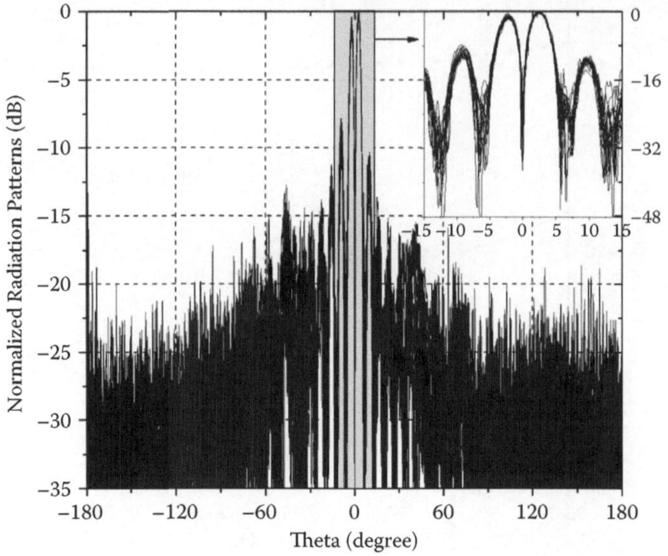

FIGURE 6.25 Measured E-plane radiation patterns of the two-dimensional SIW monopulse antenna array at different frequencies excited at port 3.

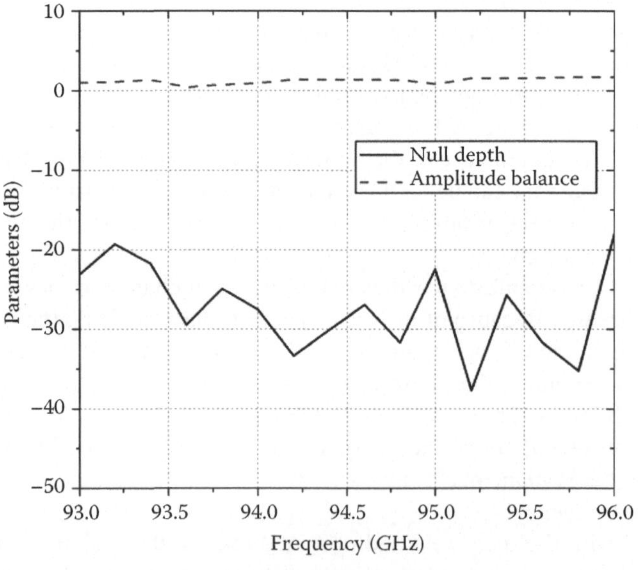

FIGURE 6.26 Measurements of the two-dimensional SIW monopulse antenna array at different frequencies excited at port 3.

fabrication tolerances. Tolerance in couplers leads to the imbalance between feeding signals. The amplitude imbalance also comes from different paths with different lengths. Tolerance in phase shifters is the main reason for the phase imbalance. Considering ±0.05 mm uniform distribution tolerances for a, d, and s, the worst phase error is up to 73.6° for the delay line and 48.2° for the equal-length unequal-width phase shifter, respectively. The antenna performance would be further deteriorated once considering tolerances in substrate such as the thickness and the relative permittivity. The input ports are open-ended in the measurement owing to lack of enough W-band matching loads, which also deteriorate antenna performance and increase SLL.

7 Substrate Integrated Multibeam Antennas

Multibeam systems have been used in many applications, such as communication systems, multiple target radars, etc. In the millimeter-wave frequency band, the power budget is usually costly and difficult to meet the requirement for long-distance applications. Therefore, high-gain, high-efficiency, and potentially multibeam schemes are necessary for millimeter-wave systems. In view of the rapid development of multibeam antenna systems, the beamforming network (BFN) has become a hotspot of research in recent years [187–189].

Substrate integrated waveguide (SIW) technology has been found to be very appropriate to construct high-frequency BFNs and integrated multibeam antenna arrays. In this chapter, recent progress of this emerging technology in applications of BFNs and multibeam antennas is overviewed and introduced. They include the SIW Butler matrix, SIW Blass matrix, SIW Nolen matrix, SIW Rotman lens, SIW modified R-KR lens, SIW reflector lens, and some two-dimensional multibeam antenna arrays.

7.1 SIW CIRCUIT BEAMFORMING TECHNOLOGY

Circuits BFN is a multiple-port network, which is constructed by transmission lines, phase shifters, connecting power dividers, and couplers. On the one hand, different types of fixed phase shifter are employed to provide required phase shifts to produce beam scanning. On the other hand, aperture amplitude distributions are controlled by power division ratios to achieve required radiation patterns. Generally speaking, the beam crossover level of a multibeam antenna fed by a circuit BFN is independent of the operating frequency, but beamwidths and directions are related to the operating frequency [188]. There are two typical circuit BFNs: the Blass matrix and the Butler matrix. Besides, the Nolen matrix is a generalization of both the Blass and Butler matrices.

7.1.1 Substrate Integrated Multibeam Antenna Array Fed by SIW Butler Matrix

The conventional Butler matrix [190] has 2^n input ports and 2^n output ports. It connects to an equal number of radiating elements. As shown in Figure 7.1, a four-element Butler matrix consists of 3 dB and crossing couplers, as well as fixed phase shifters. Integrated with an antenna array, it can generate four beams with different

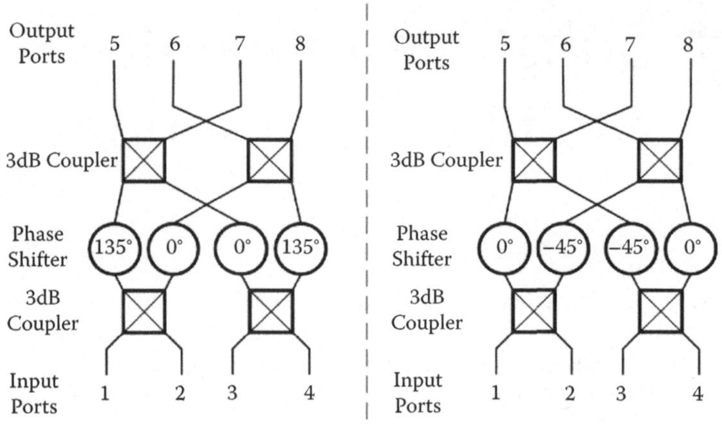

FIGURE 7.1 Topologies of different Butler matrices.

beam pointing, as presented in Figure 7.2. For two topologies with different fixed phase shifting values, the beam directions are just inverted.

The first SIW Butler matrix was designed in 2002 [191]. This multibeam antenna has a cosecant pattern in the vertical plane and a beam scanning ability in the horizontal plane. Four beams are switched to cover a sector of 90° in the horizontal plane. Another 8 × 8 SIW Butler matrix was developed at 25.6 GHz in [192]. Besides, a V-band SIW Butler matrix was fabricated in [193]. Its measured return loss and isolation coefficients are better than 13.5 dB, and the insertion losses are below 2.5 dB for the operating bandwidth from 58 to 62 GHz.

The research of sidelobe level (SLL) reduction of the multibeam antenna fed by the Butler matrix was reported in [194, 195]. These Butler circuits enable one to get low SLL by using four 2-way dividers with unequal power division to obtain a tapered amplitude distribution. Another optimized low SLL SIW Butler matrix was proposed in [196]. For this Ku-band prototype, the length of the substrate integrated multibeam antenna is 329 mm, and the length of the SIW longitudinal slot array antenna is 154 mm. The 4 × 8 Butler matrix, which consists of 3 dB directional couplers, crossing couplers, phase shifters, and power dividers with unequal division ratio, occupies the other 175 mm. The steering beam with low SLL can be generated in the E-plane. The beams generated by the multibeam antenna excited at input ports 1 and 4 have an SLL less than −21.5 dB at 16 GHz, and the beams excited at input ports 2 and 3 have an SLL less than −14.5 dB. The SLL characteristics of this multibeam antenna are better than those of most multibeam antennas fed by conventional Butler matrices. Measured return loss is better than 14 dB from 15.6 to 16.6 GHz.

Moreover, the conventional Butler matrix always has crossover, resulting in problems of additional loss, coupling, and mismatching. To solve this problem, an SIW BFN based on the eight-port hybrid was proposed in [197] at the center frequency of 30 GHz. The eight-port hybrid has been widely used in a monopulse radar system to determine the azimuth and evaluation of a target. It is based on the use of four 90° directional couplers connected together in a symmetrical arrangement. Based on

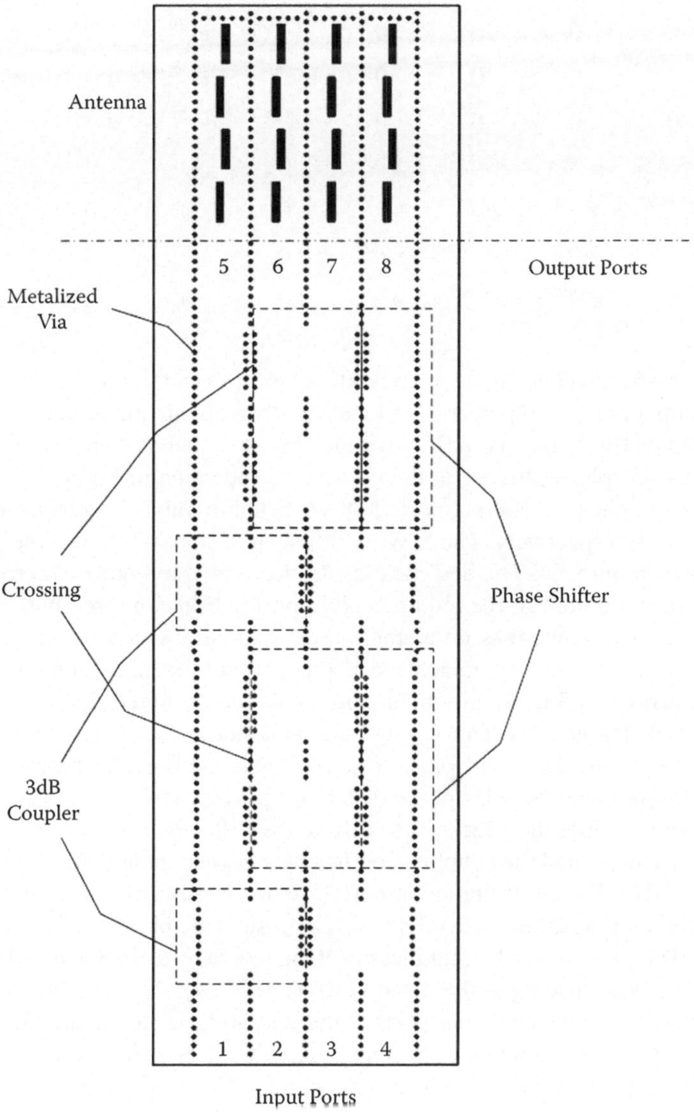

FIGURE 7.2 Configuration of the substrate integrated multibeam antenna array fed by the SIW Butler matrix.

properties of the 90° coupler, the S-matrix of the eight-port hybrid has the following form at an appropriate reference plane:

$$[S]=\frac{1}{2}\begin{bmatrix} 0 & 0 & 0 & 0 & e^{j0} & e^{-j\pi/2} & e^{-j\pi/2} & e^{-j\pi} \\ 0 & 0 & 0 & 0 & e^{-j\pi/2} & e^{-j0} & e^{-j\pi} & e^{-j\pi/2} \\ 0 & 0 & 0 & 0 & e^{-j\pi/2} & e^{-j\pi} & e^{-j0} & e^{-j\pi/2} \\ 0 & 0 & 0 & 0 & e^{-j\pi} & e^{-j\pi/2} & e^{-j\pi/2} & e^{-j0} \\ e^{-j0} & e^{-j\pi/2} & e^{-j\pi/2} & e^{-j\pi} & 0 & 0 & 0 & 0 \\ e^{-j\pi/2} & e^{-j0} & e^{-j\pi} & e^{-j\pi/2} & 0 & 0 & 0 & 0 \\ e^{-j\pi/2} & e^{-j\pi} & e^{-j0} & e^{-j\pi/2} & 0 & 0 & 0 & 0 \\ e^{-j\pi} & e^{-j\pi/2} & e^{-j\pi/2} & e^{-j0} & 0 & 0 & 0 & 0 \end{bmatrix} \tag{7.1}$$

All ports are matched, there is mutual isolation between the sets of four input and four output ports, and power is divided equally from any input port to the output ports. Based on this S-matrix, a BFN without crossover becomes possible only when adding two –45° phase shifters at the corners, as indicated in Figure 7.3. The ideal amplitudes and phases of output ports 5–8 are listed in Table 7.1 when excited from input ports 1–4, respectively. The SIW modified eight-port hybrid has the configuration shown in Figure 7.4. The designed results are shown in Figures 7.5 to 7.7.

As shown in Figure 7.4, the modified eight-port hybrid can directly be integrated with four slot array antennas on a single-layer substrate with a permittivity of 2.2 and a thickness of 0.508 mm to achieve a compact multibeam antenna design. Each linear slot array has four longitudinal slots. It should be noted that these antennas are fed from different sides. This configuration maintains the compact size and symmetry of the circuit, but contributes to a 180° phase difference between adjacent antennas. In this case, the ±45° phase difference provided by the BFN will generate the edge beams, while the ±135° phase difference will generate the center beams.

The return losses and the coupling coefficients between input ports were measured over 28–32 GHz. For an excitation at port 1, return losses and isolations with other ports are greater than 20 dB at 30 GHz. Similar results are observed for excitations at ports 2–4. The patterns of the multibeam antenna were measured at a series of equidistant angles with spacing of 0.1° from –180° to 180° over 29–32 GHz, which do not change much with frequency. For ports 1 and 2 at 30 GHz, the beam directions are 47° and 15°, the 3 dB beamwidths are approximately 32° and 30°, the gains are 13 and 12.2 dBi, and the radiation efficiencies are 44 and 46%, respectively.

7.1.2 Substrate Integrated Multibeam Antenna Array Fed by Miniaturized SIW Butler Matrix

7.1.2.1 SIW Multilayered Butler Matrix

Although the SIW Butler matrix possesses a number of advantages, its practical use is still limited considering its relatively large size. Some researchers began to use the

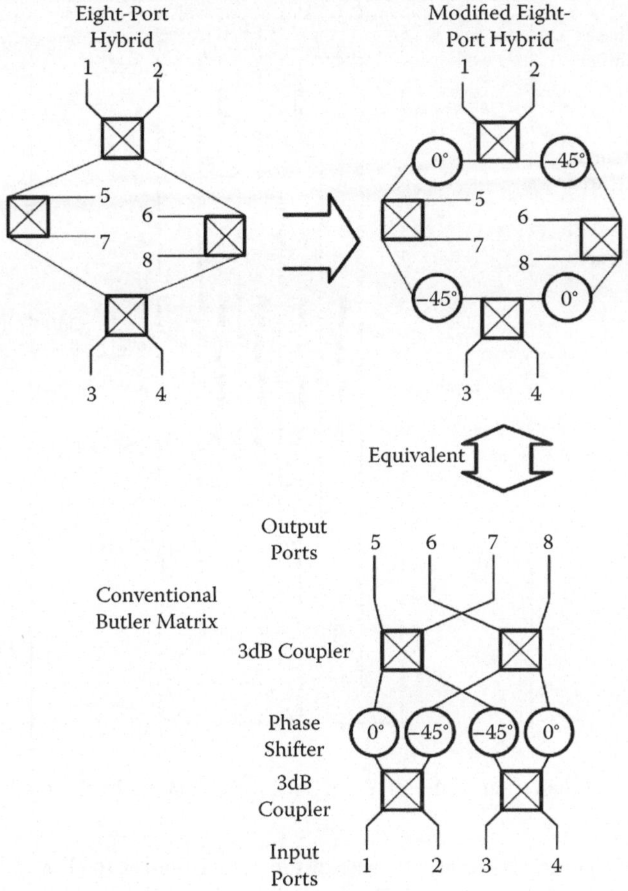

FIGURE 7.3 Topologies of the conventional Butler matrix and the modified eight-port hybrid.

TABLE 7.1
Ideal Outputs of the Modified Eight-Port Hybrid

		5	6	7	8
1	Amplitude	0.5	0.5	0.5	0.5
	Phase	0°	–45°	–90°	–135°
2	Amplitude	0.5	0.5	0.5	0.5
	Phase	–90°	45°	180°	–45°
3	Amplitude	0.5	0.5	0.5	0.5
	Phase	–45°	180°	45°	–90°
4	Amplitude	0.5	0.5	0.5	0.5
	Phase	–135°	–90°	–45°	0°

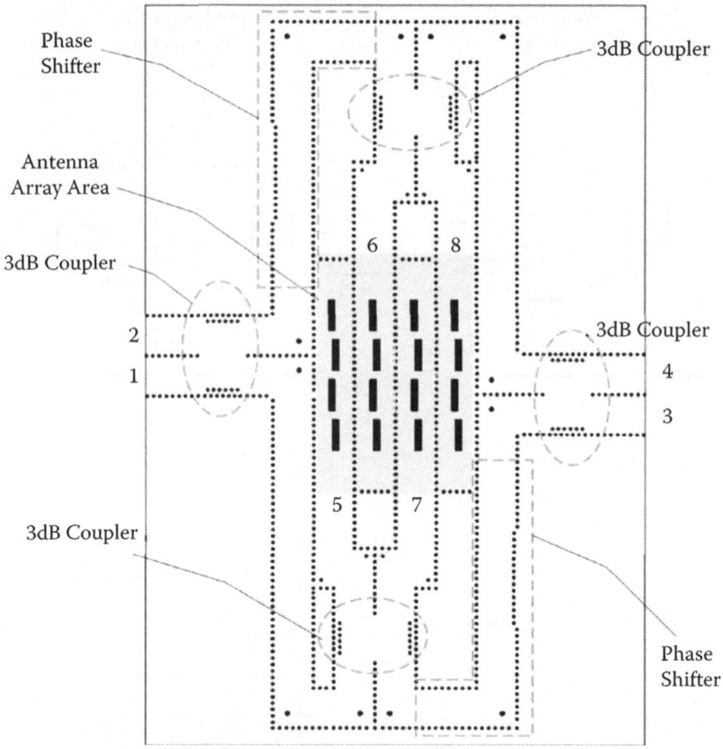

FIGURE 7.4 Configuration of the SIW multibeam antenna fed by the modified eight-port without crossover.

multilayer technology to reduce the size of the Butler matrix in [198–200]. However, it should be noted that the used BFN and the radiator array are placed side by side in a two-dimensional plane taking up a large area. In this situation, there is still a possibility of reducing the total size of the substrate integrated multibeam antenna through a three-dimensional approach. Based on low-temperature cofired ceramic (LTCC) technology, a substrate integrated multibeam array antenna fed by a folded Butler matrix was designed in [201].

First, a 4 × 4 Butler matrix was folded several times to miniaturize the circuit area and developed as shown in Figure 7.8. As revealed in Figure 7.9, the proposed folded Butler matrix, derived from a more appropriate Butler matrix topology, consists of H/E-plane couplers, layer-to-layer transitions and phase shifters to realize the optimized architecture, which suppresses crossover and improves the overall performance. This 4 × 4 Butler matrix has nine stacked ceramic tapes. Three ceramic tapes are laminated as an SIW. Four input ports are distributed in the bottom layer and the middle layer, respectively. Four output ports are all placed in the top layer. The first two couplers are of the E-plane type to conveniently connect two input ports placed in different layers while dividing the power into two equal parts. Then, two H-plane couplers are overlapped at adjacent layers, and their input ports are connected to the output ports of two E-plane couplers. The outputs of the bottom H-plane coupler

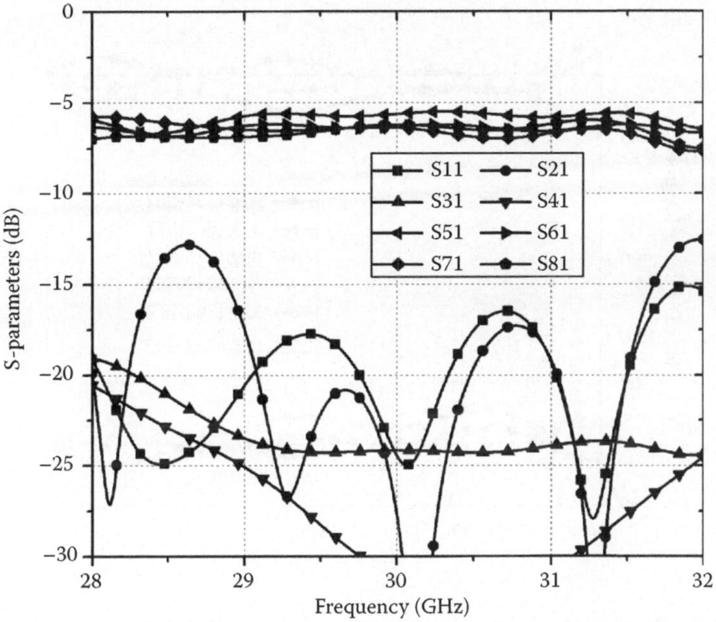

FIGURE 7.5 S-parameters of the SIW modified eight-port hybrid excited at port 1.

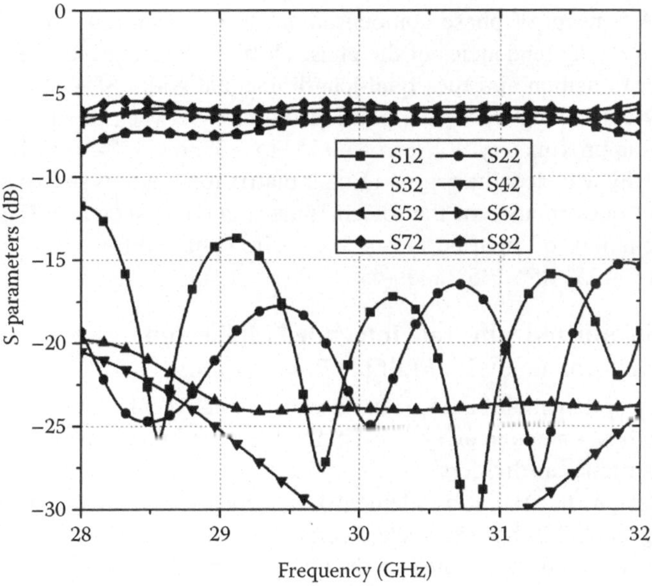

FIGURE 7.6 S-parameters of the SIW modified eight-port hybrid excited at port 2.

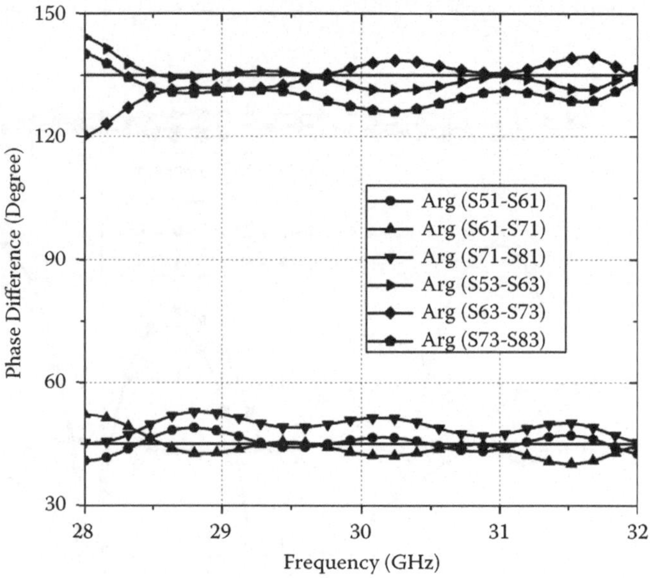

FIGURE 7.7 Phase difference between output ports of the SIW modified eight-port hybrid.

are transferred to the top layer through 1-to-3 transitions, while the outputs of the H-plane coupler in the middle layer are transferred to the top layer through 1-to-2 transitions. The distance between two neighboring output ports is $0.58\lambda_0$ @ 60 GHz.

The fixed phase shifters are positioned properly to achieve the desired phase characteristics. A concept of phase compensation can be employed here, based on the fact that the varying tendencies of the phase shift versus frequency obtained by the layer-to-layer transition and the equal-length unequal-width SIW structure are just opposite. By combining them, the phase difference between the 1-to-2 and 1-to-3 transitions is approximately 90° (error −0.75° to 4.3°) over 50–70 GHz. This is the very reason why we select the original Butler matrix topology, as shown in Figure 7.9. The excellent performance of the folded Butler matrix is shown in Figures 7.10 to 7.12. For excitations of input ports 1–4, the desired phase differences between adjacent ports are −135°, 45°, 135°, and −45°.

7.1.2.2 Miniaturized Substrate Integrated Multibeam Array Antenna with Multiple Polarizations

Then, an integrated multibeam antenna can be realized by the proposed Butler matrix and a 4 × 4 antenna array consisting of six-layer tapes used as radiators and other six-layer tapes as dividers.

In this design, the radiating element is an open-ended square SIW cavity, as shown in Figure 7.13. It has six stacked ceramic tapes and is fed by an SIW feedline through a 45° inclined slot located on the bottom conductor cladding of the cavity in order to generate a 45° linear polarization (LP) beam. A square radiating aperture is cut on the top conductor cladding.

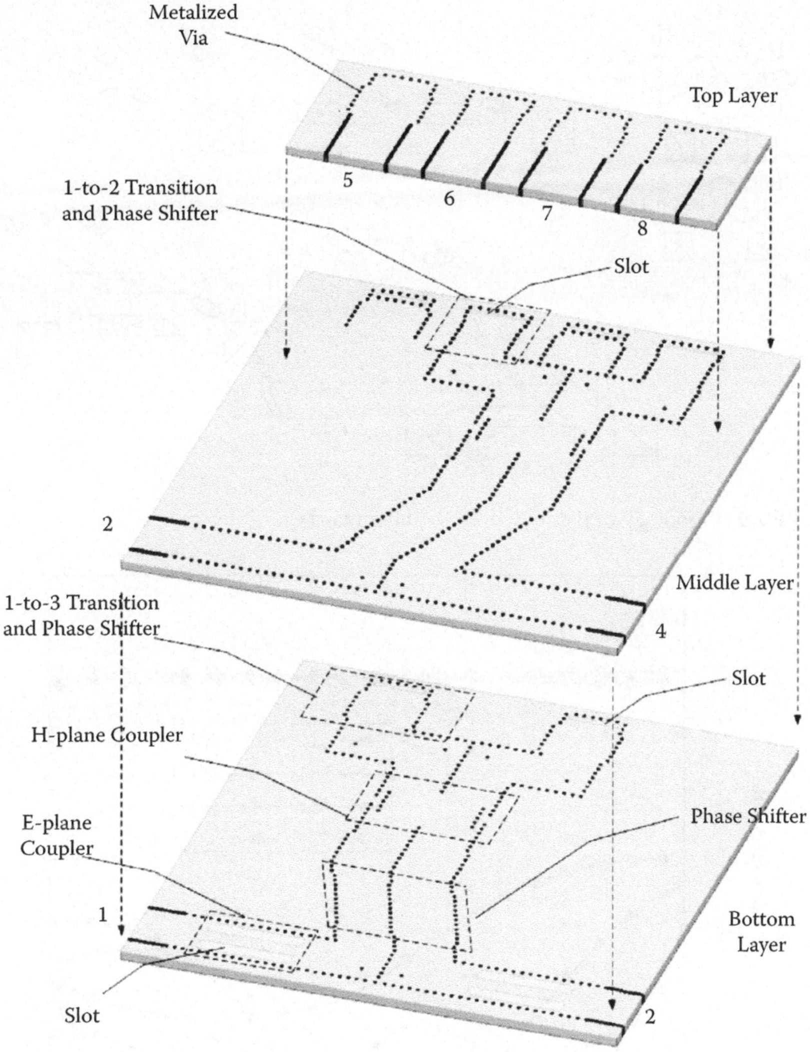

FIGURE 7.8 Configuration of the folded 4 × 4 SIW Butler matrix.

An E-plane divider is used to excite a four-element linear array vertically in a multilayer configuration. Four such linear arrays can be grouped as a 4 × 4 45° LP array antenna. Two identical folded SIW Butler matrices described above are placed back-to-back to excite two 4 × 4 antenna arrays with 45° and −45° polarizations, and then to construct a dual LP multibeam array antenna. The used ceramic material is Ferro A6-M with a relative permittivity of 5.9 and a loss tangent of 0.002. It has 21 stacked ceramic tapes and is excited by coplanar waveguides (CPWs). The total circuit area is 32.7 × 14.6 mm^2, excluding the additional feeding structures. The feeding Butler matrix is completely hidden underneath the radiators. The total size of the proposed multibeam array antenna is only equal to that of the radiating aperture, leading to a

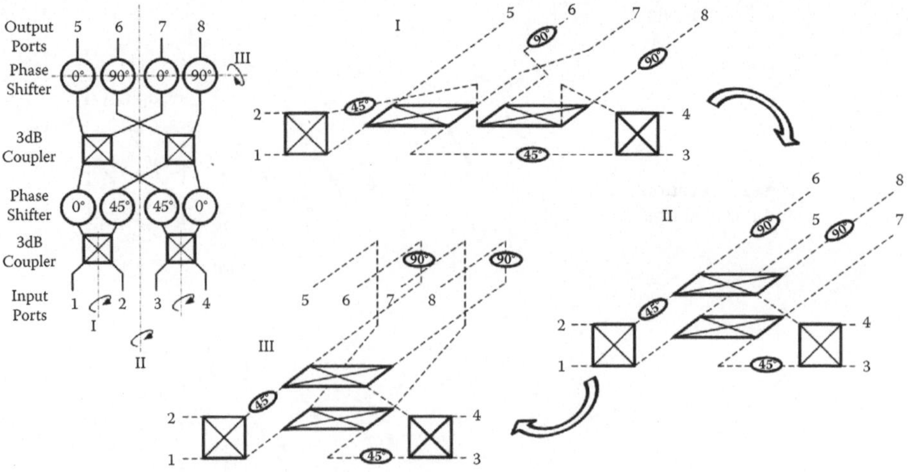

FIGURE 7.9 Topology of the folded 4 × 4 Butler matrix.

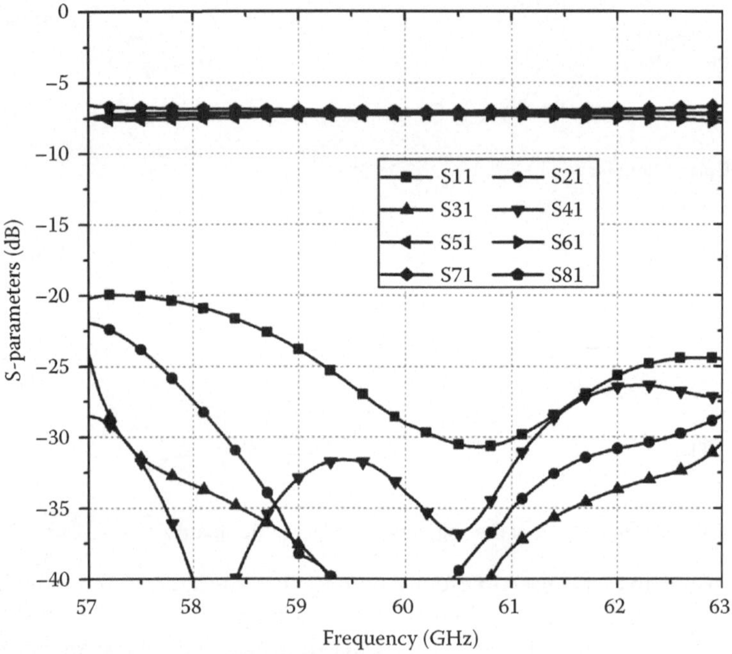

FIGURE 7.10 S-parameters of the folded 4 × 4 SIW Butler matrix excited at port 1.

remarkable miniaturization. Reflection and coupling coefficients are below –10 dB for all ports within the whole tested frequency band from 55 to 65 GHz. Measured gains at beam directions are 12.4, 13.4, and 13.5 dBi at 58, 60, and 62 GHz excited at port 1, while 14.9, 15.3, and 15.7 dBi at 58, 60, and 62 GHz excited at port 2. The overall loss of the fabricated miniaturized multibeam antenna is approximately 3 dB.

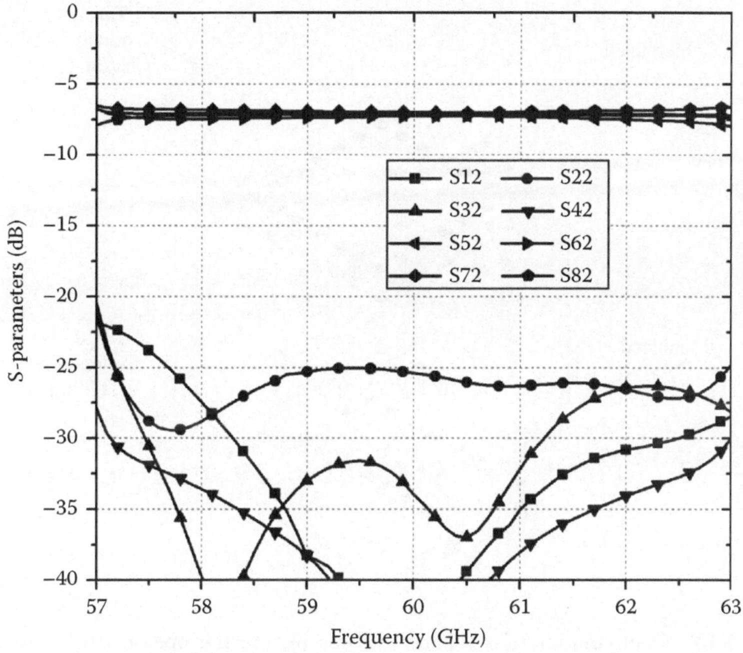

FIGURE 7.11 S-parameters of the folded 4 × 4 SIW Butler matrix excited at port 2.

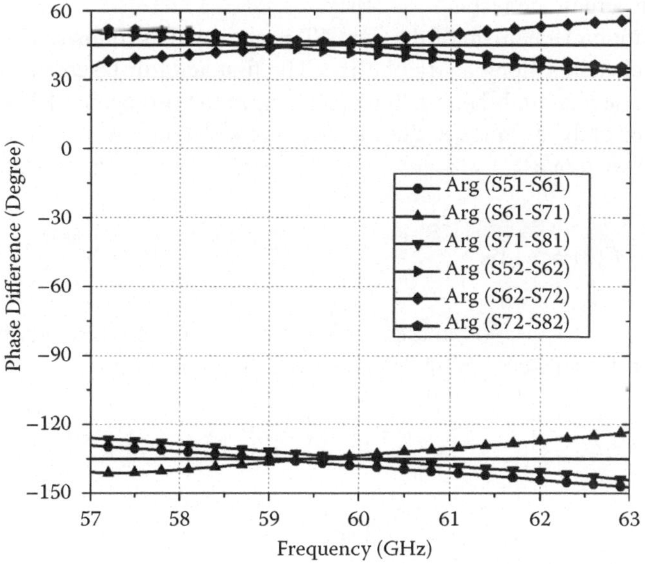

FIGURE 7.12 Phase difference between output ports of the folded 4 × 4 SIW Butler matrix.

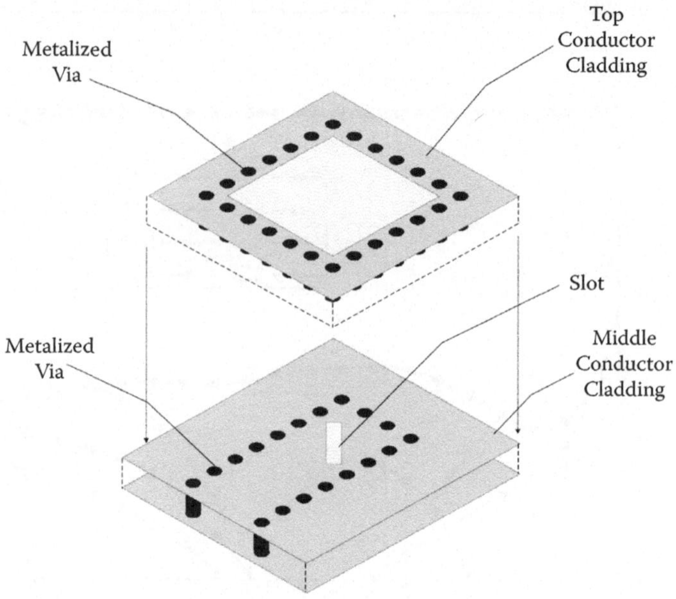

FIGURE 7.13 Configuration of a V-band SIW-fed rectangular open-ended cavity antenna with 45° LP.

Furthermore, two 1 × 4 45° and two 1 × 4 –45° LP arrays are arranged alternately to construct a 4 × 4 circular polarization (CP) array. Such a CP array antenna can generate two orthogonal CP beams, i.e., LHCP and RHCP beams, pointing at different directions simultaneously.

The operation mechanism of the dual CP is analyzed as follows. The model consists of four elements along a straight line. The first and third elements have the 45° LP, while the second and the fourth elements have the orthogonal LP. Elements are equally spaced at the distance *d*. They are excited with constant amplitude. The array factor can be written as

$$\vec{F}(\theta) = \hat{\alpha} + \hat{\beta} e^{j\left(\frac{2\pi d}{\lambda}\sin\theta - \varphi_1\right)} + \hat{\alpha} e^{j\left(\frac{4\pi d}{\lambda}\sin\theta - \varphi_2\right)} + \hat{\beta} e^{j\left(\frac{6\pi d}{\lambda}\sin\theta - \varphi_3\right)} \tag{7.2}$$

When $\varphi_1 = -45°$, $\varphi_2 = 90°$, and $\varphi_3 = 45°$, which can be realized only by using two –45° phase shifters instead of 45° phase shifters in Figure 7.9,

$$\begin{aligned}\vec{F}(\theta) &= \hat{\alpha}\left[1 + e^{j\left(\frac{4\pi d}{\lambda}\sin\theta - \frac{\pi}{2}\right)}\right] + \hat{\beta}\left[e^{j\left(\frac{2\pi d}{\lambda}\sin\theta + \frac{\pi}{4}\right)} + e^{j\left(\frac{6\pi d}{\lambda}\sin\theta - \frac{\pi}{4}\right)}\right] \\ &= \left[1 + e^{j\left(\frac{4\pi d}{\lambda}\sin\theta - \frac{\pi}{2}\right)}\right]\left[\hat{\alpha} + e^{j\left(\frac{2\pi d}{\lambda}\sin\theta + \frac{\pi}{4}\right)}\hat{\beta}\right]\end{aligned} \tag{7.3}$$

In this case, the main beam direction, θ, of this array can be calculated by

$$\frac{4\pi d}{\lambda}\sin\theta - \frac{\pi}{2} = 0 \tag{7.4}$$

or

$$\frac{4\pi d}{\lambda}\sin\theta - \frac{\pi}{2} = -2\pi \tag{7.5}$$

Considering $d = 0.58\lambda_0$, θ can be calculated at 12° or −40°. In the beam direction of 12°, Equation (7.3) becomes

$$\vec{F}(\theta) = 2[\hat{\alpha} + j\hat{\beta}] \tag{7.6}$$

That is a LHCP beam. At the beam pointing of −40°, Equation (7.3) becomes

$$\vec{F}(\theta) = 2[\hat{\alpha} - j\hat{\beta}] \tag{7.7}$$

That is an RHCP beam. The theoretical analysis is the same with simulated results, as shown in Figures 7.14 and 7.15.

When a modified folded SIW Butler matrix is employed to feed the 4 × 4 CP array, each port supports two operation modes with different polarizations and beam

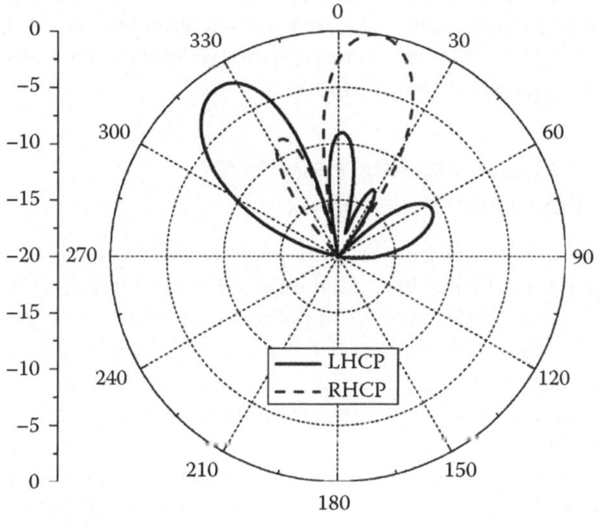

FIGURE 7.14 Simulated patterns of the 4 × 4 dual CP multibeam antenna (phase for ports 1–4: 0°, 45°, 270°, and 315°).

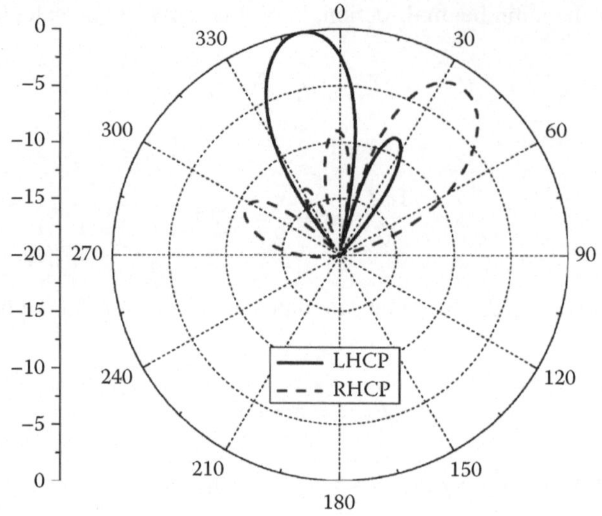

FIGURE 7.15 Simulated patterns of the 4 × 4 dual CP multibeam antenna (phase for ports 1–4: 0°, –45°, 90°, and 45°).

directions. When the polarization mode is selected, the beam pointing for each port is determined. The total circuit area of the fabricated prototype is 16.5 × 14.6 mm^2, excluding the additional CPW feedlines. This size is approximately equal to that of a 4 × 4 array antenna. For an excitation at port 3, the axial ratios in the beam direction are 2.3 and 2.62 dB for 60 GHz LHCP and RHCP beams, respectively. A similar performance can be found excited at port 1. Polarization purity is deteriorated when excited at ports 2 and 4. There exists an offset between the main beam direction and the beam direction with the best axial ratio.

7.1.3 Substrate Integrated Multibeam Antenna Array Fed by SIW Blass Matrix

The Blass matrix consists of a set of array port lines intersecting a set of beam port lines. A directional coupler is located at each intersection [202]. This matrix is far more flexible than the Butler matrix. In addition, it is usually more lossy considering the presence of line terminations. Because of its inherent loss, the Blass matrix has limited application compared to the Butler matrix. However, it has a particular advantage: the capability of generating arbitrary beams at arbitrary positions with arbitrary shapes [203].

A Ku-band double-layer 4 × 16 SIW Blass matrix was proposed in [204]. This matrix is used to feed 16 SIW slot antennas. A broad-wall to broad-wall slot coupler was employed to achieve a 26–5.52 dB coupling factor range. Owing to the series feeding architecture, the Nolen matrix has frequency-dependent phase behavior, which makes the phase difference between neighboring outputs vary with the operating frequency. It is bad news for wideband applications. However, this gradient for each output can be employed to weaken the beam squint effect.

7.1.4 Substrate Integrated Multibeam Antenna Array Fed by SIW Nolen Matrix

The Nolen matrix [205] is a special case of the Blass matrix without the termination loads. Similar to the Butler matrix, this lossless characteristic of the Nolen matrix imposes the excitation laws to be orthogonal.

In [206], a conventional 4 × 4 SIW Nolen matrix was designed at the center frequency of 12.5 GHz with a configuration as shown in Figure 7.16. It was fabricated on a substrate with a relative permittivity of 2.33. The entire circuit size is 137 × 105 mm^2 without the transitions. SIW H-plane cruciform directional couplers of 3, 4.77, and 6.02 dB are selected, as well as phase shifters, to construct a 4 × 4 Nolen matrix. Measured power-split unbalance stays within 0.5 dB over the frequency band from 12.25 to 12.75 GHz. Similar to the Blass matrix, the Nolen matrix also introduces significant phase dispersion in its standard form, which still needs to be improved and optimized.

To overcome this weakness, a wideband Nolen matrix in the SIW technology was investigated in [207]. In this work, adequate coupler delay compensation is employed

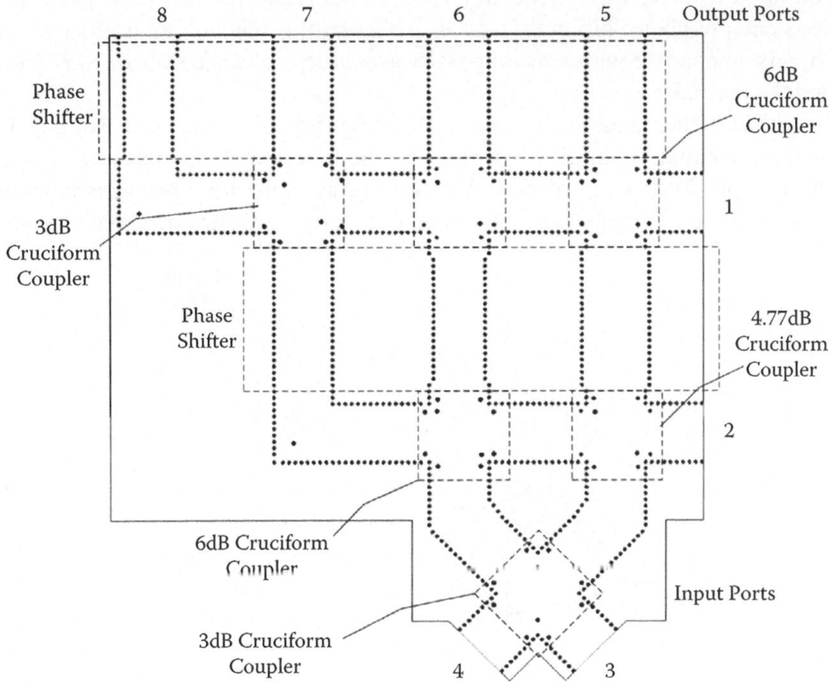

FIGURE 7.16 Configuration of the SIW Nolen matrix.

to achieve a wide operating frequency band. H-plane couplers with continuous aperture are used in this design instead of the cross-couplers previously used in the conventional serial configuration. That means a parallel matrix topology is obtained. This 4 × 4 Nolen matrix was designed at 77 GHz. It is able to achieve excellent measured results over an 11.7% relative bandwidth.

7.2 SIW LENS BEAMFORMING TECHNOLOGY

Circuit BFN technology has the disadvantages of bulky and prohibitive flexibility, especially for applications with a number of beams. Compared with circuit BFN counterparts, lens-based BFNs are easier for generating tens of beams with simple configuration and high efficiency.

7.2.1 Substrate Integrated Multibeam Antenna Array Fed by SIW Rotman Lens

The most famous lens-based BFN is the Rotman lens [208]. In [209], a Rotman lens was developed in SIW technology. A similar design is described in [210]. The original Rotman lens operates with the TEM mode and exhibits a relatively wide band. However, losses and small physical dimensions complicate a TEM lens implementation at the millimeter-wave frequency band. A non-TEM parallel plate propagation mode is one possible solution [211–213]. Although this technology limits the bandwidth, it is compatible with a rectangular waveguide, dielectric slab, or SIW feeders around the lens edge.

Figure 7.17 illustrates the geometrical configuration of a typical Rotman lens. It has three contours with different shapes. The first one is a circular focal arc, on which multiple feeders are placed. A pickup array along the inner lens contour is defined by the set of points $P(x, y)$ relating to the origin O. The antenna elements are

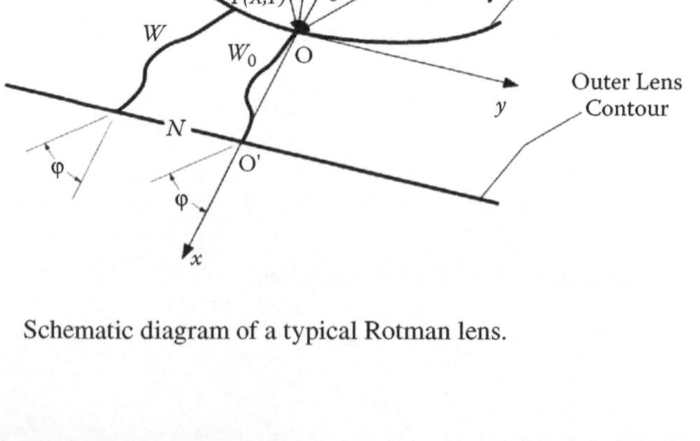

FIGURE 7.17 Schematic diagram of a typical Rotman lens.

placed on the straight outer lens contour. The position is determined by the point N relative to the horizontal axis O'.

To determine the lens periphery, four independent conditions are set on the lens, including a straight front face, an on-axis focal point G_0, and two symmetrical off-axis focal points F_1 and F_2. They are for radiation at angles of 0°, α, and $-\alpha$, respectively. Therefore, the coordinates of G_0, F_1, and F_2 are $(-G, 0)$, $(-F\cos\alpha, F\sin\alpha)$, and $(-F\cos\alpha, -F\sin\alpha)$, respectively.

To simplify analysis, a set of parameters are normalized to F:

$$\eta=\frac{N}{F},\quad x=\frac{X}{F},\quad y=\frac{Y}{F},\quad g=\frac{G}{F},\quad \omega=\frac{W-W_0}{F}$$

Also,

$$a_0=\cos\alpha,\quad b_0=\sin\alpha,\quad a_1=\cos\varphi,\quad b_1=\sin\varphi$$

Considering the SIW is a fully enclosed structure, there exists only one relative permittivity, namely, the permittivity of the used dielectric substrate, without any other permittivity appearing in the former equations as described in [214, 215]. As such, electrical constraints should be modified as

$$(\overline{F_1P})+W+\frac{N\sin\varphi}{\sqrt{\varepsilon_r}}=F+W_0 \tag{7.8}$$

$$(\overline{F_2P})+W-\frac{N\sin\varphi}{\sqrt{\varepsilon_r}}=F+W_0 \tag{7.9}$$

$$(\overline{G_1P})+W=G+W_0 \tag{7.10}$$

Geometrical constraints should be

$$(\overline{F_1P})^2=F^2+X^2+Y^2+2FX\cos\alpha-2FY\sin\alpha \tag{7.11}$$

$$(\overline{F_2P})^2=F^2+X^2+Y^2+2FX\cos\alpha+2FY\sin\alpha \tag{7.12}$$

$$(\overline{G_1P})^2=(G+X)^2+Y^2 \tag{7.13}$$

For given parameters of α, g, and the scanning angle φ, the lens periphery can be calculated by

$$y=\frac{b_1}{b_0\sqrt{\varepsilon_r}}\eta(1-\omega) \tag{7.14}$$

$$x^2 + y^2 + 2gx = \omega^2 - 2g\omega \tag{7.15}$$

$$a\omega^2 + b\omega + c = 0 \tag{7.16}$$

In Equation (7.16),

$$a = 1 - \frac{\eta^2}{\sqrt{\varepsilon_r}}\left(\frac{b_1}{b_0}\right)^2 - \left(\frac{g-1}{g-a_0}\right)^2 \tag{7.17}$$

$$b = 2g\frac{g-1}{g-a_0} - \frac{b_1^2\eta^2(g-1)}{(g-a_0)^2\varepsilon_r} + \frac{2}{\varepsilon_r}\left(\frac{\eta b_1}{b_0}\right)^2 - 2g \tag{7.18}$$

$$c = \frac{(b_1\eta)^2 g}{(g-a_0)\varepsilon_r} - \frac{(b_1\eta)^4}{4\varepsilon_r^2(g-a_0)^2} - \left(\frac{b_1}{b_0}\right)^2\frac{\eta^2}{\varepsilon_r} \tag{7.19}$$

For a given set of α, φ, and g, the required ω can be calculated at different η according to Equation (7.16). These values of ω and η are then substituted into Equations (7.14) and (7.15) to evaluate x and y. Now, the design of the SIW Rotman lens is finished.

As an example, a prototype was designed to operate at a center frequency of 28.5 GHz, as shown in Figure 7.18. This SIW lens was fabricated on Rogers 5880 substrate with a thickness of 0.508 mm and a permittivity of 2.2. Other parameters are $\alpha = 30°$, $\varphi = 30°$, $R = 19$ mm, and $G = 32$ mm. The width of the SIW is 5 mm, and the diameter of the metalized via is 0.4 mm. This structure has seven input ports and nine output ports. Dummy ports terminated with 50 Ω loads can be employed to optimize the performance.

Because the TEM mode is not supported in the SIW, the TE_{10} mode has to be selected to feed the BFN. As is well known, the non-TEM mode has transverse or longitudinal components. In the SIW structure, the TE_{10} mode has not only space dependency, but also frequency dependency. The latter results in the beam direction varying with the operating frequency. When the lens is excited in low-order TE modes, this beam deviation is not very obvious.

In the original Rotman lens, some cables with different electric lengths were used to connect the ports distributed on the inner lens contour to the ports on the outer lens contour. For integration, a moderate bandwidth SIW phase shifting network was used to replace the initial fixed cables. The required phase shift can be realized by variable SIW width.

The developed 7 × 9 SIW Rotman lens was employed to feed nine linear SIW slot arrays. Each one has four longitudinal slots. At 28.5 GHz, measured beam directions are −41°, −28°, −14°, and 0°, and gains are 13.8, 17.0, 17.8, and 18.5 dBi corresponding to the input ports 1–4, respectively. Comparing the signal levels from these beams, the optimal beam can be selected. As shown in Figure 7.19, seven E-plane beams are able to cover 95° with their 5 dB beamwidths. For the planar configuration, this

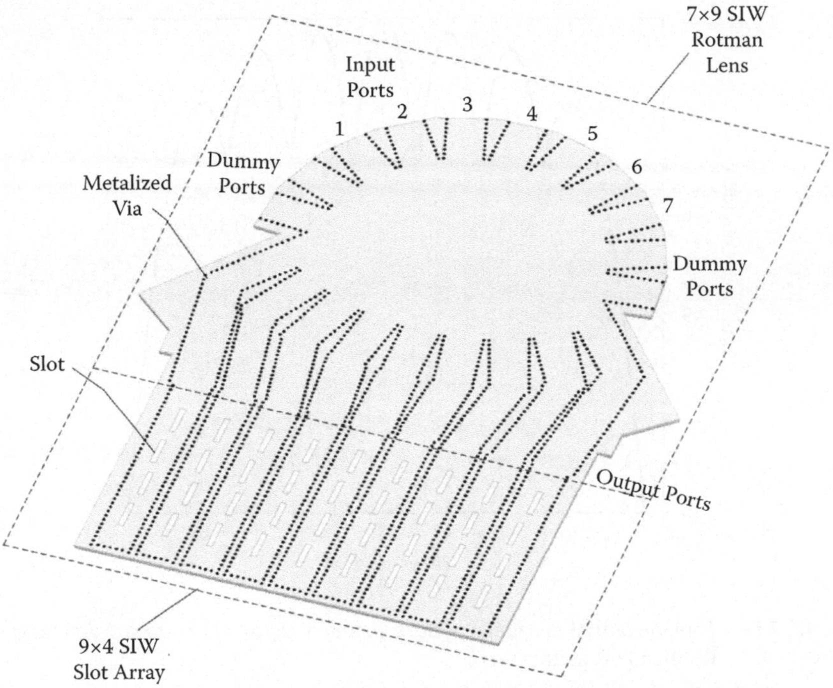

FIGURE 7.18 Configuration of the substrate integrated multibeam antenna array fed by the SIW Rotman lens.

substrate integrated multibeam antenna can be easily conformal to the curved platforms or surfaces without notable deterioration of performances.

7.2.2 Substrate Integrated Multibeam Antenna Array Fed by SIW Modified R-KR Lens

A typical R-KR lens [187] can generate tens or even hundreds of beams for large angle scanning. Such a lens can provide perfect focusing for all feeding points.

The traditional R-KR lens has several feeding ports placed on the perimeter of the lens fringe, as shown in Figure 7.20. The radius of the lens is *KR*. On the opposite side of the lens, several picking ports are also distributed on the inner lens contour. Each picking port is connected to the circular radiating array placed on the outer lens contour with a radius of *R* by equal-length cables. Moving the feeding point along the perimeter of the lens fringe, the beam is scanned. It has the following relationships [217, 218]:

$$d = R(1 - \cos\alpha) \tag{7.20}$$

$$d_1 = 2KR \tag{7.21}$$

$$d_2 = 2KR\cos\alpha \tag{7.22}$$

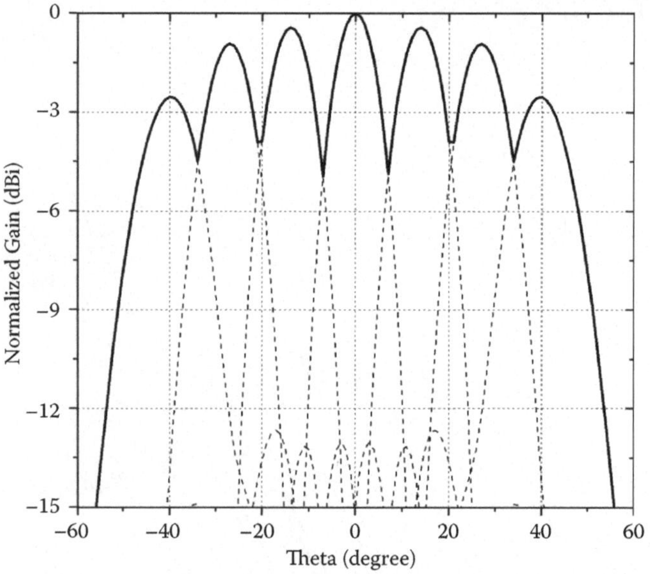

FIGURE 7.19 E-plane radiation pattern of the substrate integrated multibeam antenna array fed by the SIW Rotman lens at 28.5 GHz.

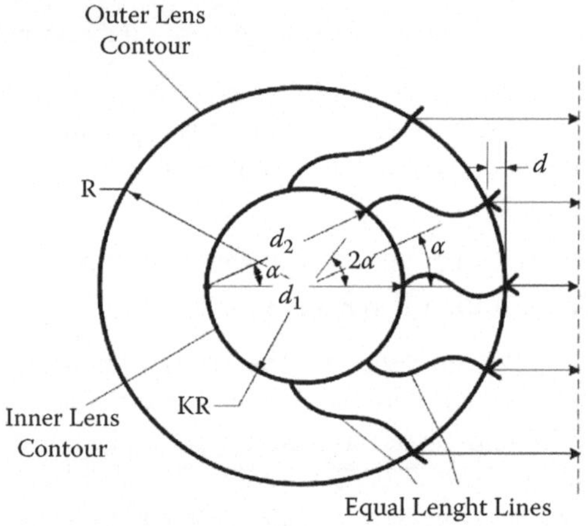

FIGURE 7.20 Schematic diagram of the conventional R-KR lens.

If $d + d_2 = d_1$, the perfect focusing is provided. For an air-filled lens, K is larger than 1. That means the diameter of the outer lens contour is smaller than that of the inner lens contour. The lens is not inside the radiating array unless a material with a relative permittivity larger than 4 is filled in the lens.

In [216], a modified R-KR lens was developed to excite the linear array, as shown in Figure 7.21. The feeding ports and picking ports are distributed on the inner contour of the lens. These internal ports are connected to the input ports and output

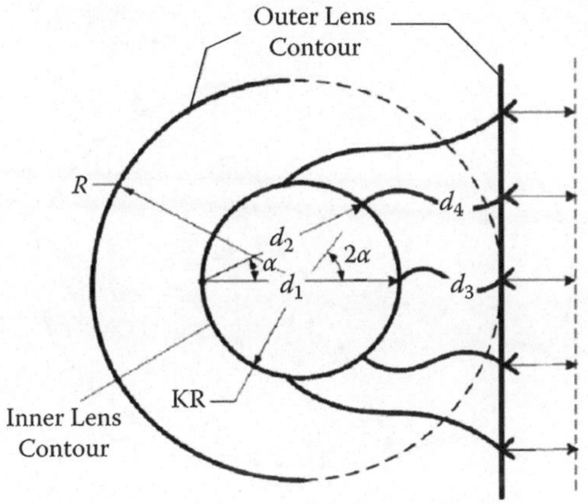

FIGURE 7.21 Schematic diagram of the modified R-KR lens.

ports, respectively. The input ports are distributed on the outer contour of the lens. The output ports are placed in line and connected to radiating elements.

If

$$d_1 + d_3 = d_2 + d_4 \tag{7.23}$$

this proposed lens will perform similar to the conventional one. Several sets of desired phase coefficients can be provided. To satisfy Equation (7.23), additional phase shifters need to be employed.

This lens has only one perfect focus, while the classical Rotman lens has three focal points. However, the design of this R-KR lens is much simpler and easier to model and optimize than its Rotman lens counterpart. Therefore, this modified R-KR lens is an attractive BFN without complicated calculation and arrangement.

The feeding port is distributed diametrically opposite to the picking port. To simplify the problem, the *i*th feeding port or *j*th picking port is supposed to be located on the inner contour of the lens with the central angle of $2i\alpha$ or $2j\alpha$, as shown in Figure 7.22. l_{ji} is the path length from feeding port i to picking port j.

When the lens is excited at center feeding port 0,

$$l_{00} = l_{10} + \delta_1 = l_{20} + \delta_2 = \ldots = l_{j0} + \delta_j \tag{7.24}$$

In Equation (7.24), δ_1, δ_2, … are path lengths implemented by the phase shifters.

$$\begin{aligned}
\delta_1 &= l_{00} - l_{10} = 2KR - 2KR\cos\alpha \\
\delta_2 &= l_{00} - l_{20} = 2KR - 2KR\cos 2\alpha \\
&\ldots \\
\delta_j &= l_{00} - l_{j0} = 2KR - 2KR\cos j\alpha
\end{aligned} \tag{7.25}$$

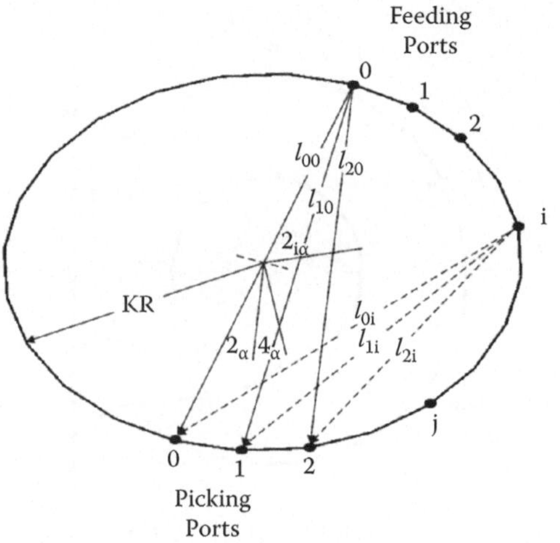

FIGURE 7.22 Geometric relationship of the modified R-KR lens.

When the lens is excited at the feeding port i, the path length differences Δl_j between adjacent paths can be expressed as

$$\begin{aligned}
\Delta l_1 &= l_{0i} - (l_{1i} + \delta_1) \\
&= 2KR\cos i\alpha - 2KR\cos(i+1)\alpha - 2KR + 2KR\cos\alpha \\
\Delta l_2 &= (l_{1i} + \delta_1) - (l_{2i} + \delta_2) \\
&= 2KR\cos(i+1)\alpha - 2KR\cos(i+2)\alpha - 2KR\cos\alpha + 2KR\cos 2\alpha \\
&\cdots \\
\Delta l_j &= (l_{(j-1)i} + \delta_{j-1}) - (l_{ji} + \delta_j) \\
&= 2KR\cos(i+j-1)\alpha - 2KR\cos(i+j)\alpha - 2KR\cos(j-1)\alpha \\
&\quad + 2KR\cos j\alpha
\end{aligned} \tag{7.26}$$

The relative phase differences between adjacent picking ports can be calculated as follows:

$$\begin{aligned}
\Delta\xi_1 &= \frac{4\pi KR}{\lambda}\left[\cos i\alpha - \cos(i+1)\alpha - 1 + \cos\alpha\right] \\
\Delta\xi_2 &= \frac{4\pi KR}{\lambda}\left[\cos(i+1)\alpha - \cos(i+2)\alpha - \cos\alpha + \cos 2\alpha\right] \\
&\cdots \\
\Delta\xi_j &= \frac{4\pi KR}{\lambda}\begin{bmatrix}\cos(i+j-1)\alpha - \cos(i+j)\alpha - \cos(j-1)\alpha \\ +2KR\cos j\alpha\end{bmatrix}
\end{aligned} \tag{7.27}$$

In Equation (7.27), λ denotes the wavelength in the substrate. If $\Delta\xi_0 = \Delta\xi_1 = \ldots$, constant phase coefficients will be produced between adjacent elements and make the beam scan. However, phase error $(\Delta\xi_1 - \Delta\xi_j)$ is zero only at $i\alpha = 0$ or $i = 0$. That means this lens has only a single focus. When the feeding port has a width of pw the same as that of the picking port, Equation (7.27) can be rewritten as

$$\begin{aligned}
\Delta\xi_1 &= \frac{4\pi KR}{\lambda}\begin{bmatrix}\cos\left(i\frac{pw}{2KR}\right)-\cos\left((i+1)\frac{pw}{2KR}\right)\\ -1+\cos\left(\frac{pw}{2KR}\right)\end{bmatrix}\\
\Delta\xi_2 &= \frac{4\pi KR}{\lambda}\begin{bmatrix}\cos\left((i+1)\frac{pw}{2KR}\right)-\cos\left((i+2)\frac{pw}{2KR}\right)\\ -\cos\left(\frac{pw}{2KR}\right)+\cos\left(\frac{2pw}{2KR}\right)\end{bmatrix}\\
&\cdots\\
\Delta\xi_j &= \frac{4\pi KR}{\lambda}\begin{bmatrix}\cos\left((i+j-1)\frac{pw}{2KR}\right)-\cos\left((i+j)\frac{pw}{2KR}\right)\\ -\cos\left((j-1)\frac{pw}{2KR}\right)+\cos\left(j\frac{pw}{2KR}\right)\end{bmatrix}
\end{aligned} \tag{7.28}$$

Supposing $pw/\lambda = 1.2$, the curves of phase errors as the function of i and j are presented in Figures 7.23 to 7.25. In these figures, KR/λ equals 8, 10, and 12.5,

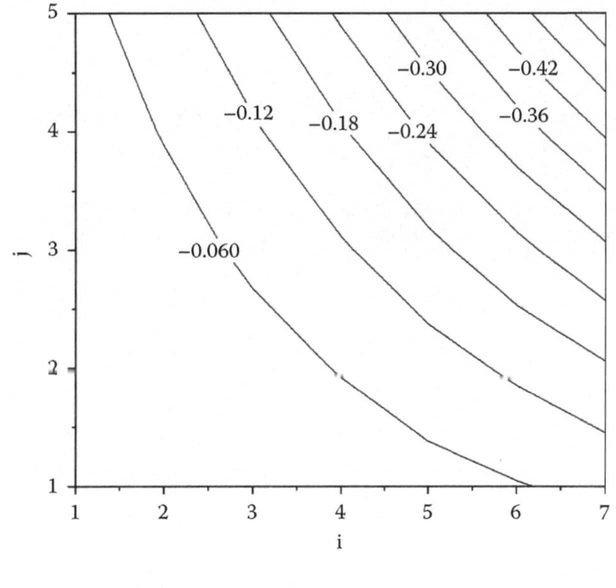

FIGURE 7.23 Phase errors of the modified R-KR lens versus i and j with $KR/\lambda = 8$.

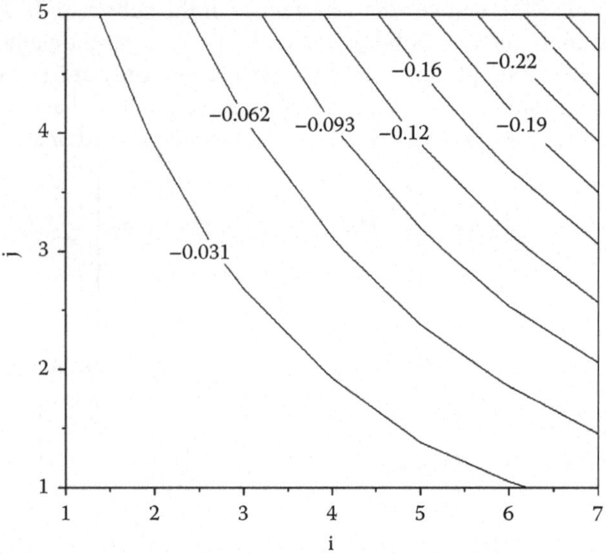

FIGURE 7.24 Phase errors of the modified R-KR lens versus *i* and *j* with $KR/\lambda = 10$.

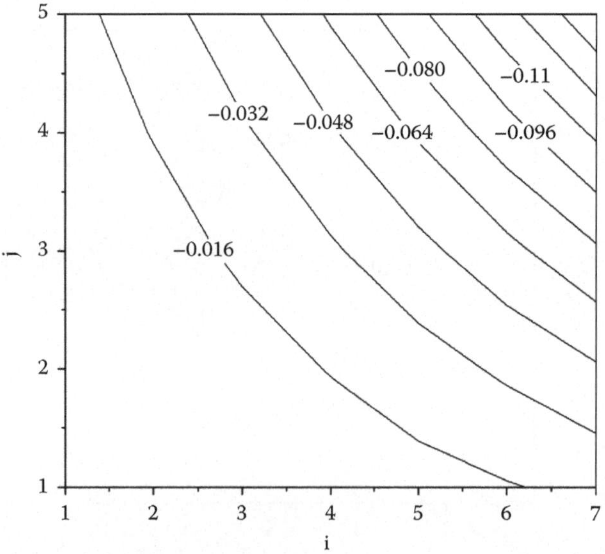

FIGURE 7.25 Phase errors of the modified R-KR lens versus *i* and *j* with $KR/\lambda = 12.5$.

respectively. In general, phase error increases when j or i is increased. Besides, a larger lens has a smaller phase error.

The phase error results in the deterioration of radiation characteristics, such as high SLL and decreasing gain. Thus, all of the phase errors should not exceed the allowable error ε_{max} to guarantee reliable performance. In this case, the minimum KR_{min} can be calculated by

$$\begin{aligned} \varepsilon_{max} &= \frac{4\pi KR_{min}}{\lambda}[\cos\alpha - \cos(i+1)\alpha - 1 + \cos i\alpha] \\ &- \frac{4\pi KR_{min}}{\lambda}[\cos j\alpha - \cos(i+j)\alpha - \cos(j-1)\alpha + \cos(i+j-1)\alpha] \end{aligned} \tag{7.29}$$

Element spacing d controls the appearance of grating lobes. For a maximum scanning angle of θ_0, the spacing between adjacent radiating elements should satisfy the condition as described in Equation (7.30) [208].

$$d \le \frac{\lambda_0}{1+|\cos\theta_0|} \tag{7.30}$$

In Equation (7.30), λ_0 is the wavelength in vacuum. This condition restricts the locations of output ports.

For a linear array, the excitation currents are assumed to have a uniform progressive phase, i.e., $\varphi_n = n\sigma$. When $\sigma = 0$, the main beam direction $\theta = \pi/2$. When $\sigma \neq 0$, the main beam points at an angle θ_i [79]:

$$\theta_i = -\arcsin\frac{\sigma}{kd} \tag{7.31}$$

In Equation (7.20), $i = -M\sim M$. The total number of feeding points is $2M + 1$. Thus, σ can be used to control the beam direction. Besides, when the lens is excited at the ith feeding port, σ_i is the phase difference between neighboring picking ports. Combining Equation (7.27) with Equation (7.31), the location of the ith feeding port can be determined to generate a beam with the desired pointing θ_i.

$$\begin{aligned} -kd\sin\theta_i &= \frac{4\pi KR}{\lambda}[\cos i\alpha - \cos(i+1)\alpha - 1 + \cos\alpha] \\ -kd\sin\theta_i &= \frac{4\pi KR}{\lambda}[\cos(i+1)\alpha - \cos(i+2)\alpha - \cos\alpha + \cos 2\alpha] \\ &\cdots \\ -kd\sin\theta_i &= \frac{4\pi KR}{\lambda}\begin{bmatrix} \cos(i+j-1)\alpha - \cos(i+j)\alpha - \cos(j-1)\alpha \\ +2KR\cos j\alpha \end{bmatrix} \end{aligned} \tag{7.32}$$

Generally speaking, Equation (7.32) is inconsistent and has no accurate solutions. The method of least squares can be used to find the approximate value. As described before, picking ports are distributed diametrically opposite to feeding points. Thus, the position of the picking port is also determined.

To achieve the desired phase distributions at the radiating elements front, a phase-shifting network is required in the modified R-KR lens to implement the phase conversion between picking ports and output ports. The phase shift can be realized just by changing SIW widths. First, when the lens is excited at the center feeding port, the phase of each picking port can be recorded through a full-wave simulation. Then, an SIW phase shifting network is designed to make every output port in phase, when excited at the center feed port.

After detailed introductions of the design processes, a prototype lens was designed and fabricated together with 11 SIW linear slot array antennas on Rogers 5880 substrate to operate at the center frequency of 30 GHz, as shown in Figure 7.26. It has 15 input ports and 11 output ports, and the radius of the inner circular is 28 mm. Each output port connects with an eight-slot antenna. The remaining circular sidewall is lined with dummy ports terminated with 50 Ω loads. The calculated E-plane

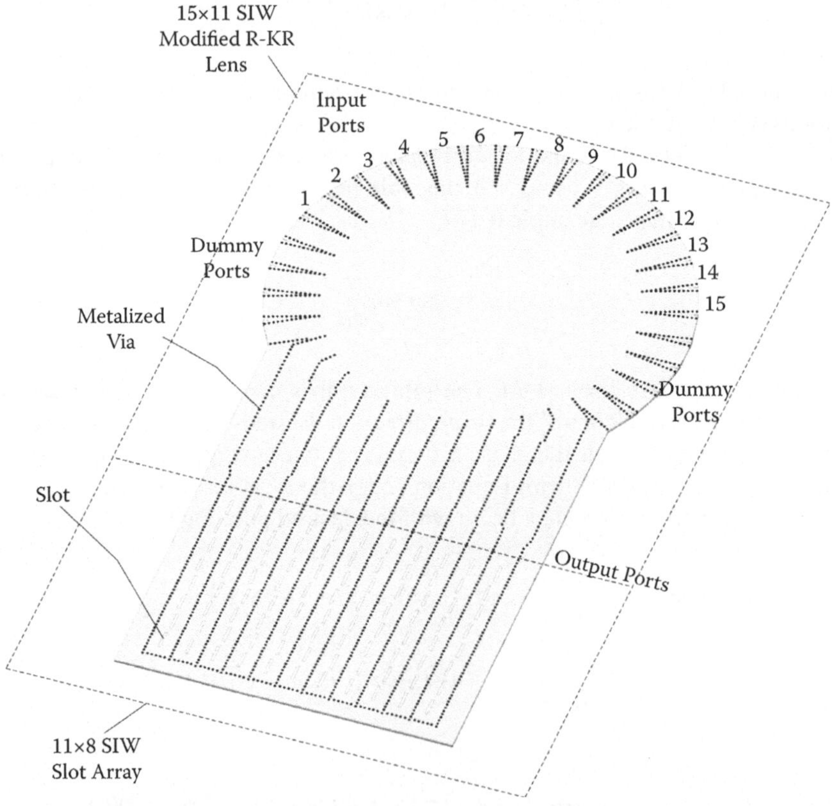

FIGURE 7.26 Configuration of the substrate integrated multibeam antenna array fed by the modified SIW R-KR lens.

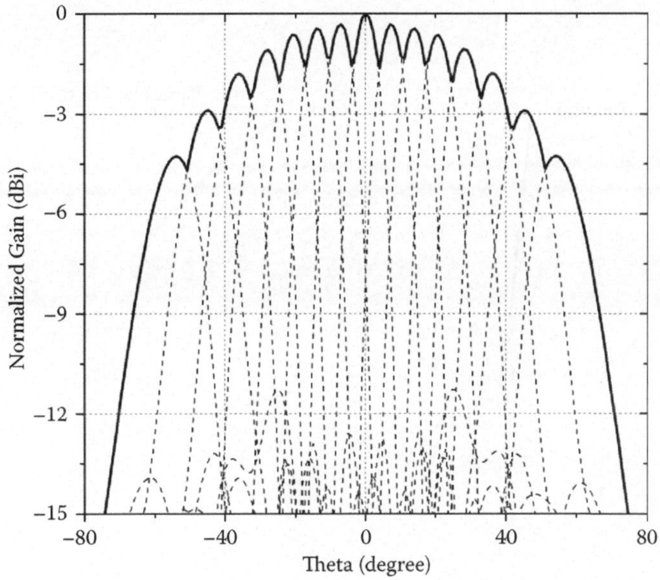

FIGURE 7.27 E-plane radiation pattern of the substrate integrated multibeam antenna array fed by the modified SIW R-KR lens at 30 GHz.

radiation pattern of the multibeam antenna array fed by the modified SIW R-KR lens is shown in Figure 7.27. After selecting the optimal beam, 15 beams are able to cover 117° with their 5 dB beamwidth. At 30 GHz, measured beam directions are 53.7°, 45.2°, 35.1°, 28.8°, 20.2°, 12.9°, 7.1°, and 0.0° corresponding to input ports 1–8. Measured gains excited at different ports range from 17.44 to 20.2 dBi, while radiation efficiencies range from 21.6 to 40.8%.

7.2.3 Substrate Integrated Multibeam Antenna Array Fed by SIW Reflector Lens

The geometry of a parabolic reflector is illustrated in Figure 7.28. When a point source feed is located at the focus F of the paraboloid, a spherical incident wave is produced, reflected, and converted to a plane wave, which travels from the reflector surface to the picking contour along the parabolic axis. For any direction of the wave, the total path length is equal. Therefore, a uniform phase distribution can be realized on the picking contour. The offset-feed topology can eliminate the aperture blocking, as well as the mismatch at the feeder because of the intercepted radiation.

The geometry of the paraboloid can be described by a simple relationship as follows:

$$y^2 = 4fx \tag{7.33}$$

In Equation (7.33), f is the focal length. When the feeder is away from the focal point, waves reflected to the picking contour are no longer in phase. In this case, the

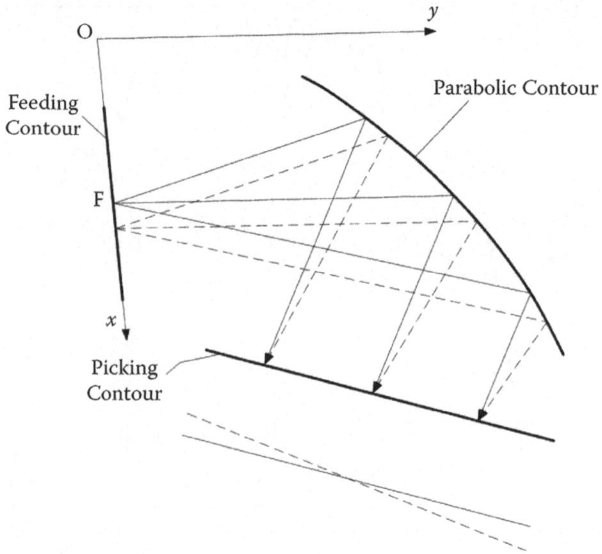

FIGURE 7.28 Schematic diagram of the reflector lens and the principle of multibeam generation.

beam produced by a parabolic reflector can be moved. This concept can be used to generate multiple beams by placing multiple feeders at different locations near the focal point. When the shape of the reflector is changed from the paraboloid to the spherical surface, it is possible to achieve a wider scanning angle [219].

The SIW technology can be easily employed to simulate the shape of a planar parabolic contour with arrays of metalized vias in a planar substrate. A SIW reflector lens prototype was developed in [220] on Rogers 5880 substrate with a thickness of 0.508 mm and a relative permittivity of 2.2, as shown in Figure 7.29. It has a focal length of 30 mm and is designed to operate at 37.5 GHz. This 8 × 11 SIW reflector lens was directly integrated with 11 SIW linear slot array antennas without any phase shifter. Measured beam directions are –30°, –21°, –13°, –5°, 5°, 14°, and 23° for the input ports 2–8, respectively. Gains excited at different ports range from 15.8 to 19.1 dBi. The calculated E-plane radiation pattern of the multibeam antenna array fed by the SIW reflector lens is shown in Figure 7.30. After selecting the optimal beam, seven beams are able to cover 62° from –34.5° to 27.5° with their 4 dB beamwidth.

Besides, an SIW multibeam pillbox antenna was introduced in [221]. The antenna consists of integrated feeding horns, quasi-optical system, and radiating array. The radiating array and feeding horns are placed in two different stacked dielectric substrates. They are connected by an optimized quasi-optical system. This multibeam antenna concept has been validated by experiment, showing a scanning capability range over ±30° at 24 GHz.

A dual-layer pillbox antenna with mechanical scanning of the beam was presented in [222]. Moving an integrated SIW feeding horn in the focal plane of an SIW parabolic reflector can steer the beam. This SIW reflector behaves as an E-plane

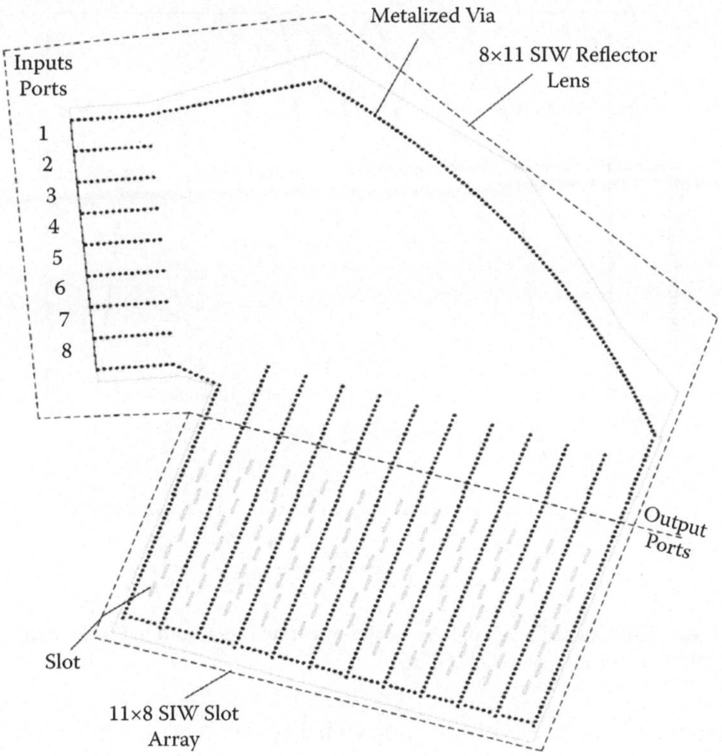

FIGURE 7.29 Configuration of the substrate integrated multibeam antenna array fed by the SIW reflector lens.

coupler type bent between the lower substrate and the upper one. An SIW slot array antenna is integrated in the upper substrate as the radiating part. The measured scan range is ±35° at 24.1 GHz.

7.2.4 Substrate Integrated Multibeam Antenna Array Fed by SIW Miniaturized Multimode Lens

A dual-layer multibeam antenna was introduced in [223] with a folded Rotman lens used as a compact BFN in SIW technology. The overall size of the multibeam antenna can be reduced by folding the Rotman lens on two stacked dielectric substrates along the inner lens contour. A transition between the lower substrate and the upper one is based on a reflector and several coupling holes.

Besides, a miniaturized SIW multibeam antenna with endfire radiation was described in [224]. The BFN can be viewed as an asymmetrical multimode lens with four input ports, but only one output port. The width of the output port is four times the width of the input SIW. To reduce the circuit size as much as possible, the output port of the BFN is directly connected to a four-element tapered slot antenna (TSA) array without connection structures.

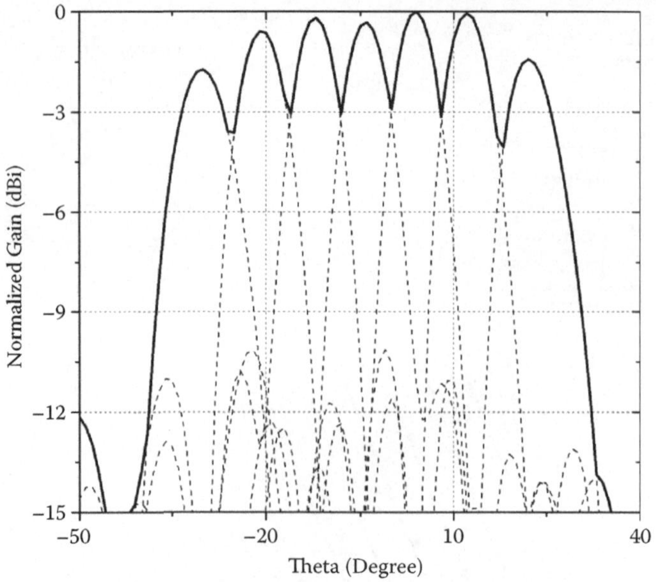

FIGURE 7.30 E-plane radiation pattern of the substrate integrated multibeam antenna array fed by the SIW reflector lens at 37.5 GHz.

This type of BFN consists of two parts with different functionalities. The first part is used to excite several different modes within the multimode SIW and realize the amplitude transformations of fields. It can be considered the conventional H-plane moder. The second part is a multimode SIW used to transform the phase of the fields. The amplitude of each mode propagated along a lossless network is unchanged.

Different types of radiating elements, such as horn antenna, TSA, etc., may be employed in this design. The horn antenna is a very popular endfire antenna with wide bandwidth. Because of the large width-to-height ratio of a conventional SIW, it is very difficult to radiate energy out effectively with the horn implemented in SIW technology. To solve this problem, the thickness of the substrate has to be increased much more in order to improve the radiation efficiency. In this case, the substrate has excessive thickness and is not compatible with other active circuits, which usually require thin substrates for compactness and thermal dissipation. The TSA fed by an SIW is another choice. The aperture width of each TSA should be reduced approximating to the width of the SIW feedline. This compact TSA can be optimized by adjusting its taper length and cross-width.

The configuration of the Ka-band SIW multimode multibeam antenna is shown in Figure 7.31. The width of the feeding SIW is 5 mm, the distance between neighboring metalized vias is 0.8 mm, and the diameter of the metalized via is 0.4 mm. The thickness and permittivity of the substrate are 0.508 mm and 2.2, respectively. The circuit size of the asymmetric multimode network is only 40% of that of a conventional 4 × 4 SIW Butler matrix.

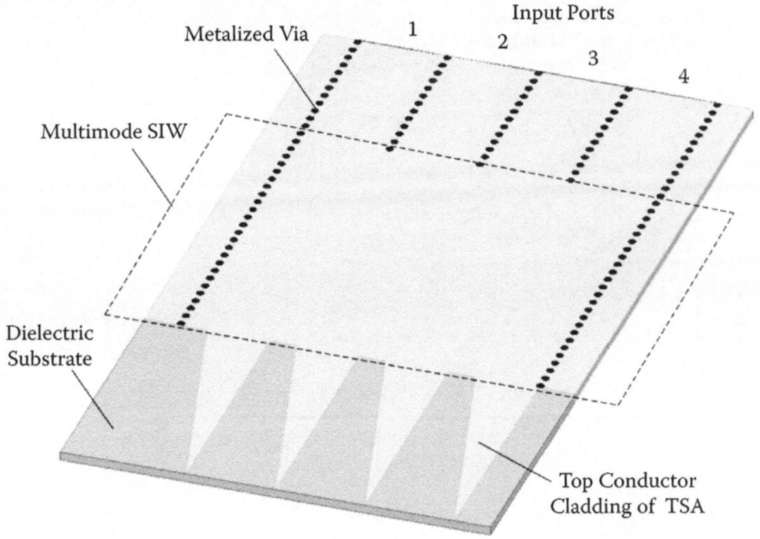

FIGURE 7.31 Configuration of the SIW miniaturized multibeam antenna array.

7.3 TWO-DIMENSIONAL SIW MULTIBEAM ANTENNA ARRAY

The realization of multiple beams in two orthogonal planes seems difficult owing to its complexity. Only a few investigations have been reported until now. For example, two-dimensional scanning was realized by cross-connecting two stacked Rotman lenses or Butler matrices [188]. How to synthesize a two-dimensional BFN was discovered in [225] based on the odd hexagonal lattice. In this topology, the center is not a single element but rather a triad of elements. In this section, several two-dimensional SIW multibeam antenna arrays are introduced.

7.3.1 Mechanism Setting

In [209], several one-dimensional multibeam antennas fed by SIW Rotman lenses were grouped to realize the two-dimensional beam scanning. This approach is shown in Figure 7.32. When excited at input ports 2–6 of the feeding Rotman lens, the substrate integrated multibeam antenna can generate five beams in the θ_y-direction. Then, four identical multibeam antennas are mounted with different pointings, such as 30°, 10°, –10°, and –30°, to offer four different beams in the θ_x-direction. After that, each multibeam antenna is shifted with a small angle of 3.75° or –3.75° in the θ_y-direction to remove the blind spots in the coverage.

Absolutely, influences of adjacent antennas, as well as the framework, deteriorate radiation patterns, but slightly. This type of two-dimensional SIW multibeam array is able to generate 4 × 5 beams, and the overall coverage is around (–40°, 40°) × (–35°, 35°) with 5 dB beamwidth.

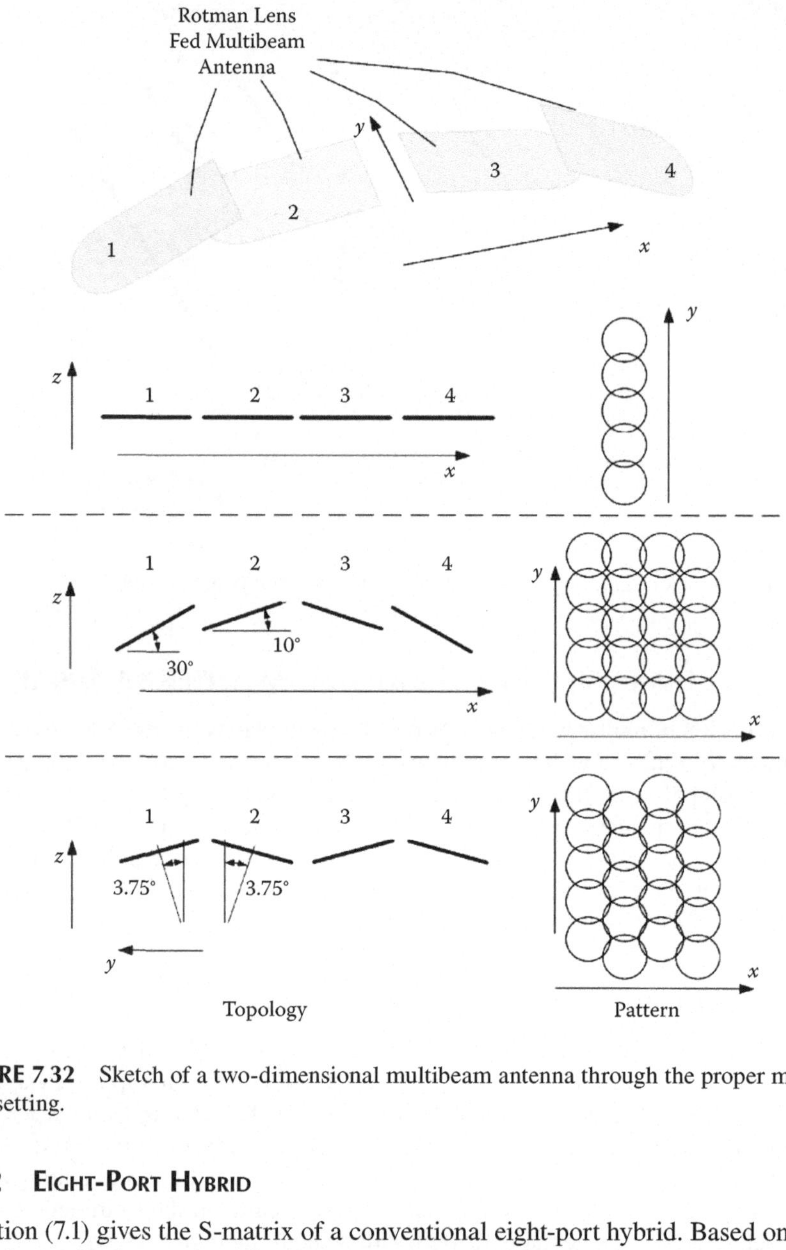

FIGURE 7.32 Sketch of a two-dimensional multibeam antenna through the proper mechanism setting.

7.3.2 Eight-Port Hybrid

Equation (7.1) gives the S-matrix of a conventional eight-port hybrid. Based on this, it can be applied to excite a 2 × 2 antenna array and then produce 2 × 2 beams, as shown in Figure 7.33 [197]. It has no connectors and cables in the completely integrated configuration, which is superior in the design of a small multibeam antenna. Simulated –3 dB contour lines of input ports 1–4 are shown in Figure 7.34. Measured beam directions are (–30°, –48°), (30°, –48°), (–30°, 48°), and (30°, 48°), respectively. The 3 dB beamwidth is approximately 40° in the E-plane and 32° in the H-plane. A similar Ka-band two-dimensional scan design can be found in [226].

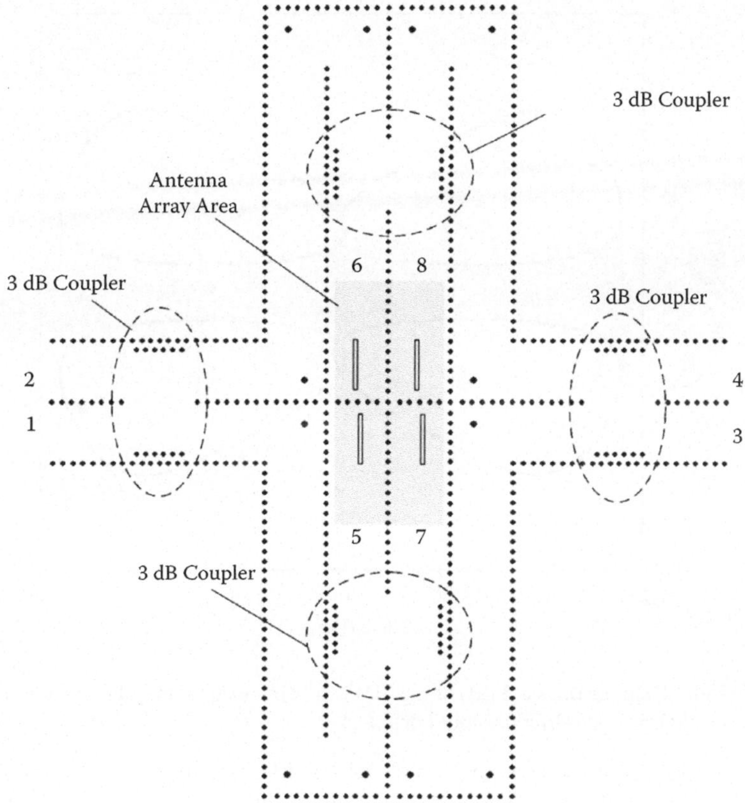

FIGURE 7.33 Configuration of the SIW two-dimensional multibeam antenna fed by the eight-port hybrid.

7.3.3 Quasi-Traveling-Wave Antenna

Most of the slot array antennas are designed based on standing-wave excitation. In this case, the distance of neighboring longitudinal slots is half of the guided wavelength at the center frequency. If the slot spacing is not just half of the guided wavelength, the main beam will be at a given angle other than normal. Based on this principle, a quasi-traveling-wave array was developed in [209], which characters the controlled beam direction. Here, a standing-wave array (antenna 1) and two quasi-traveling-wave arrays (antennas 2 and 3) were developed. Their beam directions are 0°, 11°, and 20°, respectively. Then, they were fed by SIW Rotman lenses to construct substrate integrated multibeam antenna arrays.

Figure 7.35 illustrates this way for accomplishing two-dimensional multibeam coverage. Five SIW multibeam antennas were mounted on a curved surface, offering different beam directions, such as –20°, –11°, 0°, 11°, and 20° in the θ_x-direction. When excited at input ports 2–6 of the feeding Rotman lens, each substrate integrated multibeam antenna can generate five beams in the θ_y-direction. The curved

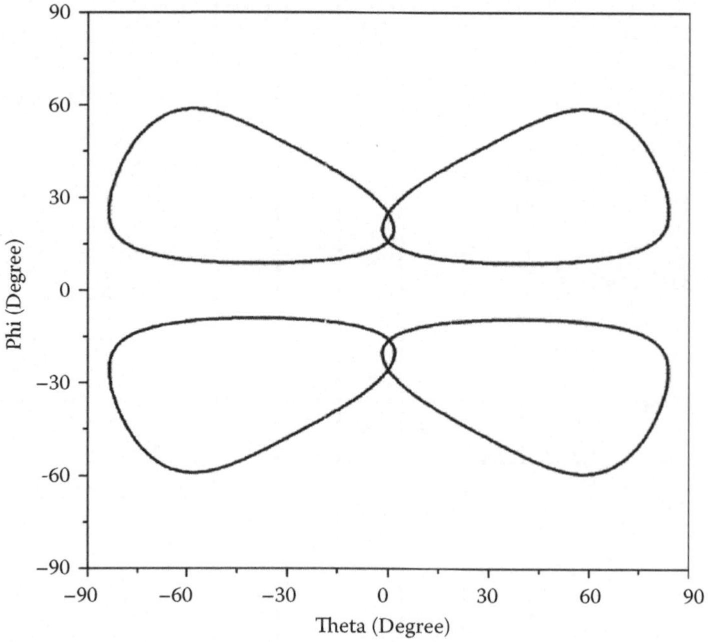

FIGURE 7.34 Contour lines (–3 dB) of the SIW two-dimensional multibeam antenna fed by the eight-port hybrid according to input ports 1–4.

surface was used to provide a small angle of 7.5° likewise to cross the adjacent multiple beams to remove the blind spots in the coverage.

7.3.4 Leaky-Wave Array

Another method of the two-dimensional beam scanning array was proposed in [227]. A one-dimensional phased array of leaky-wave antennas can steer a pencil beam in both elevation and azimuth directions. In [138], a two-dimensional multibeam antenna was realized by combining the technologies of the multibeam antenna and the leaky-wave antenna. As shown in Figure 7.36, 14 low SLL long slot leaky-wave antennas were combined with a broadband SIW 10 × 14 reflector lens. Scanning in one plane is achieved by the use of a passive BFN arranged in the feeding structure of the one-dimensional leaky-wave array to avoid using expensive millimeter-wave phase shifters. Scanning in the orthogonal plane is obtained in usual leaky-wave fashion by varying the operating frequency. The whole circuit was fabricated on a single-layer Rogers 5880 substrate. All input ports are fed by the standard WR-28 waveguide.

Excited at port 6, measured gains are 17.1, 17.5, 18.3, 19.1, and 18.9 dBi, and measured beam directions are –27.2°, –33.6°, –39.1°, –45.0°, and –48.2° at 33, 34, 35, 36, and 37 GHz. When the frequency is changed from 33 GHz to 37 GHz, the angular region of 37.5° in the *yoz*-plane can be covered by a 3 dB beamwidth of those continuous scanning beams. Excited at port 8, measured gains are 17.9, 18.1, 18.9,

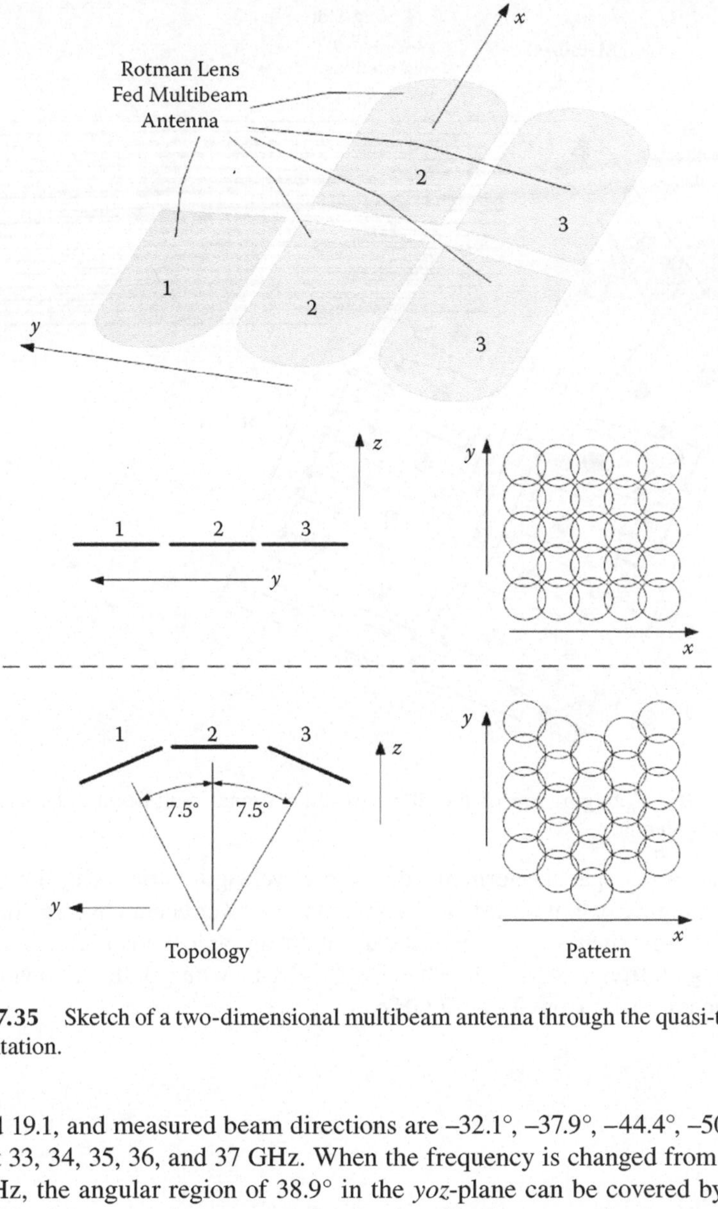

FIGURE 7.35 Sketch of a two-dimensional multibeam antenna through the quasi-traveling-wave excitation.

19.2, and 19.1, and measured beam directions are –32.1°, –37.9°, –44.4°, –50.3°, and –56.6° at 33, 34, 35, 36, and 37 GHz. When the frequency is changed from 33 GHz to 37 GHz, the angular region of 38.9° in the *yoz*-plane can be covered by a 3 dB beamwidth of those continuous scanning beams. Similar ranges in elevation can be covered and excited at other ports.

Besides, at 35 GHz, measured gains are 17.5, 18.5, 18.9, 18.3, and 17.3 dBi, and measured beam directions are –33.5°, –18.9°, 2.0°, 22.6°, and 41.8° excited at ports 2, 4, 6, 8, and 10. Excited at input ports 1–10 of such a two-dimensional multibeam antenna at 35 GHz, the angular region of 86.6° in azimuth can be covered by a 3 dB beamwidth of 10 pencil beams. Similar results in the *xoz*-plane can be achieved at other frequencies.

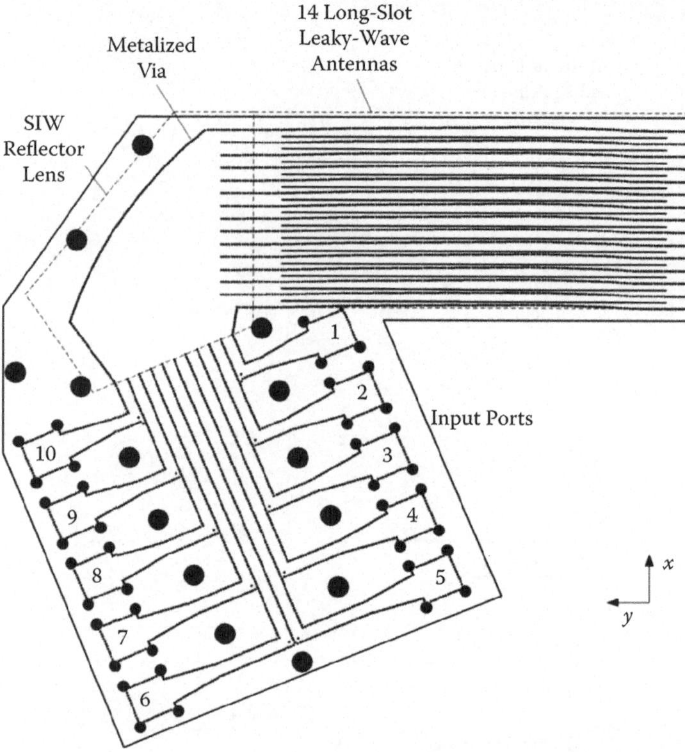

FIGURE 7.36 Configuration of the SIW two-dimensional multibeam antenna using the leaky-wave antenna.

Because array spacing normalized to the wavelength varies with the operating frequency, higher operating frequency leads to narrower beam coverage in the *xoz*-plane. However, such a two-dimensional multibeam antenna can almost cover the angular region from (–45.4°, –20.3°) to (41.2°, –58.4°) with a 3 dB beamwidth, when the frequency varies from 33 to 37 GHz.

References

1. H. Uchimura, T. Takenoshita, and M. Fujii. Development of a laminated waveguide. *IEEE Transactions on Microwave Theory and Techniques*, 46(12): 2438–2443, 1998.
2. H. Yong and K.L. Wu. A broad-band LTCC integrated transition of laminated waveguide to air-filled waveguide for millimeter-wave applications. *IEEE Transactions on Microwave Theory and Techniques*, 51(5): 1613–1617, 2003.
3. H. Yong, K.L. Wu, and M. Ehlert. An integrated LTCC laminated waveguide to microstrip line T-junction. *IEEE Microwave and Wireless Components Letters*, 13(8): 338–339, 2003.
4. J.A. Cruz and K.A. Zaki. Canonical ridge waveguide filters in LTCC or metallic resonators. *IEEE Transactions on Microwave Theory and Techniques*, 53(1): 174–182, 2005.
5. B.J. Chen, T.M. Shen, and R.B. Wu. Dual-band vertically stacked laminated waveguide filter design in LTCC technology. *IEEE Transactions on Microwave Theory and Techniques*, 57(6): 1554–1562, 2009.
6. J. Hirokawa and M. Ando. Efficiency of 76-GHz post-wall waveguide-fed parallel-plate slot arrays. *IEEE Transactions on Antennas and Propagation*, 48(11): 1742–1745, 2000.
7. J. Hirokawa and M. Ando. Sidelobe suppression in 76-GHz post-wall waveguide-fed parallel-plate slot arrays. *IEEE Transactions on Antennas and Propagation*, 48(11): 1727–1732, 2000.
8. J. Hirokawa and M. Ando. 45° linearly polarised post-wall waveguide-fed parallel-plate slot arrays. *IEE Proceedings—Microwaves, Antennas and Propagation*, 147(6): 515–519, 2000.
9. T. Kai, J. Hirokawa, and M. Ando. Analysis of a feeding structure for TEM wave excitation in an oversized rectangular waveguide. In *IEEE Antennas and Propagation Society International Symposium*, 2003, pp. 1177–1180.
10. J.R. Bray and L. Roy. Resonant frequencies of post-wall waveguide cavities. *IEE Proceedings—Microwaves, Antennas and Propagation*, 150(5): 365–368, 2003.
11. S. Park, Y. Okajima, J. Hirokawa, and M. Ando. A slotted post-wall waveguide array with inter-digital structure for 45-deg linear and dual polarization. *IEEE Transactions on Antennas and Propagation*, 53(9): 2865–2871, 2005.
12. Y. Cassivi, L. Perregrini, P. Arcioni, M. Bressan, and K. Wu. Dispersion characteristics of substrate integrated rectangular waveguide. *IEEE Microwave and Wireless Components Letters*, 12(1): 333–335, 2002.
13. F. Xu and K. Wu. Guided-wave and leakage characteristics of substrate integrated waveguide. *IEEE Transactions on Microwave Theory and Techniques*, 53(1): 66–73, 2005.
14. L. Yan, W. Hong, K. Wu, et al. Investigations on the propagation characteristics of the substrate integrated wave-guide based on the method of lines. *IEE Proceedings—Microwaves, Antennas and Propagation*, 152(1): 35–42, 2005.
15. D. Deslandes and K. Wu. Accurate modeling, wave mechanisms, and design considerations of a substrate integrated waveguide. *IEEE Transactions on Microwave Theory and Techniques*, 54(6): 2516–2526, 2006.
16. R. Garg. The effect of tolerances on microstripline and slotline performances. *IEEE Transactions on Microwave Theory and Techniques*, 26(1): 16–19, 2010.

17. K.C. Gupta, R. Garg, I. Bahl, et al. *Microstrip lines and slotlines*, 2nd ed. Artech House, London, U.K., 1996.
18. Y.J. Cheng, W. Hong, and K. Wu. Investigation on tolerances of substrate integrated waveguide (SIW). In *Proceedings of Asia-Pacific Microwave Conference*, Bangkok, Thailand, December 2007, pp. 2305–2308.
19. Y.J. Cheng, K. Wu, and W. Hong. Power handling capability of substrate integrated waveguide interconnects and related transmission line systems. *IEEE Transactions on Advanced Packaging*, 31(4): 900–909, 2008.
20. I.J. Bahl. Average power handling capability of multiplayer microstrip lines. *International Journal of RF and Microwave Computer-Aided Engineering*, 11(6): 385–395, 2001.
21. W.Y. Yin, X.T. Dong, J.F. Mao, and L.W. Li. Average power handling capability of finite-ground thin-film microstrip lines over ultra-wide frequency ranges. *IEEE Microwave and Wireless Components Letters*, 15(10): 715–717, 2005.
22. W.Y. Yin, Y.J. Zhang, X.T. Dong, and Y.B. Gan. Average power handling capability of the signal line in coplanar waveguides on polyimide/GaAs substrates including the irregular line edge shape effects. *International Journal of RF and Microwave Computer-Aided Engineering*, 15(1): 156–163, 2005.
23. R.E. Collin. *Field theory of guided waves*, 2nd ed. Piscataway, NJ: IEEE Press, 1990.
24. J.P. Holman. *Heat transfer*, 5th ed. New York: McGraw-Hill, 1981.
25. D.G. Jablonski. Power-handling capabilities of circular dielectric waveguide at millimeter wavelengths. *IEEE Transactions on Microwave Theory and Techniques*, 33(2): 85–89, 1985.
26. High frequency circuit materials product selector guide. Rogers Corporation. www.rogerscorporation.com.
27. http://www.infrared-thermography.com/material.htm.
28. High frequency laminates—Product selector guide. Rogers Corporation. www.rogerscorporation.com.
29. High frequency circuit materials properties guide. Rogers Corporation. www.rogerscorporation.com.
30. R. Hornung and J. Frankosky. Microwave laminate material considerations for multilayer military applications. Arlon, Inc. www.rfglobalnet.com.
31. P. Kabacik and M.E. Bialkowski. The temperature dependence of substrate parameters and their effect on microstrip antenna performance. *IEEE Transactions on Antennas and Propagation*, 47(6): 1042–1049, 1999.
32. A. Patrovsky and K. Wu. Substrate integrated image guide (SIIG)—A planar dielectric waveguide technology for millimeter-wave applications. *IEEE Transactions on Microwave Theory and Techniques*, 54(6): 2872–2879, 2006.
33. A. Patrovsky and K. Wu. 94-GHz broadband transition from coplanar waveguide to substrate integrated image guide (SIIG). In *Proceedings of IEEE International Microwave Symposium*, June 2007, pp. 1551–1554.
34. Y.J. Cheng, X.Y. Bao, and Y.X. Guo. LTCC-based substrate integrated image guide and its transition to conductor-backed coplanar waveguide. *IEEE Microwave and Wireless Components Letters*, 23(9): 450–452, 2013.
35. R.M. Knox. Dielectric waveguide microwave integrated circuits—An overview. *IEEE Transactions on Microwave Theory and Techniques*, 24(11): 806–814, 1976.
36. D. Lioubtchenko, S. Tretyakov, and S. Dudorov. *Millimeter-wave waveguides*. Boston: Kluwer Academic, 2003.
37. Y. Cassivi and K. Wu. Substrate integrated nonradiative dielectric waveguide. *IEEE Microwave and Wireless Components Letters*, 14(3): 89–91, 2004.

38. Y. Cassivi and K. Wu. Substrate integrated circuits concept applied to the nonradiative dielectric guide. *IEE Proceedings—Microwaves, Antennas and Propagation*, 152(6): 424–433, 2005.
39. F. Xu and K. Wu. Substrate integrated nonradiative dielectric waveguide structures directly fabricated on printed circuit boards and metallized dielectric layers. *IEEE Transactions on Microwave Theory and Techniques*, 59(12): 3076–3086, 2011.
40. Y. Cassivi, and K. Wu. Substrate integrated NRD (SINRD) guide in high dielectric constant substrate for millimetre wave circuits and systems. In *IEEE International Microwave Symposium*, June 2004, pp. 1639–1642.
41. P. Mondal and K. Wu. Single mode operation of substrate integrated non-radiative dielectric waveguide and an excitation scheme of LSM_{11} mode. *IEEE Microwave and Wireless Components Letters*, 23(8): 418–420, 2013.
42. J. Hirokawa, K. Sakurai, M. Ando, et al. An analysis of a waveguide T junction with an inductive post. *IEEE Transactions on Microwave Theory and Techniques*, 39(3): 563–566, 1991.
43. Z. Ma and E. Yamashita. Efficient fullwave analysis of a waveguide T-junction with an inductive post. In *1994 Asia-Pacific Microwave Conference Proceedings*, December 1994, pp. 259–262.
44. S. Germain, D. Deslandes, and K. Wu. Development of substrate integrated waveguide power dividers. In *2003 IEEE Canadian Conference on Electrical and Computer Engineering*, May 2003, pp. 1921–1924.
45. X. Xu, R. G. Bosisio, and K. Wu. A new six-port junction based on substrate integrated waveguide technology. *IEEE Transactions on Microwave Theory and Techniques*, 53(7): 2267–2273, 2005.
46. K. Sakakibara, Y. Kimura, A. Akiyama, et al. Alternating phase-fed waveguide slot arrays with a single-layer multiple-way power divider. *IEE Proceedings—Microwaves, Antennas and Propagation*, 144(6): 425–430, 1997.
47. B.M. Schiffman. A new class of broad-band microwave 90-degree phase shifters. *IRE Transactions on Microwave Theory and Techniques*, 6(4): 232–237, 1958.
48. Schiek and J. Kohler. A method for broad-band matching of microstrip differential phase shifters. *IEEE Transactions on Microwave Theory and Techniques*, 25(8): 666–671, 1977.
49. J.L.R. Quirarte and J.P. Starski. Novel Schiffman phase shifters. *IEEE Transactions on Microwave Theory and Techniques*, 41(1): 9–14, 1993.
50. S.Y. Eom, S.I. Jeon, J.S. Chae, et al. Broadband 180° bit phase shifter using a $\lambda/2$ coupled line and parallel $\lambda/8$ stubs. *IEEE Microwave and Wireless Components Letters*, 14(5): 228–231, 2004.
51. J. Dittloff, F. Amdt, and D. Grauerholtz. Optimum design of waveguide E-plane stub-loaded phase shifters. *IEEE Transactions on Microwave Theory and Techniques*, 36(3): 582–587, 1988.
52. F. Alessandri, M. Mongiardo, and R. Sorrentino. A technique for the fullwave automatic synthesis of waveguide components: Application to fixed phase shifters. *IEEE Transactions on Microwave Theory and Techniques*, 40(7): 1484–1495, 1992.
53. F. Arndt and R. Frye. Double dielectric-slab-filled waveguide phase shifter. *IEEE Transactions on Microwave Theory and Techniques*, 33(5): 373–381, 1985.
54. B. Liu, W. Hong, Z.C. Hao, et al. Substrate integrated waveguide 180-degree narrow-wall directional coupler. In *2005 Microwave Conference Proceedings*, December 2005, pp. 4–7.
55. K. Sellal, L. Talbi, T.A. Denidni, et al. Design and implementation of a substrate integrated waveguide phase shifter. *IET Microwaves, Antennas and Propagation*, 2(2): 194–199, 2008.

56. Y.J. Cheng, W. Hong, and K. Wu. Novel substrate integrated waveguide fixed phase shifter for 180-degree directional coupler. In *2007 IEEE International Microwave Symposium*, HI, June 2007, pp. 189–192.
57. K. Morimoto, J. Hirokawa, and M. Ando. Design of a 180-degree single-layer divider to control sidelobe and crossover levels in Butler-matrix beam-switching antenna. In *Proceedings of 2007 Asia-Pacific Microwave Conference*, Bangkok, Thailand, December 2007, pp. 1–4.
58. E. Sbarra, L. Marcaccioli, R.V. Gatti, et al. A novel Rotman lens in SIW technology. In *2007 Microwave Conference*, October 2007, pp. 1515–1518.
59. Y.J. Cheng, W. Hong, and K. Wu. Broadband self-compensating phase shifter combining delay line and equal-length unequal-width phaser. *IEEE Transactions on Microwave Theory and Techniques*, 58(1): 203–210, 2010.
60. W.J. Liu. *Foundation for microwave technology.* Chengdu: UESTC Publishing Company, 1989.
61. D. Deslandes and K. Wu. Integrated microstrip and rectangular waveguide in planar form. *IEEE Microwave and Wireless Components Letters*, 11(2): 68–70, 2001.
62. D. Deslandes and K. Wu. Integrated transition of coplanar to rectangular waveguides. In *2001 IEEE International Microwave Symposium*, May 2001, pp. 619–622.
63. D. Deslandes and K. Wu. Analysis and design of current probe transition from grounded coplanar to substrate integrated rectangular waveguide. *IEEE Transactions on Microwave Theory and Techniques*, 53(8): 2487–2494, 2005.
64. Y. Huang and K.L. Wu. A broad-band LTCC integrated transition of laminated waveguide to air-filled waveguide for millimeter-wave applications. *IEEE Transactions on Microwave Theory and Techniques*, 51(5): 1613–1617, 2003.
65. T. Shimura, Y. Ohashi, and Y. Kato. Waveguide substrate and high-frequency circuit module. U.S. Patent 20060274992A1, September 12, 2005.
66. Schary, G.C. Dalman, and C.A. Lee. Dielectric waveguide-to-co-planar transmission line transitions. U.S. Patent 5225797, July 6, 1993.
67. H. Tehrani, M. Li, and K. Chang. Broadband microstrip to dielectric image line transitions. *IEEE Microwave and Wireless Components Letters*, 10(10): 409–411, 2000.
68. J.H. Schaffner and W.B. Bridges. Integrated ultra wideband element card for array antennas. U.S. Patent 7012572B1, July 2004.
69. N. Dolatsha and J. Hesselbarth. Propagation and excitation of the higher-order E_x^{11} mode in an insulated image guide. *Microwave and Optical Technology Letters*, 54(1): 179–181, 2012.
70. N. Dolatsha and J. Hesselbarth. Millimeter-wave chip-to-chip transmission using an insulated image guide excited by an on-chip dipole antenna at 90 GHz. *IEEE Microwave and Wireless Components Letters*, 22(5): 266–268, 2012.
71. A.F. Stevenson. Theory of slots in rectangular waveguides. *Journal of Applied Physics*, 19(1): 24–38, 1948.
72. A.A. Oliner. The impedance properties of narrow radiating slots in the broad face of rectangular waveguide. *IRE Transactions on Antennas and Propagation*, 5(1): 4–20, 1957.
73. H.Y. Yee. Impedance of a narrow longitudinal shunt slot in a slotted waveguide array. *IEEE Transactions on Antennas and Propagation*, 22(4): 589–592, 1974.
74. T.V. Khac and C.T. Carson. Impedance properties of a longitudinal slot antenna in the broad face of a rectangular waveguide. *IEEE Transactions on Antennas and Propagation*, 21(5): 708–710, 1973.
75. G.T. Stern and R.S. Elliott. Resonant length of longitudinal slots and validity of circuit representation: Theory and experiment. *IEEE Transactions on Antennas and Propagation*, 33(11): 1264–1271, 1985.

76. R.S. Elliott. An improved design procedure for small arrays of shunt slots. *IEEE Transactions on Antennas and Propagation*, 31(1): 48–53, 1983.
77. R.S. Elliott and W.R. O'Loughlin. The design of slot arrays including internal mutual coupling. *IEEE Transactions on Antennas and Propagation*, 34(9): 1149–1154, 1986.
78. R.S. Elliott and L. Kurtz. Main-line/branch-line couplers for waveguide slot arrays. *IEEE Transactions on Antennas and Propagation*, 27(2): 256–259, 1979.
79. R.S. Elliott. *Antenna Theory and Design*, rev. ed. John Wiley & Sons, 2003.
80. J.J. Gulick and R.S. Elliott. The design of linear and planar arrays of waveguide-fed longitudinal slots. *Electromagnetics*, 10(4), 327–347, 1990.
81. H.Y. Yee. The design of large waveguide arrays of shunt slots. *IEEE Transactions on Antennas and Propagation*, 40(7): 775–781, 1992.
82. J.C. Coetzee, J. Joubert, and D.A. McNamara. Off-center-frequency analysis of a complete planar slotted-waveguide array consisting of subarrays. *IEEE Transactions on Antennas and Propagation*, 48(11): 1746–1755, 2000.
83. A. Morini, T. Rozzi, and G. Venanzoni. On the analysis of slotted waveguide arrays. *IEEE Transactions on Antennas and Propagation*, 54(7): 2016–2021, 2006.
84. G. Montisci and G. Mazzarella. Full-wave analysis of a waveguide printed slot. *IEEE Transactions on Antennas and Propagation*, 52(8): 2168–2171, 2004.
85. J.J. Gonzalez Picazo, M. Calvo, J.L. Besada, et al. On the design of nonuniformly spaced slot arrays. *IEEE Transactions on Antennas and Propagation*, 38(11): 1780–1783, 1990.
86. S.S. Oh, J.W. Lee, M.S. Song, et al. Two-layer slotted-waveguide antenna array with broad reflection/gain bandwidth at millimetre-wave frequencies. *IEE Proceedings—Microwaves, Antennas and Propagation*, 151(5): 393–398, 2004.
87. J.A. Rodriguez, F. Ares, E. Moreno, and G. Franceschetti. Design of shunt slot arrays without weak excitations. *Electronics Letters*, 35(17): 1396–1397, 1999.
88. Y. Oh, J.H. Hwang, and J. Choi. Single-layer waveguide slot array antenna with diaphragms. *Electronics Letters*, 41(16): 5–6, 2005.
89. S. Park, Y. Tsunemitsu, J. Hirokawa, et al. Center feed single layer slotted waveguide array. *IEEE Transactions on Antennas and Propagation*, 54(5): 1474–1480, 2006.
90. M. Zhang, J. Hirokawa, and M. Ando. A four-way divider for partially-corporate feed in an alternating-phase fed single-layer slotted waveguide array. *IEEE Transactions on Antennas and Propagation*, 56(6): 1790–1794, 2008.
91. K. Sakakibara, Y. Kimura, A. Akiyama, et al. Alternating phase-fed waveguide slot arrays with a single-layer multiple-way power divider. *IEE Proceedings—Microwaves, Antennas and Propagation*, 144(6): 425–430, 1997.
92. J. Hirokawa, M. Ando, and N. Goto. Waveguide-fed parallel plate slot array antenna. *IEEE Transactions on Antennas and Propagation*, 40(2): 218–223, 1992.
93. M. Ando, J. Hirokawa, T. Yamamoto, et al. Novel single-layer waveguides for high-efficiency millimeter-wave arrays. *IEEE Transactions on Antennas and Propagation*, 46(6): 792–799, 1998.
94. K. Sakakibara, J. Hirokawa, M. Ando, et al. High-gain and high-efficiency single-layer slotted waveguide array for use in 22 GHz band. *Electronics Letters*, 32(4): 283–284, 1996.
95. Y. Kimura, J. Hirokawa, and M. Ando. Low sidelobe single-layer slotted waveguide arrays at 76 GHz band. In *Proceedings of IEEE International Antennas and Propagation Symposium*, July 2000, vol. 1, pp. 86–89.
96. J.F. Xu, W. Hong, P. Chen, et al. Design and implementation of low sidelobe substrate integrated waveguide longitudinal slot array antennas. *IET Microwaves, Antennas and Propagation*, 3(5): 790–797, 2009.

97. H.C. Lu and T.H. Chu. Equivalent circuit of radiating longitudinal slots in substrate integrated waveguide. In *Proceedings of IEEE International Antennas and Propagation Symposium*, June 2004, vol. 3, pp. 2341–2344.
98. S.R. Rengarajan and D.D. Nardi. On internal higher order mode coupling in slot arrays. *IEEE Transactions on Antennas and Propagation*, 39(5): 694–698, 1991.
99. http://www.etsi.org Harmonized European Standard Telecommunications series.
100. D.M. Pozar. The active element pattern. *IEEE Transactions on Antennas and Propagation*, 42(8): 1176–1178, 1994.
101. X.S. Yang, H. Qian, B.Z. Wang, et al. Radiation pattern computation of pyramidal conformal antenna array with active element pattern technique. *IEEE Antennas and Propagation Magazine*, 53(1): 28–37, 2011.
102. R.H. Otten and L.P. Ginneken. *The annealing algorithm*. Boston: Kluwer Academic, 1989.
103. J. Martinez-Fernandez, J.M. Gil, and J. Zapata. Ultrawideband optimized profile monopole antenna by means of simulated annealing algorithm and the finite element method. *IEEE Transactions on Antennas and Propagation*, 55(6): 1826–1832, 2007.
104. P. Chen, Y.Y. Zheng, and W. Zhu. Optimized simulated annealing algorithm for thinning and weighting large planar arrays in both far-field and near-field. *IEEE Journal of Oceanic Engineering*, 36(4): 658–664, 2011.
105. J. Hirokawa and M. Ando. Single-layer feed waveguide consisting of posts for plane TEM wave excitation in parallel plates. *IEEE Transactions on Antennas and Propagation*. 46(5): 625–630, 1998.
106. K. Hashimoto, J. Hirokawa, and M. Ando. A post-wall waveguide center-feed parallel plate slot array antenna in the millimeter-wave band. *IEEE Transactions on Antennas and Propagation*. 58(11): 3522–3538, 2010.
107. D.Y. Kim, W.S. Chung, C.H. Park, et al. A series slot array antenna for 45°-inclined linear polarization with SIW technology. *IEEE Transactions on Antennas and Propagation*, 60(4): 1785–1795, 2012.
108. D.Y. Kim and S. Nam. Excitation control method for a low sidelobe SIW series slot array antenna with 45° linear polarization. *IEEE Transactions on Antennas and Propagation*, 61(11): 5807–5812, 2013.
109. Y.J. Cheng, W. Hong, and K. Wu. Millimeter-wave half mode substrate integrated waveguide frequency scanning antenna with quadri-polarization. *IEEE Transactions on Antennas and Propagation*. 58(6): 1848–1855, 2010.
110. A.J. Simmons. Circularly polarized slot radiators. *IRE Transactions on Antennas and Propagation*. 5(1): 31–36, 1957.
111. G. Montisci, M. Musa, and G. Mazzarella. Waveguide slot antennas for circularly polarized radiated field. *IEEE Transactions on Antennas and Propagation*. 52(2): 619–623, 2004.
112. J. Hirokawa, M. Ando, N. Goto, et al. A single layer slotted leaky waveguide array antenna for mobile reception of direct broadcast from satellite. *IEEE Transactions on Vehicle Technology*, 44(4): 749–755, 1995.
113. Y. Songnan, S.H. Suleiman, and A.E. Fathy. Ku-band slot array antennas for low profile mobile DBS applications: Printed vs. machined. In *Proceedings of IEEE International Antennas and Propagation Symposium*, July 2006, pp. 3137–3140.
114. P. Chen, Q.Z. Kuai, J.X. Chen, et al. Substrate integrated waveguide (SIW) traveling-wave circular polarized antennas. In *China Microwave Conference*, Xian, China, May 2009, pp. 821–824.
115. G. Montisci. Design of circularly polarized waveguide slot linear arrays. *IEEE Transactions on Antennas and Propagation*. 54(10): 3025–3029, 2006.

116. P. Chen, W. Hong, Z.Q. Kuai, et al. A substrate integrated waveguide circular polarized slot radiator and its linear array. *IEEE Antennas and Wireless Propagation Letters*, 8: 120–123, 2009.
117. IEEE 145. IEEE standard definitions of terms for antennas. 1993.
118. L. Josefsson and P. Persson. *Conformal array antenna theory and design*. Piscataway, NJ: IEEE Press, 2006.
119. K.L. Wong. *Design of nonplanar microstrip antennas and transmission lines*. New York: Wiley, 1999.
120. J.L. Volakis. *Antenna engineering handbook*, 4th ed. New York: McGraw-Hill, 2007.
121. Y.J. Cheng, H. Xu, D. Ma, et al. Millimeter-wave shaped-beam substrate integrated conformal array antenna. *IEEE Transactions on Antennas and Propagation*, 61(9): 4558–4566, 2013.
122. O. Bayraktar and O.A. Civi. Circumferential traveling wave slot array on cylindrical substrate integrated waveguide (CSIW). *IEEE Transactions on Antennas and Propagation*, 62(7): 3557–3566, 2014.
123. Z. Hao, W. Hong, J. Chen, et al. A novel feeding technique for antipodal linearly tapered slot antenna array. In *Proceedings of IEEE International Microwave Symposium*, June 2005, pp. 1641–1643.
124. T. Djerafi and K. Wu. Corrugated substrate integrated waveguide (SIW) antipodal linearly tapered slot antenna array fed by quasi-triangular power divider. *Progress in Electromagnetics Research*, 26: 139–151, 2012.
125. W. Hong, J.X. Chen, P.P. Yan, et al. CMOS millimeter wave and THz ICs. In *China-Japan Joint Microwave Conference*, April 2011.
126. S. Kobayashi, R. Mittra, and R. Lampe. Dielectric tapered rod antennas for millimeter-wave applications. *IEEE Transactions on Antennas and Propagation*, 30(1): 54–58, 1982.
127. D.V. Lioubtchenko, S.N. Dudorov, J.A. Mallat, et al. Dielectric rod waveguide antenna for W band with good input match. *IEEE Microwave Wireless Component Letters*, 15(1): 4–6, 2005.
128. S. Hanham, T.S. Bird, A. Hellicar, et al. A 600 GHz dielectric rod antenna. In *3rd European Conference on Antennas and Propagation*, April 2009, pp. 1645–1647.
129. A. Patrovsky and K. Wu. 94-GHz planar dielectric rod antenna with substrate integrated image guide (SIIG) feeding. *IEEE Antennas and Wireless Propagation Letters*, 5(1): 435–437, 2006.
130. N. Ghassemi and K. Wu. Planar dielectric rod antenna for gigabyte chip-to-chip communication. *IEEE Transactions on Antennas and Propagation*, 60(10): 4924–4928, 2012.
131. Z.Y. Zheng, K. Wu, and N. Yang. Broadband millimeter-wave quasi-Yagi antenna using substrate integrated waveguide technique. In *IEEE Radio and Wireless Symposium*, January 2008, pp. 671–674.
132. G.H. Zhai, W. Hong, K. Wu, et al. Printed quasi-Yagi antenna fed by half mode substrate integrated waveguide. In *2008 Asia-Pacific Microwave Conference*, December 2008.
133. X. Zou, C.M. Tong, J.S. Bao, et al. SIW-fed Yagi antenna and its application on monopulse antenna. *IEEE Antennas and Wireless Propagation Letters*, 13: 1035–1038, 2014.
134. W. Hong, C. Yu, G.H. Zhai, et al. Frequency notched wideband printed directional antennas. In *2010 International Workshop on Antenna Technology*, March 2010.
135. G.S. Scharp. Continuous slot antennas. U.S. Patent 4328502, May 4, 1982.
136. F. Whetten and C.A. Balanis. Meandering long slot leaky-wave antennas. *IEEE Transactions on Antennas and Propagation*, 39(11): 1553–1559, 1991.

137. J.L. Gómez-Tornero, A.T. Martínez, D.C. Rebenaque, et al. Design of tapered leaky-wave antennas in hybrid waveguide-planar technology for millimeter wave band applications. *IEEE Transactions on Antennas and Propagation*, 53(8): 2563–2578, 2005.
138. Y.J. Cheng, W. Hong, K. Wu, et al. Millimeter-wave substrate integrated waveguide long slot leaky-wave antennas and two-dimensional multibeam applications. *IEEE Transactions on Antennas and Propagation*, 59(1): 40–47, 2011.
139. P. Mondal and K. Wu. A leaky-wave antenna in substrate integrated non-radiative dielectric (SINRD) waveguide with controllable scanning rate. *IEEE Transactions on Antennas and Propagation*, 61(4): 2294–2297, 2013.
140. J.F. Xu, W. Hong, H.J. Tang, et al. Half-mode substrate integrated waveguide (HMSIW) leaky-wave antenna for millimeter-wave applications. *IEEE Antennas and Wireless Propagation Letters*, 7: 85–88, 2008.
141. D. Deslandes and K. Wu. Substrate integrated waveguide leaky-wave antenna: Concept and design considerations. In *2005 Asia-Pacific Microwave Conference*, December 2005.
142. A.J. Martinez-Ros, J.L. Gómez-Tornero, and F. Quesada-Pereira. Efficient analysis and design of novel SIW leaky-wave antenna. *IEEE Antennas and Wireless Propagation Letters*, 12: 496–499, 2013.
143. F. Xu, K. Wu, and X. Zhang. Periodic leaky-wave antenna for millimeter wave applications based on substrate integrated waveguide. *IEEE Transactions on Antennas and Propagation*, 58(2): 340–347, 2010.
144. A.J. Martinez-Ros, J.L. Gómez-Tornero, and G. Goussetis. Holographic pattern synthesis with modulated substrate integrated waveguide line-source leaky-wave antennas. *IEEE Transactions on Antennas and Propagation*, 61(7): 3466–3474, 2013.
145. J. Liu, D.R. Jackson, and Y. Long. Substrate integrated waveguide (SIW) leaky-wave antenna with transverse slots. *IEEE Transactions on Antennas and Propagation*, 60(1): 20–29, 2012.
146. Y. Mohtashami and J. Rashed-Mohassel. A butterfly substrate integrated waveguide leaky-wave antenna. *IEEE Transactions on Antennas and Propagation*, 62(6): 3384–3388, 2014.
147. C. Jin and A. Alphones. Leaky-wave radiation behavior from a double periodic composite right/left-handed substrate integrated waveguide. *IEEE Transactions on Antennas and Propagation*, 60(4): 1727–1735, 2012.
148. A. Patrovsky and K. Wu. Substrate integrated image guide array antenna for the upper millimeter-wave spectrum. *IEEE Transactions on Antennas and Propagation*, 55(11): 2994–3001, 2007.
149. Y.J. Cheng, Y.X. Guo, X.Y. Bao, et al. Millimeter-wave low temperature co-fired ceramic leaky-wave antenna and array based on the substrate integrated image guide technology. *IEEE Transactions on Antennas and Propagation*, 62(2): 669–676, 2014.
150. O. Kramer, T. Djerafi, and K. Wu. Very small footprint 60 GHz stacked Yagi antenna array. *IEEE Transactions on Antennas and Propagation*, 59(9): 3204–3210, 2011.
151. N. Ghassemi and K. Wu. High-efficient patch antenna array for E-band gigabyte point-to-point wireless services. *IEEE Antennas and Wireless Propagation Letters*, 11: 1261–1264, 2012.
152. N. Ghassemi, K. Wu, S. Claude, et al. Low-cost and high-efficient W-band substrate integrated waveguide antenna array made of printed circuit board process. *IEEE Transactions on Antennas and Propagation*, 60(4): 1648–1653, 2012.
153. W.M. Abdel-Wahab and S. Safavi-Naeini. Wide-bandwidth 60-GHz aperture-coupled microstrip patch antennas (MPAs) fed by substrate integrated waveguide (SIW). *IEEE Antennas and Wireless Propagation Letters*, 10: 1003–1005, 2011.

154. Y.J. Cheng, Y.X. Guo, and Z.Q. Liu. W-Band large-scale high-gain planar integrated antenna array. *IEEE Transactions on Antennas and Propagation*, 62(6): 3370–3373, 2014.
155. A. Borji, D. Busuioc, and S. Safavi-Naeini. Efficient, low-cost integrated waveguide-fed planar antenna array for Ku-band applications. *IEEE Antennas and Wireless Propagation Letters*, 8: 336–339, 2009.
156. M.H. Awida, S.H. Suleiman, and A.E. Fathy. Substrate-integrated cavity-backed patch arrays: A low-cost approach for bandwidth enhancement. *IEEE Transactions on Antennas and Propagation*, 59(4): 1155–1163, 2011.
157. Y.J. Li and K.M. Luk. Low-cost high-gain and broadband substrate-integrated-waveguide-fed patch antenna array for 60-GHz band. *IEEE Transactions on Antennas and Propagation*, 62(11): 5531–5538, 2014.
158. T.Y. Yang, W. Hong, and Y. Zhang. Wideband millimeter-wave substrate integrated waveguide cavity-backed rectangular patch antenna. *IEEE Antennas and Wireless Propagation Letters*, 13: 205–208, 2014.
159. Y. Li, Z.N. Chen, X.M. Qing, et al. Axial ratio bandwidth enhancement of 60-GHz substrate integrated waveguide-fed circularly polarized LTCC antenna array. *IEEE Transactions on Antennas and Propagation*, 60(10): 4619–4626, 2012.
160. A.B. Guntupalli and K. Wu. 60-GHz circularly polarized antenna array made in low-cost fabrication process. *IEEE Antennas and Wireless Propagation Letters*, 13: 864–867, 2014.
161. Q.H. Lai, C. Fumeaux, W. Hong, et al. Comparison of the radiation efficiency for the dielectric resonator antenna and the microstrip antenna at Ka band. *IEEE Transactions on Antennas and Propagation*, 56(11): 3589–3592, 2008.
162. S.A. Long, M.W. McAllister, and L.C. Shen. The resonant cylindrical dielectric cavity antenna. *IEEE Transactions on Antennas and Propagation*, 31(3): 406–412, 1983.
163. W.M.A. Wahab, D. Busuioc, and S.S. Naeini. Low cost planar waveguide technology-based dielectric resonator antenna (DRA) for millimeter-wave applications: Analysis, design, and fabrication. *IEEE Transactions on Antennas and Propagation*, 58(8): 2499–2507, 2010.
164. W.M.A. Wahab, D. Busuioc, and S.S. Naeini. Millimeter-wave high radiation efficiency planar waveguide series-fed dielectric resonator antenna (DRA) array: Analysis, design, and measurements. *IEEE Transactions on Antennas and Propagation*, 59(8): 2834–2843, 2011.
165. Q.H. Lai, C. Fumeaux, W. Hong, et al. 60 GHz aperture-coupled dielectric resonator antennas fed by a half-mode substrate integrated waveguide. *IEEE Transactions on Antennas and Propagation*, 58(6): 1856–1864, 2010.
166. J.F. Xu, Z.N. Chen, X.M. Qing, et al. Bandwidth enhancement for a 60 GHz substrate integrated waveguide fed cavity array antenna on LTCC. *IEEE Transactions on Antennas and Propagation*, 59(3): 826–832, 2011.
167. Y. Zhang, Z.N. Chen, X.M. Qing, et al. Wideband millimeter-wave substrate integrated waveguide slotted narrow-wall fed cavity antennas. *IEEE Transactions on Antennas and Propagation*, 59(5): 1488–1496, 2011.
168. Y. Lang, S.W. Qu, and J.X. Chen. Wideband circularly polarized substrate integrated cavity-backed antenna array. *IEEE Antennas and Wireless Propagation Letters*, 13: 1513–1516, 2014.
169. L. Wang, Y.J. Cheng, D. Ma, et al. Wideband and dual-band high-gain substrate integrated antenna array for E-band multi-gigahertz capacity wireless communication systems. *IEEE Transactions on Antennas and Propagation*, 62(9): 4602–4611, 2014.
170. U. Nickel. Overview of generalized monopulse estimation. *IEEE Aerospace and Electronic Systems Magazine*, 21(6): 27–56, 2006.

171. J.T. Branigan, and M.W. Wronski. Amplitude monopulse slotted array. U.S. Patent 4958166, September 1990.
172. J.T. Nemit. Multibeam slot array. U.S. Patent 4164742, August 1979.
173. R.R. Kinsey. An edge-slotted waveguide array with dual-plane monopulse. *IEEE Transactions on Antennas and Propagation*, 47(3): 474–481, 1999.
174. M.S. Ferez, P.R. Fernandez, J.L.M. Campos, et al. Monopulse waveguide patch array antenna in 37 GHz band. In *2005 IEEE Antennas and Propagation Society International Symposium*, July 2005, pp. 689–690.
175. H. Wang, D.G. Fang, and X.G. Chen. A compact single layer monopulse microstrip antenna array. *IEEE Transactions on Antennas and Propagation*, 54(2): 503–509, 2006.
176. Y.J. Cheng, W. Hong, and K. Wu. Millimetre-wave monopulse antenna incorporating substrate integrated waveguide phase shifter. *IET Microwaves, Antennas and Propagation*, 2(1): 48–52, 2008.
177. H. Wang, D.G. Fang, B. Zhang, et al. Dielectric loaded substrate integrated waveguide (SIW) H-plane horn antennas. *IEEE Transactions on Antennas and Propagation*, 58(3): 640–647, 2010.
178. S.W. Drabowitch. Multimode antennas. *Microwave Journal*, 9(1): 41–51, 1966.
179. K.M. Lee and R.S. Chu. Design and analysis of a multimode feed horn for a monopulse feed. *IEEE Transactions on Antennas and Propagation*, 36(2): 171–181, 1988.
180. Y.J. Cheng, W. Hong, and K. Wu. Design of a monopulse antenna using a dual V-type linearly tapered slot antenna (DVLTSA). *IEEE Transactions on Antennas and Propagation*, 56(9): 2903–2909, 2008.
181. P.W. Hannan. Optimum feeds for all three modes of a monopulse antenna. I. Theory. *IRE Transactions on Antennas and Propagation*, 9(9): 444–452, 1961.
182. P.W. Hannan. Optimum feeds for all three modes of a monopulse antenna. II. Practice. *IRE Transactions on Antennas and Propagation*, 9(9): 453–461, 1961.
183. Y.J. Cheng, W. Hong, and K. Wu. Multimode substrate integrated waveguide H-plane monopulse feed. *Electronic Letters*, 44(2): 78–79, 2008.
184. B. Liu, W. Hong, Z.Q. Kuai, et al. Substrate integrated waveguide (SIW) monopulse slot antenna array. *IEEE Transactions on Antennas and Propagation*, 57(1): 275–279, 2009.
185. H.D. Chen, W.Q. Che, Q.Q. He, et al. Compact substrate integrated waveguide (SIW) monopulse network for Ku-band tracking system applications. *IEEE Transactions on Microwave Theory and Techniques*, 62(3): 472–480, 2014.
186. Y.J. Cheng, W. Hong, and K. Wu. 94 GHz substrate integrated monopulse antenna array. *IEEE Transactions on Antennas and Propagation*, 60(1): 121–128, 2012.
187. P.S. Hall and S.J. Vetterlein. Review of radio frequency beam-forming techniques for scanned and multiple beam antennas. *IEE Proceedings—Microwave Antennas and Propagation*, 137(5): 293–303, 1990.
188. R.C. Hansen. *Phased array antennas*. John Wiley & Sons, 1998.
189. Y.T. Lo and S.W. Lee. *Antenna handbook*, vol. 3. New York: Van Nostrand Reinhold, 1993.
190. J. Butler and R. Lowe. Beamforming matrix simplifies design of electronically scanned antennas. *Electronic Design*, 1961(4): 170–173, 1961.
191. S. Yamamoto, J. Hirokawa, and M. Ando. A beam switching slot array with a 4-way Butler matrix installed in a single layer post-wall waveguide. In *2002 IEEE Antennas and Propagation Society International Symposium*, June 2002, pp. 138–141.
192. S. Yamamoto, J. Hirokawa, and M. Ando. A single-layer hollow-waveguide 8-way Butler matrix. In *2005 IEEE Antennas and Propagation Society International Symposium*, June 2005, pp. 647–650.

193. C.J. Chen and T.H. Chu. Design of a 60-GHz substrate integrated waveguide Butler matrix—A systematic approach. *IEEE Transactions on Microwave Theory and Techniques*, 58(7): 1724–1733, 2010.
194. K. Morimoto, J. Hirokawa, and M. Ando. Single-layer waveguide circuit to control sidelobe and crossover levels in Butler-matrix beam-switching antenna. In *2007 IEEE International Symposium on Antennas and Propagation*, August 2007, pp. 1106–1109.
195. K. Morimoto, J. Hirokawa, and M. Ando. Design of a 180-degree single-layer divider to control sidelobe and crossover levels in Butler-matrix beam-switching antenna. In *Proceedings of Asia-Pacific Microwave Conference*, Bangkok, Thailand, December 2007.
196. P. Chen, W. Hong, Z.Q. Kuai, et al. A multibeam antenna based on substrate integrated waveguide technology for MIMO wireless communications. *IEEE Transactions on Antennas and Propagation*, 57(6): 1813–1821, 2009.
197. Y.J. Cheng, W. Hong, and K. Wu. Millimeter-wave multibeam antenna based on eight-port hybrid. *IEEE Microwave and Wireless Components Letters*, 19(4): 212–214, 2009.
198. A.A.M. Ali, N.J.G. Fonseca, F. Coccetti, et al. Design and implementation of two-layer compact wideband Butler matrices in SIW technology for Ku-band applications. *IEEE Transactions on Antennas and Propagation*, 59(2): 503–512, 2011.
199. Y.J. Cheng, C.A. Zhang, and Y. Fan. Miniaturized multilayer folded substrate integrated waveguide butler matrix. *Progress in Electromagnetics Research C*, 21: 45–58, 2011.
200. T. Djerafi and K. Wu. Multilayered substrate integrated waveguide 4×4 Butler matrix. *International Journal of RF and Microwave Computer-Aided Engineering*, 22(3): 336–344, 2012.
201. Y.J. Cheng, X.Y. Bao, and Y.X. Guo. 60-GHz LTCC miniaturized substrate integrated multibeam array antenna with multiple polarizations. *IEEE Transactions on Antennas and Propagation*, 61(12): 5958–5967, 2013.
202. J. Blass. Multidirectional antenna—A new approach to stacked beams. *IRE International Convention Record*, 8: 48–50, 1960.
203. Bonnedal, I. Karlsson, and K.V. Klooster. A dual beam slotted waveguide array antenna for SAR applications. In *7th IEEE International Symposium on Antennas and Propagation*, April 1991, pp. 559–562.
204. P. Chen, W. Hong, Z.Q. Kuai, et al. A double layer substrate integrated waveguide Blass matrix for beamforming applications. *IEEE Microwave and Wireless Components Letters*, 19(6): 374–376, 2009.
205. J. Nolen. Synthesis of multiple beam networks for arbitrary illuminations. PhD dissertation, Radio Division, Bendix Corporation, Baltimore, MD, 1965.
206. T. Djerafi, N.J.G. Fonseca, and K. Wu. Planar 4×4 Ku-band Nolen matrix in SIW technology. *IEEE Transactions on Microwave Theory and Techniques*, 58(2): 259–266, 2010.
207. T. Djerafi, N.J.G. Fonseca, and K. Wu. Broadband substrate integrated waveguide 4×4 Nolen matrix based on coupler delay compensation. *IEEE Transactions on Microwave Theory and Techniques*, 59(7): 1740–1745, 2011.
208. W. Rotman and R.F. Turner. Wide-angle microwave lens for line source application. *IEEE Transactions on Antennas and Propagation*, 11(11): 623–632, 1963.
209. Y.J. Cheng, W. Hong, K. Wu, et al. Substrate integrated waveguide (SIW) Rotman lens and its Ka-band multibeam array antenna applications. *IEEE Transactions on Antennas and Propagation*, 56(8): 2504–2513, 2008.
210. E. Sbarra, L. Marcaccioli, R.V. Gatti, et al. A novel Rotman lens in SIW technology. In *Proceedings of 37th European Microwave Conference*, October 2007, pp. 1515–1518.
211. E.O. Rausch, A.F. Peterson, and W. Wiebach. A low cost, high performance electronically scanned MMW antenna. *Microwave Journal*, 40(1): 20–32, 1997.

212. A.F. Peterson and E.O. Rausch. Scattering matrix integral equation analysis for the design of a waveguide Rotman lens. *IEEE Transactions on Antennas and Propagation*, 47(5): 870–878, 1999.
213. J. Kim, C.S. Cho, and F.S. Barnes. Dielectric slab Rotman lens for microwave millimeter-wave applications. *IEEE Transactions on Microwave Theory and Techniques*, 53(8): 2622–2627, 2005.
214. D.R. Gagnon. Procedure for correct refocusing of the Rotman lens according to Snell's law. *IEEE Transactions on Antennas and Propagation*, 37(3): 390–392, 1989.
215. J. Kim and F.S. Barnes. Scaling and focusing of the Rotman lens. In *2011 IEEE Antennas and Propagation Society International Symposium*, June 2001, pp. 773–776.
216. Y.J. Cheng, W. Hong, and K. Wu. Design of a substrate integrated waveguide modified R-KR lens for millimetre-wave application. *IET Microwave Antennas and Propagation*, 4(4): 484–491, 2010.
217. D. Archer. Lens-fed multiple-beam arrays. *Microwave Journal*, 10: 37–42, 1975.
218. D. Archer. Lens-fed multiple beam array. *Microwave Journal*, 19: 171–195, 1984.
219. K.F. Lee. *Principles of antenna theory*. John Wiley & Sons, 1984.
220. Y.J. Cheng, W. Hong, and K. Wu. Millimeter-wave substrate integrated waveguide multibeam antenna based on the parabolic reflector principle. *IEEE Transactions on Antennas and Propagation*, 56(9): 3055–3058, 2008.
221. M. Ettorre, R. Sauleau, and L. Le Coq. Multi-beam multi-layer leaky-wave SIW pillbox antenna for millimeter-wave applications. *IEEE Transactions on Antennas and Propagation*, 59(4): 1093–1100, 2011.
222. E. Gandini, M. Ettorre, M. Casaletti, et al. SIW slotted waveguide array with pillbox transition for mechanical beam scanning. *IEEE Antennas and Wireless Propagation Letters*, 11: 1572–1575, 2012.
223. K. Tekkouk, M. Ettorre, R. Sauleau, et al. Folded Rotman lens multibeam antenna in SIW technology at 24 GHz. In *Proceedings of 6th European Conference of Antennas and Propagation*, March 2012, pp. 2308–2310.
224. Y.J. Cheng and Y. Fan. Millimeter-wave miniaturized substrate integrated multibeam antenna. *IEEE Transactions on Antennas and Propagation*, 59(12): 4840–4844, 2011.
225. J.L. McFarland. The RN* multiple beam-array family and the beam-forming matrix. In *1979 IEEE Antennas and Propagation Society International Symposium*, June 1979, pp. 728–731.
226. A.B. Guntupalli, T. Djerafi, and K. Wu. Two-dimensional scanning antenna array driven by integrated waveguide phase shifter. *IEEE Transactions on Antennas and Propagation*, 62(3): 1117–1124, 2014.
227. A.A. Oliner. A new class of scannable millimeter wave antennas. In *Proceedings of 20th European Microwave Conference*, September 1990, pp. 95–104.

Index

T

U

V

W

X

Y

Z